WITHDRAWN
WRIGHT STATE UNIVERSITY LIBRARIES

Current Research in Alzheimer Therapy

Cholinesterase Inhibitors

Edited by Ezio Giacobini and Robert Becker

Taylor & Francis
New York • Philadelphia • Washington D.C. • London

USA	Publishing Office:	Taylor & Francis • New York 3 East 44th St., New York, NY 10017
	Sales Office:	Taylor & Francis • Philadelphia 242 Cherry St., Philadelphia, PA 19106-1906
UK		Taylor & Francis Ltd. 4 John St., London WC1N 2ET

Current Research in Alzheimer Therapy:
Cholinesterase Inhibitors

Copyright © 1988 Taylor & Francis • New York

All rights reserved. No part of this publication may be reproduced, stored in a retrieval system, or transmitted, in any form or by any means, electronic, electrostatic, magnetic tape, mechanical, photocopying, recording or otherwise, without the prior permission of the copyright owner.

First published 1988
Printed in the United States of America

Library of Congress Cataloging in Publication Data

Current research in Alzheimer therapy : cholinesterase inhibitors /
   edited by Ezio Giacobini & Robert Becker.
       p.   cm.
    Includes bibliographies and index.
    ISBN 0-8448-1555-1
    1. Alzheimer's disease—Chemotherapy. 2. Cholinesterase
inhibitors—Therapeutic use. I. Giacobini, Ezio. II. Becker,
Robert E.
    [DNLM: 1. Alzheimer's Disease—drug therapy. 2. Cholinesterase
Inhibitors—pharmacology. 3. Cholinesterase Inhibitors—therapeutic
use. WM 220 C976]
RC523.C87 1988
618.98'683—dc19
DNLM/DLC
for Library of Congress   88-20044
                                                                  CIP

Contents

Introduction by the Editors: Advances in the Therapy of
Alzheimer's Disease: Issues in the Study and Development of
Cholinomimetic Therapies 1
R. Becker and E. Giacobini

Part I. Preclinical Studies and Mechanisms of Action of Cholinesterase Inhibitors in the Brain

Effects of Anticholinesterases Pertinent for SDAT Treatment but Not
Necessarily Underlying Their Clinical Effectiveness 15
A. Karczmar and N.J. Dun

Use of Long-Acting Organophosphates in Alzheimer's Disease 31
P.G. Waser and C. Streichenberg

Accumulation and Turnover of Acetylcholine after Administration of
Acetylcholinesterase Inhibitors in Rat Brain 43
A. Enz

Senescence-Like Learning Retention Deficits in Autoimmune Mice:
Reversal By Physostigmine 53
M.J. Forster, K.C. Retz and H. Lal

Pharmacological Consequences of Cholinergic Plus Noradrenergic
Lesions 63
V.H. Haroutunian, G.K. Tsuboyama, P.D. Kanof and K.L. Davis

Blood Cholinesterase Inhibition as a Guide to the Efficacy of
Putative Therapies for Alzheimer's Dementia: Comparison of
Tacrine and Physostigmine 73
K.A. Sherman and E. Messamore

Choice Reaction Time in Aged Rhesus Monkeys as a Model of AD 87
W.A. Ashford and J. Bice

Cholinesterase Inhibitors Increase the Brain's Need for Free Choline 95
R. Wurtman, J.K. Blusztajn, J.H. Growdon and I.H. Ulus

Part II. Physostigmine Treatment in Alzheimer's Disease

Physostigmine Treatment in SDAT: Type of Administration, Dose
and Duration 103
L.J. Thal, B.R. Lasker, D.M. Masur, A.D. Blau, and S. Knapp

Intracerebroventricular Administration of Cholinergic Drugs: Preclinical
Trials and Clinical Experience in Alzheimer Patients 113
E. Giacobini, R. Becker, M. Mcilhany, and V. Kumar

Treatment of Alzheimer Dementia with Steady-State Infusion of
Physostigmine 123
R. Elble, E. Giacobini, R. Becker, R. Zec, S. Vicari, C. Womack,
E. Williams, and C. Higgins

Measuring Changes in Memory and Cognitive Functioning in Alzheimer's
Disease with Administration of Oral Physostigmine 141
D.M. Masur, A.D. Blau, L.J. Thal, and P.A. Fuld

Effects of Physostigmine on Recognition Memory in AD Patients:
A Comparative Review 153
R. Zec

Cognitive and Neuroendocrine Changes with Physostigmine in Alzheimer's
Disease Patients 163
V. Kumar, J. Murphy, K.A. Sherman, J.W. Ashford, and J. Markwell

High Performance Liquid Chromatography-Electrochemical Detection: Effect of
Electrode Surface Area on Sensitivity of Physostigmine Determination. 173
L.K. Unni, R.E. Becker, and E. Giacobini

Part III. THA: Preclinical Studies and Mechanism of Action of Cholinesterase Inhibitors

Effects of Cholinergic and Adrenergic Enhancing Drugs on
Memory in Aged Monkeys 179
R. Bartus and R.L. Dean

Actions of THA, 3,4-Diaminopyridine, Physostigmine and Galanthamine
on Neuronal K + at a Cholinergic Nerve Terminal 191
A.L. Harvey and E.G. Rowan

Effects of Cholinergic Drugs Used in Alzheimer Therapy at the
Mammalian Neuromuscular Junction 199
R.J. Bradley, M.T. Edge, S.G. Moran, and A.M. Freeman

THA-Memory and Cognitive Enhancement in Mouse, Monkey and Man 211
L.J. Fitten, K. Perryman and P. Gross

Tacrine: Levels and Effects on Biogenic Amines and Their Metabolites
in Specific Areas of the Rat Brain 217
K.H. Tachiki, K. Spidell, L. Samuels, R.F. Ritzmann, A. Steinberg,
R.L. Lloyd, W.K. Summers and A. Kling

Part IV. Clinical Experience with Cholinesterase Inhibitors

Efficacy and Side Effects of THA in Alzheimer's Disease Patients 225
V. Kumar

Preliminary Experiences and Results with THA for the Amelioration
of Symptoms of Alzeimer's Disease 231
H. Nyback, H. Nyman, G. Ohman, I. Nordgren and B. Lindstrom

Tetrahdroaminoacridine and Lecithin in Alzheimer's Disease 237
S. Gauthier, H. Masson, L. Gauthier, R. Bouchard, B. Collier,
Y. Bacher, P. Bailey, R. Becker, H. Bergman, R. Charbonneau,
D. Dastoor, D. Gayton, J. Kennedy, C. Kissel, M. Krieger, S. Kushnir,
A. Lamontagne, M. St-Martin, J. Morin, N.P.V. Nair, L. Neirinik,
J. Ratner, S. Suissa, Y. Tesfaye and S. Vida

Effect of THA on Acetylcholine Release and Cholinergic Receptors
in Alzheimer Brains 247
A. Nordberg, L. Nilsson, A. Adem, J. Hardy and B. Winblad

Tacrine, Back to the Future? 259
W.K. Summers, L.V. Majovski, G.M. Marsh, K. Tachiki
and A. Kling

The History of THA 267
J.E. Thornton and S. Gershon

Part V. New Approaches to Pharmacotherapy of Alzheimer's Disease

Metrifonate: A Review 281
I. Nordgren and B. Holmstedt

Studies on the Nootropic Effects of Huperzine A and B: Two Selective
Acetylcholinesterase Inhibitors 289
X.C. Tang, X.D. Zhu, and W.H. Lu

Galanthamine: Another Look at an Old Cholinesterase Inhibitor 295
E.F. Domino

Methanesulfonyl Fluoride: A CNS Selective Cholinesterase Inhibitor 305
D.E. Moss, H. Kobayashi, G. Pacheco, R. Palacios, and
R.G. Perez

Intra-cerebro-ventricular Bethanechol: Dose and Response 315
S.L. Read

Intraventricular Bethanechol Infusion for Alzheimer's Disease:
Results of Double-Blind and Escalating Dose Trials 325
R.D. Penn, E.M. Martin, R.S. Wilson, J.H. Fox, S.M. Savoy, and
D.A. Groose

Cholinergic Agonists in Alzheimer's Disease Patients 333
M. Davidson, E. Hollander, Z. Zemishlany, L.J. Cohen, R.C. Mohs, and
K.L. Davis

4-Aminopydine (4-AP) - Derivatives as Central Cholinergic Agents 337
P. Waser, S. Berger, H.L. Haas, and A. Hofmann

From Physostigmine to Physostigmine Derivatives as New
Inhibitors of Cholinesterase 343
M. Brufani, C. Castellano, M. Marta, F. Murroni, A. Oliverio,
P.G. Pagella, F. Pavone, M. Pomponi, and P.L. Rugarli

Part VI. Neurochemical Correlates of Alzheimer's Disease: Neural and Non-Neural Markers

Peripheral Cholinergic Changes and Pharmacological Aspects
in Alzheimer's Disease 355
L. Ravizza, P. Ferrero, C. Eva, P. Rocca, L. Tarenzi, and P. Benna

Subject Index 365

INTRODUCTION BY THE EDITORS --
ADVANCES IN THE THERAPY OF ALZHEIMER'S DISEASE:
Issues in the Study and Development of Cholinomimetic Therapies

Robert E. Becker[1,2] and Ezio Giacobini[1]
[1]Dept. of Pharmacology, [2]Dept. of Psychiatry,
S.I.U. School of Medicine, P.O. Box 19230, Springfield, IL

Alzheimer's disease (AD) is the fourth highest cause of death among the elderly and produces a long period of severe and progressive impairment of mental functions that leads to final extreme or complete dependency. The emotional burden on families is sharpened by the progressive destruction of the personality of the affected family members. The financial burden for families with an affected member can be impoverishing due to the need for medical and nursing care, respite care while the member is at home and a final prolonged institutionalization that is usually not reimbursed from third party sources.

A cure or prevention for the disease would be most desirable and evidence is accruing that at least some non-sporadic forms are possibly genetic or have familial features (Mohs et al., 1987). Knowledge about the gene locus and the role of associated genetic features such as amyloid protein will continue to grow but preventive or therapeutic applications of this information are in the future (Selkoe, 1987). Therefore there remains a need for the development of palliative treatments or especially for the development of any measures that might arrest or slow the expression of the disease process.

One strategy that is being widely investigated is the development of a drug with therapeutic effects in AD. An alternative but interrelated approach is the identification of the biochemical or physiological defect(s) with the intention that a therapeutic agent could then be found. In the main four areas of research have been explored, the acetylcholine system of the brain originating in the nucleus basalis of Meynert and radiating widely to cortex and especially hippocampus and adjacent structures, the catecholamine and indoleamine systems of the brain which exert widespread neuromodulatory effects in the same areas and in core brain structures, somatostatin activity and concentrations and the effects of and control of the expression of nerve growth or other neurotropic factors.

Steady state drug effects and behavioral response

This volume addresses the function and pharmacology of the central cholinergic system in AD. Many authors share the concern that the cholinergic deficit in AD may be too severe to be treated. Our inability to diagnose early AD may depend on the fact that our patients are already too deteriorated to respond to treatment. Similarly treatment response to manipulation of the cholinergic system may be limited by deficits in other, i.e. monoaminergic (Haroutunian et al in this volume), systems. Since the monoaminergic deficits have been proposed to be typical of early onset Alzheimer's disease (Adolfsson et al 1979) subtyping may prove to be important to development of effective therapies. The neurochemical deficiencies in AD may be heterogeneous as may the response to drug therapies. These issues are discussed in this volume by Thal et al and others who have noted a variable response to the administration of cholinesterase inhibitors. Another focus of discussion is the complexity of the cholinergic system. The functions of the central cholinergic system may be too complex to respond to presently available pharmacological agents. This could account for the variability of response to different drugs of the same class i.e. physostigmine (PHY) (Thal et al), tetrahydroaminocridine (THA) (Summers et al, Gauthier et al, Nyback et al in this volume) and to indirect and direct agonists such as physostigmine and bethanicol (Read et al in this volume).

We have argued that a long term steady state modification of brain acetylcholinesterase (AChE) activity may be needed to demonstrate the efficacy of this approach to AD. We have assumed that acute single dose administration of drugs that change neurotransmitter or neuromodulator concentrations may not allow sufficient time for adaptation or a stable enough environment for the brain to use its improved neurotransmission in the organization of a complex behavioral response such as memory performance. Similarly multiple dose administration of short acting drugs such as PHY may provide a continually changing environment within which the brain cannot organize an improved behavioral response. This volume addressed these issues of cholinomimetic therapy.

Early diagnosis of Alzheimer's Disease

A second issue, that is of equal importance to the effectiveness of any palliative treatment, is early detection and early treatment of the disease. The mode of onset of diagnosed AD suggests that in most patients a long downhill course has gone unnoticed until a dramatic and obvious impairment in social functioning forces recognition of the presence of a problem by the family and brings a relatively advanced patient to be diagnosed (Kuhl et al., 1985). This mode of clinical presentation strongly suggests that the lesion has been present for a prolonged period and that excessive losses of brain cells or damage to brain systems may have already occurred and created an irreversible process by the time of diagnosis (Kuhl et al., 1985). A model for this exists is Parkinson's disease where it has been estimated that an 80% loss in substantia nigra dopaminergic neurons occurs before the disorder is clinically manifest and diagnosable (McGeer and McGeer, 1984). In order to determine the true efficacy of any long-term treatment that might slow or arrest deterioration it may be very important to be able to reliably diagnose AD at a much earlier stage than is now possible.

We and others find that the diagnostic standards recommended by the interdisciplinary group of the National Institute of Neurological and Communicative Disorders and Stroke (NINCDS) and the AD and Related Disorders Association (ADRDA) (Neurology 34(7), July 1984) can achieve over 80% accuracy confirmed by subsequent autopsy (Sulkava et al., 1983). Routine clinical and laboratory examinations specific for non-Alzheimer's dementia are required by these standards. As a result criteria developed from these standards effectively screen out patients with dementias originating from etiologies other than Alzheimer's. One exception to this is that these standards will not screen out patients with Pick's disorders, a relatively rare condition. Using these standards, diagnosis of the demented state requires a moderate to severe dementia to be present in order for the diagnosis to be confirmed. Memory disorders and intellectual deterioration must be noted to be progressive and present for over six months. In addition, the cognitive disorder must be of sufficient severity to interfere with social or occupational functioning (DSM IIIR). Social and occupational functioning are generally routinized overlearned behaviors that often do not call for new learning (Kaszniak, 1986). Therefore it is understandable that disproportionately severe impairments of the ability to retain newly learned facts and procedures are frequently found among newly diagnosed AD patients (Kaszniak, 1986). The ability to assimilate new information and incorporate it in a working store of knowledge is an early loss in Alzheimer's, the performance of routinized behaviors is impaired only in later stages. DSM IIIR, diagnoses the presence of AD using behaviors that are impaired only after the more specific but clinically silent pathology impairments of memory and intellect have become severe and irreversible. This same observation can be applied to the criteria that require the presence of impaired abstract thinking, apraxia, aphasia and personality change (DSM IIIR). It has been reported that 100 grams or more of cerebral cortex must be lost prior to the loss of abstracting ability, the appearance of concrete thinking and the appearance of organic personality change (Tomlinson et al. 1970).

Similarly aphasia, apraxia and agnosia are associated with localized dysfunction or damage to specific receptive, motor or associational cortex (Huff et al., 1986; Cummings et al., 1985; Flekkoy, 1976; Foster et al., 1986). These requirements restricts the diagnosis to individuals with considerable loss of cerebral cortical tissues or functions.

Even anatomical examination of brain obtained by biopsy does not offer a reliable means for early diagnostic identification of AD patients. Using expanded DSM IIIR criteria that meet NINCDS and ADRDA standards for diagnosis we and others have found biopsy confirmed the clinical diagnosis (Sulkava et al., 1983). But in earlier studies it was reported that about 30% of brain biopsies in suspected AD patients are nondiagnostic (Smith, 1966). This is not unexpected if "earlier" cases do not yet meet DSM IIIR criteria are included since it is apparent from several studies that AD is not simply the result of a diffuse and unpatterned loss of neurons. The senile plaques and neurofibrillary tangles occur in the "normal" aged, as well as in AD. Both exhibit highest density in the mesial temporal lobes and posterior temporoparietal areas (Cummings and Benson, 1983, Chapter 3) and biopsy of these areas carries the greatest potential surgical risk. In addition, these histologic findings may lateralize to one hemisphere, generally the frontal lobe or the temporal lobe pole (Crystal et al., 1982). The anterior frontal lobe and temporal lobe pole are commonly the preferred biopsy sites, because of the lower associated surgical risk. However, these areas may not always reveal the true frequency of tangles and plaques. Therefore, even if cerebral biopsy could be considered as a useful method of detection it is not surprising that it is not generally accepted as a clinical tool in the diagnosis of AD (see Cummings and Benson, 1983 for a review). Also brain biopsy can carry a 13% risk of major morbidity (epilepsy, infection, subdural hematoma, pulmonary embolism) and mortality (Kaufman and Catalano, 1979). In view of these risks and the nonspecificity of diagnosis based on only anatomical examination of biopsy material the use of biopsy in early diagnosis will require the identification in brain biopsy of diagnostic features, possibly neurochemical, neuropharmacological or even neurohistological, that are specific for early AD. Intriguing findings are the lower levels of ACh production and the reduced numbers of synapses found in biopsies obtained from Alzheimer patients (Nilsson et al., 1986; De Kosky et al., 1985) as well as the lower density of nicotinic receptors in the cortex (De Sarno et al, 1988).

Subtyping of Alzheimer's disease
Since early AD cannot be reliably diagnosed, during early stages, investigations of the disorder are severely limited by the heterogeneity in the population. The diagnosis of Alzheimer's includes both presenile and senile forms of the disease. There has been an attempt to distinguish two types of AD (early onset vs. late onset) on the basis of clinical presentation. Heston et al. (1981) reported that in early onset AD patients the duration of the illness was shorter, there was relatively less variation in the course of illness and greater risk to siblings for the development of AD than in late onset patients. Recently Selzer and Sherwin (1983) observed a higher prevalence of language disorder, a shorter age specific life expectation, more common gait disturbance and possible selective left hemisphere involvement in early cases.

AD may also be heterogeneous in other ways. Neuropathological studies (Roth and Wischnik, 1985; MountJoy et al., 1983; Whitehouse et al., 1982) have attempted to distinguish early onset patients from the late onset patients. These studies suggest that pathological changes are more severe in early onset patients than late onset patients. Similarly, neurochemical studies of the brain of AD patients (Roth and Wischnik, 1985; Rosser et al., 1984) have suggested that cholineacetyltransferase activity and noradrenaline, GABA and somatostatin deficits were more severe in early onset patients than late onset patients.

There has been additional increasing evidence showing heterogeneity of AD (Roth, 1986). A variant of AD has been known for several years to be a heritable condition that follows the pattern of autosomal dominant transmission (Feldman et al., 1963; Cook et al., 1979). A number of studies have reported that about 1/5 to 1/3 of the patients have at least one family member suffering from dementia of Alzheimer's type. We combined the data of eight studies including 515 probands and found that 113 (21.9%) of them have at least one other affected member (Matsuyama et al., 1985). Familial AD may differ from non-familial AD. Folstein et al. (1981) reported that agraphia was significantly more frequently present in familial cases than non-familial cases. Recently Chui et al. (1985) reported a relationship between aphasia and family history of dementia but this correlation was not significant ($P=0.08$). These findings all strongly suggest that there may be subgroups within the population presently diagnosed as AD using DSM IIIR criteria or even more stringent research criteria. The presence of these subgroups may contribute sufficient heterogeneity to the AD population to confound both research into disease mechanisms and therapeutics.

In sum there is considerable evidence to suggest that not only studies of drug efficacy but also studies of diagnostic criteria, biological or behavioral markers of early AD, disease mechanisms and other therapeutics in AD may be confounded by the diagnostic heterogeneity present in AD patient populations as they are presently defined. It is widely accepted that the effectiveness of research in clinical conditions can be vitiated by the lack of homogeneity in the population chosen for study.

Testing the cholinergic hypothesis

In our view, major problems exist with tests of cholinomimetic therapies conducted to date (Becker and Giacobini, 1988). In the later stages of AD the number of cholinergic neurons surviving and capable of releasing ACh may be inadequate even with ChE inhibition. At early stages probably only 5% or 10% of neurons are affected, at later stages as many as 90% of neurons may be totally inactive (Whitehouse et al., 1982; McGeer et al., 1984; Giacobini et al., 1987). The amplification of the chemical signal as a result of 50% of ChE inhibition in the synapse and of an increase of 10-20% in vesicular ACh concentration may not be sufficiently high to stimulate cholinergic receptors at more advanced stages of AD. If the impairment of the postsynaptic cholinergic receptors in the CNS is modest and not too rapidly progressing (Mash et al., 1985; DeKosky et al., 1985) moderately reduced numbers of presynaptic nerve endings may still allow for an adequate stimulation of cholinoceptive neurons with ChE inhibition. The reduction in presynaptic muscarinic receptors seen in AD (Mash et al., 1985) may be an advantage as it may reduce the inhibitory feedback effect of the neurotransmitter on the release mechanism. On the other hand the decrease seen in nicotinic presynaptic receptors (De Sarno et al 1988) may limit the release of ACh. However it has been reported by Nilsson et al. that PHY and THA may restore ^{3}H-ACh efflux from brain slices of AD patients to a normal level (in this volume). Using concentrations of PHY that occur in vivo after dosing at therapeutic levels we could not demonstrate increased ^{3}H-ACh efflux. Therefore this stimulatory effect of Phy may be present only with high concentrations of the drug which may selectively activate nicotinic presynaptic autoreceptors and increase ACh release. Thus, the ratio of presynaptic muscarinic/nicotinic autoreceptors present in AD patients at different stages of the disease may determine the efficacy of the ChE inhibitors and the effect of increased ACh levels.

We propose that more needs to be understood about early neuronal impairment, how it is similar to the better described and studied later stages of the illness and how it differs. We do not see this as a problem of the cholinergic system alone. We suggest the need for a more exact description and understanding of the function of the brain in early Alzheimer's to aid in the identification of neurochemically homogenous groups for study and to identify candidates for therapeutic trials.

The second major limitation to effective research into the therapeutics of AD is inherent in the methodologies available for therapeutic research into Alzheimer's dementia and in our knowledge of the relevant neurochemistry and pharmacology. Most drugs, for example, PHY are short acting and inappropriate as long-term treatments. Their long-term effects are not known. The effectiveness of drugs presently available are limited. Also there is a lack of drugs or naturally occurring substances that can be administered to humans to therapeutically probe many other known dysfunctional systems, e.g., somatostatin, nerve growth factors, etc. This leaves important questions unanswerable. Finally, all studies are restricted by the absence of widely accepted behavioral measures that have a demonstrated ability to reliably distinguish drug induced changes in memory and intellectual functions.

An overview of the cholinesterase inhibitors.

Much of our work with cholinesterase inhibitors (ChEI) illustrates aspects of these problems. We have been studying PHY, administered by various routes, in order to determine the effects of ChEI on memory and intellectual function in AD patients. Physostigmine produces, at therapeutic doses, only modest ChE inhibition and ACh increase (Hallak and Giacobini, 1986; Sherman et al., 1987). This ChE inhibition is rather short-lasting, in the order of 1-2 hrs, depending on the dose and the route of administration. The drug is rapidly metabolized in the liver to inactive metabolites (Unni and Somani, 1986). Any further increase in dose to reach levels of ChE inhibitors in plasma higher than 30% results in side effects which may counteract the beneficial effect of the drug (Sherman et al., 1987). Physostigmine plasma levels have not been studied in relation to ChE inhibition in plasma and clinical behavioral effects. Aquilonius (1986) recently observed that the pharmacokinetics of PHY are not adequately understood. The lack of study of PHY kinetics is due to the fact that a sensitive and reliable assay of Phy was not yet available until recently, and assay sensitivity is reduced because of the interference by plasma constituents. At present the measurements of AChE inhibition in CSF and ChE in plasma provide an indirect indication of Phy effect in the CNS in humans. Large individual variations of ChE inhibition in both CSF and plasma (Thal et al., 1983a,b) demonstrate the difficulties of establishing a proper dosage in individual patients. We have developed a new HPLC method with .1 ng sensitivity (Somani et al., 1987).

We have also studyied other ChEI especially the long-acting compounds, metrifonate (MTF) and THA. A comparison of results from our studies of these different reversible ChEI uses derived from studies in experimental animals and humans (Hallak and Giacobini, 1986, 1987a,b).

We (Hallak and Giacobini, 1986) have proposed that MTF might offer several advantages as compared to both Phy and THA. It produces only mild or no side effects at therapeutic doses in humans (Nordgren et al., 1981). A dose of 10 mg/kg MTF given orally produces an 80% ChE inhibition which persists for 2-3 days with few or no side effects. In contrast, both Phy and THA produce moderate or severe side effects both in rats and humans even at significantly lower levels of ChE inhibition (Hallak and Giacobini, 1987,a,b). A second administration of THA in the rat does not produce any further increase in CNS ACh levels (Hallak and Giacobini, 1987b). Both release and synthesis of ACh are lower after THA than after Phy or MTF. A concentration of 1.4 mM THA completely blocks K^+ evoked release of ACh in rat caudate nucleus slices (Hallak and Giacobini, 1987b). Uptake of Ch is also decreased by both Phy and THA. No similar data are available for MTF. Toxicity (measured by LD_{50}) values are also higher for THA and Phy than for MTF.

Metrifonate, in contrast to Phy and THA, is not a directly acting inhibitor of ChE but requires non-enzymatic metabolism to form the active compound dichlorvos (2,2-dichlorvinyl dimethylphosphate, DDVP). The process of non-enzymatic metabolism of MTF is a slow mechanism causing the drug to act as a slow-release formulator, thus never allowing high levels of DDVP to accumlate in the CNS. This unique feature, represents probably the main advantage of the compound. Non-toxic but sufficiently high levels of the drugs are released to maintain long-lasting ChE inhibition in plasma and brain. Comparative clinical trials of these three drugs are presently in progress at our Center and studies in animals have contributed to the understanding of the shortcomings of present ChE-inhibiting therapy and indicate the requirements for an ideal drug of this category.

We have previously shown that very high levels of AChE inhibition can be achieved in the CSF of the dog by the intracerebroventricular (i.c.v.) administration of Phy (Mattio et al., 1986). We now report that similarly high (90+%) levels of inhibition of CSF AChE is achieved in humans following the i.c.v. administration of Phy (Becker and Giacobini, 1988). This was accompanied by a 300% increase in ACh concentration with persistence of elevated CSF ACh concentrations for over 6 hrs. No adverse effects were evident in spite of the very high levels of AChE inhibition.

We view the development of long-lasting drugs or of delivery systems that can be used to attain high levels of steady state drug effects, to be the second major methodological hurdle to successful therapeutic research into the palliation of AD. One aspect of this is illustrated by the problems inherent in our knowledge of the range of activities of the AChE inhibitors. We have argued elsewhere that these drugs have important, clinically significant activities in addition to inhibition of ChE, and that it is these other properties of these drugs that may limit the clinical effectiveness of inhibitors that are available for human use (Becker and Giacobini, 1988).

A second aspect of the problem of ChEI therapy is the lack of understanding of the relation of AChE inhibition, ACh concentration or turnover and behavioral effects. Unfortunately there are relatively few studies of the relations of kinetics of ACh concentrations to the behavioral changes that follow ChE inhibition. In general, the reported behavioral and physiological effects from Phy appear to occur during the time of peak levels of ACh concentration (Becker and Giacobini, 1988). For example, slowing of the EEG starts during the expected increase in ACh concentration (increased delta activity) and further slowing (theta activity) occurs with the expected peak of ACh concentration. The effects on memory (Muramoto, 1984) appear to occur shortly after the estimated peak of ACh concentration is reached and are lost as ACh values return to baseline. Adverse effects also coincide with predicted maximal brain levels of ACh concentration.

The timing of these behavioral effects is consistent with the timing of expected increases in ACh concentrations in brain and not with the timing of incremental changes in ACh concentration or of late tissue accommodations. We have proposed elsewhere (Becker and Giacobini, 1988) that this is of importance in the design of strategies for therapy since it suggests that increased steady state ACh concentrations (and neither incremental increases in the rates of release nor cellular accommodation) are the appropriate target of therapy. Unfortunately almost all cholinomimetic research has used short acting ChE inhibitors, predominantly Phy. Physostigmine has a $t_{1/2}$ elimination of 10-30 min and an enzyme dissociation $t_{1/2}$ of 20-30 min. This means that single or multiple dosing will result in a consistently and relatively rapidly varying environment for the brain. In our view this challenge to homeostasis may overwhelm the potential beneficial effects of decreased AChE activity.

Another problem with Phy is the level of ChE inhibition that can be achieved with peripheral routes of administration. In humans, we have determined blood ChE inhibition following oral and intravenous Phy (Sherman et al., 1987). There is very high (12 to 33%) interindividual variability in the level of inhibition associated with both oral and intravenous dosing. At 20 µg/kg i.v. and 60 µg/kg oral we have found approximately 25% mean inhibition of butylcholinesterase (BuChE) in blood plasma and approximately 11% mean inhibition of AChE in RBCs (Sherman et al., 1987). The maximum BuChE inhibition in plasma that we have achieved is 40%. This is consistent with the report of Thal (1983) that 30% or more AChE inhibition in CSF can follow repeated dosing with Phy. We find in humans that CNS AChE inhibition is approximately equal to plasma BuChE inhibition (Becker and Giacobini, 1988).

At these relatively low levels of inhibition the levels of uninhibited AChE activity found in many patients are still within the normal range of AChE activities found in the untreated population. It could be expected that this adjustment within the normal range is modest for many patients and within the homeostatic capacity of the body to accommodate. Thus potential therapeutic effects may be eliminated or blunted by the compensatory return of synaptic ACh concentrations toward pretreatment levels for many patients. A different behavioral outcome could result from more dramatic and prolonged steady state alterations in enzyme activity analogous to the use of monoamine oxidase inhibitors in depression. Monoamine oxidase (MAO) must be inhibited by 80% or more for 7-10 days or longer before therapeutic effects are seen (Ravaris et al., 1976). At lower levels of inhibition the MAO inhibitors are ineffective as antidepressants.

Principles of cholinomimetic therapy.

In AD, cholinomimetic therapy's main principles are: maintainance of synthesis of ACh, maintenance of adequate release of ACh and protection of released ACh from hydrolysis. Elevated (10-15%) ACh levels can be maintained by using repeated doses of both reversible and irreversible ChE inhibitors (Becker and Giacobini, 1987). However, the various AChE inhibitors, for example physostigmine PHY, THA, MET, significantly differ with regard to their effects on brain, ACh, ACh release, toxicity and duration of action (Hallak and Giacobini, 1987 a,b).

Increased ACh concentration in brain is the explicit rationale for the administration of ChE inhibitors for relief of the memory deficit found in AD. In a recent overview we presented evidence that the changes in ACh concentration that follow the administration of the different ChEI are not a direct result of the percentage of ChE inhibition and are in fact more strongly affected by properties of the individual drugs (Becker and Giacobini, 1988). Similarly, the toxicity and lethality of these compounds did not appear to be a primary effect of the degree of ChE inhibition or of changes in ACh concentration.

It is assumed that increased tissue concentrations of ACh occur after administration of ChE inhibitors. We proposed that changes in ACh concentration that follow ChE inhibition are not predictable, are not solely the direct result of the percentage of ChE inhibition and may in fact be strongly affected by properties of the individual drugs. Similarly, the toxicity and lethality of these compounds does not appear to be directly linked only to the degree of ChE inhibition or changes in ACh concentration but to be related to specific properties of the individual drugs.

We have reviewed evidence that supports the hypothesis that clinical tests of the efficacy of ChE inhibitors in AD have utilized drugs, such as PHY and THA, that may have prominent drug specific idiosyncratic effects. These drug specific effects may reduce the effectiveness with which these drugs produce increased ACh concentrations and may include adverse effects that limit the dose of drug administered to levels that are not therapeutic. Other ChE inhibitors may be more useful. For example, MTF appears to have a neurochemical and toxicological profile that suggests it may be useful in the treatment of Alzheimer disease. Also a number of other strategies, e.g., reduction of non-quantal release, regulation of ACh turnover rate rather than concentration, high levels of AChE inhibition, etc., may be useful alone or as combination therapies to increase ACh concentrations.

We suggest that a broad approach to manipulation of ACh brain concentrations is appropriate and that this approach is facilitated by restating the therapeutic strategy for the development of treatment of the memory deficit in AD in terms of ACh concentration or turnover rather than in terms of enzyme inhibition. Using ChE inhibition as the focus of therapeutic efforts may be too restrictive. At present testing of the ACh deficiency hypothesis has been compromised by the lack of availability of a drug sufficiently free of side effects. To more effectively increase ACh concentrations in brain an increased understanding of the dynamics of ACh concentration equilibrium in brain tissue may be needed.

The state of the art

Evidence presented here and elsewhere suggests that the cholinergic system is not too complex to manipulate in order to obtain selective memory effects in animals and man (Bartus et al, Thal et al and others in this volume). Both effects on memory (Thal et al) and on daily living activities and other cognitive functions have been reported (Gauthier et al in this volume, Summers et al in this volume).

THA is the only drug that has been associated with improvement in function to near normal levels (Summers et al in this volume) yet there is considerable evidence in the volume that side effects may interfere with the therapeutic use of THA (Kumar et al, Gauthier et al, Nyback et al, Summers et al in this volume). At present there is no therapeutic agent that has been shown to be both effective and safe in man . There is a need to search for newer compounds and to drastically modify existing substances to increase efficacy and safety of drug treatment in AD. This publication presents a number of new drugs which address this necessity.

This volume brings together reports by researchers whose works are germane to understanding the cholinergic based therapy of AD. There is preliminary evidence that various cholinesterase inhibitors may be efficacious. This evidence is sufficiently intriguing to warrant the further careful exploration of this hypothesis. Advances in more exact diagnosis, identification of subtypes, ability to reach diagnosis earlier, understanding of neuroanatomical changes, molecular neurobiology of AD and in neurochemistry will all be important in themselves for understanding AD but many are also closely related to our ability to identify and test possible drug therapies. We look forward to addressing many of these issues in subsequent volumes of this series.

REFERENCES

Adolfsson, R., Gottfries, C.Q., Roos, B.E., Winblad, B. 1979. Changes in the brain catecholamines in patients with dementia of Alzheimer's type. Brit J Psychiat 135:216-223.

Aquilonius, S-M and Hartvig, P., 1986. Clinical Pharmacokinetics of Cholinesterase Inhibitors. Clinical Pharmacokinetics 11:236-249.

Bartus, R. and Dean, R.L. Effects of Cholinergic and Adrenergic Enhancing Drugs on Memory in Aged Monkeys.

Becker, R.E. and Giacobini, E., 1988. Mechanisms of cholinesterase inhibition in senile dementia of Alzheimer's type - clinical, pharmacological and therapeutic aspects. Drug Development Research 12:163-95.

Chui, H.S., Teng, E.L., Henderson, V.W. and Moy, A.C., 1985. Clinical subtypes of dementia of the Alzheimer's type. Neurol. 35:1544-1550.

Cook, R.H., Ward, B.E. and Austin, H.M., 1979. Studies in aging of the brain IV. Familial Alzheimer's disease: relation to transmissible dementia, aneuploidy, and microtubular defects. Neurol. (N.Y.) 29:1402-12.

Crystal, H.A., Horoupian, D.S., Katzman, R. and Jotkowitz, S., 1982. Biopsyproved Alzheimer disease presenting as a right parietal syndrome. Ann. Neurol., 12:186-188.

Cummings, J.L., Benson, D.F., Hill, M.A. and Read, S. 1985. Aphasia in dementia of the Alzheimer type. Neurol. 35:394-397.

De Kosky, S.T., Scheff, S.W. and Markesbery, W.R., 1985. Laminar organization of cholinergic circuits in human frontal cortex in Alzheimer Disease and Aging. Neurol. 35:1425.

De Sarno, P., Giacobini, E. and Clark, B. 1988. Changes in nicotinic receptors in human and rat CNS, FASEB Meeting, Abstract.

Feldman, R.S., Chandler, K.A., Levy, L. and Glaser, G.H., 1963. Familial Alzheimer's disease. Neurol. 13:811-824.

Flekkoy, K. 1976. Visual agnosia and cognitive defects in a case of Alzheimer's disease. Biol. Psychiat. 11(3):333-337.

Folstein, M.F. and Breitner, J.C.S., 1981. Language disorder predicts familial Alzheimer's Disease. The Johns Hopkins Medical Journal, 149:145-147.

Foster, N.L., Chase, T.N., Patronas, N.J., Gillespie, M.M. and Fedio, P., 1986. Cerebral mapping of apraxia in Alzheimer's disease by positron emission tomography. Ann Neurol. 19:139-143.

Gauthier, S., Masson, H., Gauthier, L., Bouchard, R., Collier, B., Bacher, Y., Bailey, R., Becker, R., Bergman, H., Charbonneau, R., Dastoor, D., Gayton, D., Kennedy, J., Kissel, C., Krieger, M., Kushner, S., Lamontagne, A., St-Martin, M., Morin, J., Nair, N.P.V., Neirinck, L., Ratner, J., Suissa, S., Tesfaye, Y. and Vida, S. Tetrahydroaminoacridine and Lecithin in Alzheimer's Disease.

Giacobini, E., Becker, R., Elble, R., Mattio, T., McIlhany, M. & Scarsella, G., 1987. Brain acetylcholine-a view from the cerebrospinal fluid. IN: Neurobiology of Acetylcholine. (Eds.) N.J. Dun and R.L. Perlman. Plenum Press, New York, pp 85-101.

Hallak, M. and Giacobini, E., 1986. Relation of brain regional physostigmine concentration to cholinesterase activity and acetylcholine and choline levels in rat, Neurochem. Res., 11(7):1037-1048.

Hallak, M. and Giacobini, E., 1987a. A comparison of the effects of two inhibitors on brain cholinesterase, Neuropharmacology 26(6):521-530.

Hallak, M. and Giacobini, E., 1987b,(In press) Physostigmine, Tacrine and Metrifonate. The effect of three cholinesterase inhibitors on acetylcholine metabolism in rat brain. Neuropharmacology.

Haroutunian, V.H., Tsuboyama, G.K., Kanof, P.D. and Davis, K.L. Pharmacological Consequences of Cholinergic Plus Noradrenergic Lesions.

Heston, L.L., Masln, A.R., Anderson, E., White, J., 1981. Dementia of the Alzheimer type, clinical genetics, natural history and associated conditions. Arch Gen Psychiatry 38:1085-1090.

Kaszniak, A.W., 1986. IN: Neuropsychological Assessment of Neuropsychiatric Disorders, Edited by I. Grant and K.M. Adams, Oxford University Press, New York, pps. 172-220.

Kaufman, H.H. and Catalano, L.W., 1979. Diagnostic brain biopsy: a series of 50 cases and a review. Neurosurgery, 4:129-136.

Kuhl, D.E., Metter, E.J., Benson, D.F., Ashford, J.W., Riege, W.H., Fujikawa, D.G., Markham, C.H., Mazziotta, J.C., Maltese, A., and Dorsey, D.A. 1985. VIII-8. Similarities of Cerebral Glucose metabolism in Alzheimer's and Parkinsonian Dementia. Journal of Cerebral Blood Flow and Metabolism, Vol. 5., Supp 1, S169-170.

Mash, D.C., Flynn, D.D. and Potter, L.T., 1985. Loss of M2 muscarine receptors in the cerebral cortex in Alzheimer's disease and experimental cholinergic denervation, Science 228:1115-1117.

Matsuyama, S.S., Jarvik, L.F. and Kumar, V. 1985. Dementia Genetics. IN: Recent Advances in Psychogeriatrics ed Arie T. Churchill Livingston M.C. New York: 45-69.

Mattio, T., McIlhany, M., Giacobini, E. and Hallak, M., 1986. The effects of physostigmine on acetylcholinesterase activity of CSF, plasma and brain. A comparison of intravenous and intraventricular administration in beagle dogs, Neuropharmacology 25(10):1167-1177.

McGeer, P.L., McGeer, E.G., Suzuki, J., Dolman, C.E. and Nagai, T., Aging, Alzheimer's disease, and the cholinergic system of the basal forebrain, Neurol. 34(6):741-745, 1984.

Mohs, R.C., Breitner, J.C.S., Silverman, J.M., Davis, K.L., 1987. Alzheimer's Disease. Morbid Risk Among First-Degree Relatives Approximates 50% by 90 Years of Age. Arch. Gen Psychiatry 44:405-408.

MountJoy, C.Q., Roth, M., Evans, N.J.R., Evans, H.M., 1983. Corticol neuronal counts in normal elderly controls and demented patients. Neurobiology of Aging 4;1-11.

Muramoto, O., Sugishita, M. and Ando, K., 1984. Cholinergic system and constructional praxis: A further study of physostigmine in Alzheimer's disease. J. Neurol. Neurosurg. Psychiatry 47:485-491.

Nilsson, L., Nordberg, A., Hardy, J., Wester, P. and Winblad, B., 1986. Short Notes. Physostigmine Restores ^{3}H-Acetylcholine Efflux from Alzheimer Brain Slices to Normal Level. J. Neurol. Transm. 67:275-285.

Nordgren, I., Bengtsson, E., Holmstedt, B. and Prettersson, B.-M., 1981. Levels of metrifonate and dichlorvos in plasma and erythrocytes during treatment of schistosomiasis with bilarcil. Acta Pharmacol. et Toxicol. 49 suppl V. 79-86.

Nyback, H. Preliminary Experiences and Results with THA for the Amelioration of Symtoms of Alzheimer's Disease.

Ravaris, C.L., Nies, A., Robinson, D.S., Ives, J.O., Lamborn, K.R. and Korson, L., 1976. A multiple dose, controlled study of phenelzine in depression - anxiety states. Arch Gen Psychiatry. 33:347-350.

Read, S.L. Intra-cerebro-ventricular Bethanechol: Dose and Response.

Rosser, M.N., Iversen, L.L. and Raynolds, G.P. MountJoy C.Q., Roth M. 1984. Neurochemical characteristics of early and late onset types of Alzheimer's Disease. Brit. Med. J. 288:361-364.

Roth, M., 1986. The heterogeneity of Alzheimer's disease and its implication for scientific investigations of the disorder. Br. Med Bull, Jan:42(1):42-50.

Roth, M. and Wischik, C.M., 1985. Heterogeneity of Alzheimer's Disease and its implications for scientific invstigations of the disorder. In Recent Advances in Psychogeriatrics ed. Arie T. Churchill Livington Inc. New York 71-92.

Selkoe, D.J. 1987. Deciphering Alzheimer's disease: the pace quickens. Trends in Neuroscience vol 10, #5, 181-184.

Seltzer, B., Sherwin, I.A., 1983. Comparison of clinical features in early and late onset primary degenerative dementia. Arch. Neurology 40:143-146.

Sherman, K.A., Kumar, V., Ashford, J.W., Murphy, J.M., Elble, R.J., Smith, R. and Giacobini, E., 1987. (In press). Effect of oral physostigmine in senile dementia patients: utility of blood cholinesterase inhibition and neuroendocrine responses to define pharmacokinetics and pharm-acodynamics. IN: Central Nervous System Disorders of Aging: Strategies for Intervention, Ed. by R. Strong, Raven Press, NY (Aging Vol. 33) pp. 71-90.

Smith, W., Turner, E. and Sim, M., 1966. Cerebral biopsy in the investigation of pre-senile dementia: II pathological aspects. Brit. J. Psychiat., 112:127-133.

Somani, S.M., Unni, L.K. and McFadden, K.L., 1987. Drug Interaction for plasma protein binding: physostigmine and other drugs. International Journal of Clinical Pharmacology Therapy and Toxicology. 25, #8, 412-416.

Sulkava, R., Haltia, M., Paetau, A., et al., 1983. Accuracy of clinical diagnosis in primary degenerative dementia: correlation with neuropathological finding. J Neurol Neurosurg Psychiatry 46:9-13.

Summers, W.K., Majovski, L.V., Marsh, G.M., Tachiki, K. and Kling, A. Tacrine, Back to the Future?

Thal, L.J. and Fuld, P.A., 1983a. Memory enhancement with oral physostigmine in Alzheimer's disease, N. Engl. J. Med. 308:720.

Thal, L.J., Fuld, P.A., Masur, D.M. and Sharpless, N.S., 1983b. Oral physostigmine and lecithin improve memory in Alzheimer's disease, Ann Neurol. 13:491-496, 1983b.

Thal, L.J., Lasker, B.R., Masur, D.M., Blau, A.D. and Knapp, S. Physostigmine Treatment in SDAT: Type of Administration, Dose and Duration.

Thal, L.J., Masur, D.M., Fuld, P.A., Sharpless, N.S. and Davies, P., 1983. Memory improvement with oral physostigmine and lecithin in Alzheimer's disease. IN: Banbury Report 15: Biological Aspects of Alzheimer's Disease, Cold Spring Harbor Laboratory, pp. 461-469.

Tomlinson, B.E., Blessed, G. and Roth, M., 1970. Observations on the Brains of Demented Old People. J. Neurol. Sci. 11:205-242.

Unni, L.K. and Somani, S.M., 1986. Hepatic and muscle clearance of physostigmine in the rat, Drug Metabolism and Disposition 14(2):183-189.

Whitehouse, P.J., Price, D.L., Struble, R.G., Clark, A.W., Coyle, J.T. and DeLong, M.R., 1982. Alzheimer's Disease and senile dementia: loss of neurons in the basal forebrain. Science 215:1237-1239.

Part I

Preclinical Studies and Mechanisms of Action of Cholinesterase Inhibitors in the Brain

Part 1

Replicated Studies and Mechanisms of Action
of Psilocybin-assisted Therapy on the Brain

EFFECTS OF ANTICHOLINESTERASES PERTINENT FOR SDAT TREATMENT BUT NOT NECESSARILY UNDERLYING THEIR CLINICAL EFFECTIVENESS

Alexander G. Karczmar and Nae J. Dun
*Dept. of Pharmacology and Experimental Therapeutics,
Loyola University Medical Center, Maywood, IL*

A. INTRODUCTION

Anticholinesterases (antiChEs) constitute one of the categories of drugs employed in the difficult therapy of Senile Dementia of Alzheimer Type (SDAT). This use dates from their first employment in the treatment of memory loss in normal aged subjects (Drachman and Leavit, 1974), followed a few years later by the use of physostigmine in SDAT (Davis et al., 1979).

In spite of the past limited succsess with antiChE therapy in SDAT, these compounds continue being employed in this condition; let us hope that this Meeting will yield encouraging information in this respect.

What is this perseverance due to? First, the rationale for the use of antiChEs in SDAT is quite compelling in view of the general acceptance of the "cholinergic deficit" hypothesis of the etiology of SDAT (Perry et al., 1978; McGeer et al., 1984; and Whitehouse et al., 1983). It must be stressed, however, that abnormalities concerning other neurotransmitters and modulators (cf. Rossor and Iversen, 1986) and, in fact, energy metabolism (Sims and Bowen, 1986) may also be involved.

Second, our capacity for synthesis of clinically specific and effective drugs is, today, very high, and it may be rationally hoped that further drug development may yield ultimately an antiChE which would have few or no actions inconsistent with long term therapy of SDAT.

This paper is concerned, then, with describing sideactions and actions of antiChEs that are pertinent for the synthetic development in question. It stresses also the fact that in view of the diversity of these actions, the multiplicity of the mechanisms involved, and also because of certain other limitations pertinent in the present context, the task is very difficult indeed.

B. SIDEACTIONS OF ANTIChEs

This paper does not focus on well-known, common garden variety sideactions of antiChEs; they were reviewed by many authors. Suffices to state at this time that they range from intestinal including emetic effects to miosis and visual disturbances, from muscle fasciculations to changes in blood pressure, and from sweating and salivation to hypothermia. Mental and behavioral actions -- which constitute less mundane sideactions -- will be reviewed later.

A few comments are appropriate. First, in the case of the reversible antiChEs such as physostigmine or THA these effects may be short-lived; assuming every four hours dosing, they will vane following each administration and disappear prior to the next dose. Or, they may be long-lived and continue to be present throughout dosing interval; indeed, they may increase in the case of treatment with irreversible antiChEs such as metrifonate.

Second, while deleterious, these sideactions may serve as markers of cholinergic efficacy of the treatment; in the case of the ineffectiveness of a particular antiChE or dose in a SDAT patient their presence will assure us that this ineffectiveness is not due to the absence or low level of antiChE effect. Indeed, the readily measurable ChE inhibition in the blood of the patient may not reflect antiChE action and ACh accumulation in the nervous system. Particularly in the case of the intracerebral infusion of the drug -- a route employed sometimes -- these effects will be most pertinent for establishing cholinergic efficacy, as well as to warn that dose retrenchment is indicated.

C. TYPES OF ACTIONS OF ANTIChEs

Table 1 lists actions of OP and reversible antiChEs. Some of the effects listed concern the actions described above as "mundane" -- such as somatic and autonomic actions. Some are not "mundane", and they may depend on mechanisms other than ChE inhibition.

Synaptic Actions of AntiChEs: These actions occur at peripheral--somatic and ganglionic--as well as central synapses, and at the junctions with the autonomic effectors -- such as smooth or cardiac muscle. The features of these actions which will be described now are mainly based on the investigations of the autonomic ganglia which constitute the most studied cholinergic synapses; actions on the central synapses and skeletal neuromyal junctions will be described subsequently.

TABLE 1

EFFECTS OF ANTICHOLINESTERASES (ANTI ChEs)

Effects dependent on and independent of ChE inhibition

Synaptic, junctional and receptor effects

Actions on non-ACh transmitters and modulators

Cellular, genetic and metabolic effects

Morphopathologic including axonal effects

Behavioral and mental effects

Autonomic and somatic actions

First, there are augmenting actions that occur generally with small doses of antiChEs, yet doses that are sufficient to inhibit some 50 percent of the synaptic acetylcholinesterase and cause acetylcholine (ACh) accumulation; some of the augmenting actions may be not dependent on ChE inhibition (cf. Karczmar, 1967). Depending on the synaptic site, these augmentary effects concern fast (nicotinic) or slow (muscarinic) excitatory potentials, as well inhibitory -- hyperpolarizing -- potentials.

Blocking actions of these compounds become apparent with higher doses. The fast potentials are more readily blocked; the slow responses seem generally more resistant although they are blockable as well.

It may be conjectured that this block is due to accumulated ACh, either via its depolarising or, subsequent to depolarization, desensitizing action. Actually, antiChEs exhibit desensitizing actions independent of those due to their cummulating effect on ACh (Karczmar and Ohta, 1981). It must be emphasized that only few antiChES cause significant depolarization which may be more related to the presynaptic than postsynaptic action (see below), and, as the response to applied ACh is maintained under these circumstances there cannot be any significant desensitization. Thus, the block in question appears to be due mainly to a direct action of antiChEs on the postsynaptic receptors. The effect is reversible in the case of the ganglia, neuromyal junctions, brain or spinal slices; this is true even in the case of the irreversible OP antiChEs which, again, implies that the effect in question is not due to ChE inhibition and ACh accumulation.

Changes in the postsynaptic responses reflect changes in

ionic fluxes; several kinds of currents may be distinguished
with respect to the muscarinic and, to a lesser degree, the
nicotinic potential (Karczmar and Dun, 1986; Adams et al., 1982;
Shinnick-Gallagher et al., 1987). Additional generating force
underlying these potentials involve second messengers, i.e.
nucleotides and phosphatidylinositol cascade (see below, and
Karczmar and Dun,o.c.). While these complex processes are
evoked post-synaptically by ACh, and thus are affected by
antiChEs, they may be affected by these agents independently of
their effect on ACh.

A special type of the postsynaptic action is that on the
ionophore or the channel, the effect that was discovered at the
neuromyal junction by Bernard Katz (Katz and Miledi, 1973). As
the channel when excited by the transmitter-activated receptor,
serves to generate the current, it constitutes the sine qua non
of the synaptic transmission, whether at the neuromyal junction
(Adler et al., 1978; Albuquerque et al., 1987), ganglia (Skok,
1987), or in the CNS. In the present context, it shoud be
stressed that the channel may be directly blocked by antiChEs;
particularly the so-called closed conformation block may be
nefarious for synaptic transmission (Albuquerque et al., 1987).

The presynaptic site should be discussed now. Again, two
types of cholinoceptive receptors, muscarinic and nicotinic can
be distinguished (Pollak, 1965; Szerb, 1977; Nishi, 1970); ACh
present in the cleft will act on these receptors,and this effect
will be reinforced by antiChEs when they accummulate ACh.
Generally, the cholinoceptive presynaptic sites represent the
negative feedback as their activation decreases or inhibits ACh
release. Accordingly, nicotinic and muscarinic blockers
facilitate the release of ACh; ultimately, depletion of ACh may
occur as was demonstrated for the CNS (Szerb, o.c.; Polak, o.c.;
Wecker, 1986); on the other hand, most antiChEs will generally
block the release of ACh -- Soman is an exception (Dun et al.,
1987). It should be emphasized that this negative feedback is
present at some non-cholinergic terminals and affects the
release of other than ACh transmitters.

Still another nerve terminal effect pertinent for the
action of antiChEs is the repetitive discharge; this effect
obtains at the neuromyal junction (vide infra), the ganglia and
in the CNS (Karczmar, 1969).

Finally, antiChEs are effective at autonomic peripheral
sites, including the ocular muscles, the smooth -- bronchiolar
and gastrointestinal -- muscle, and the glands; the
gastrointestinal ad associated glandular actions involve also
the enteric plexus system. The effects in question are
generally muscarinic and excitatory in nature.

At the myenteric neurons the situation is complex, as fast and slow excitatory potentials occur, as well as hyperpolarizing inhibitory potentials; only the fast potentials seem to be cholinergic in nature (Hodgkiss and Lees, 1986).

After the excitatory effects, antiChEs should produce blocking actions via overstimulation of the smooth muscle and inhibition of the ganglia of the enteric system. Actually, what is generally reported - whether in the cases of myasthenic patients overtreated with antiChEs or in the case of exposure to war gases and OP antiChEs - is the muscarinic intoxication that includes bronchiolar constriction, abdominal cramps and diarrhea; paralytic ileus is not explicitly described, although such occurrence is sometimes hinted at (Grob, 1963).

Skeletal Neuromyal Junction: The skeletal neuromyal junction is essentially a nicotinic site, as it exhibits mainly the fast nicotinic potential. AntiChEs first augment and prolong the potential, resulting in the facilitation of the action potential spike and its conversion into a tetanic discharge, which leads to the facilitation of the muscle twitch. Second, there are the inhibitory effects which are due to several mechanisms: short-lived depolarizing block resulting from antiChE-induced ACh accumulation, desensitization (inactivation) of the receptor which occurs concurrently with repolarization (Thesleff, 1955; Karczmar, 1967), and direct actions of the antiChEs, some of which are independent of the desensitizing action of the cummulated ACh (Karczmar and Ohta, 1981). The direct block of antiChEs occurs with relatively high doses of these compounds and includes, as in the case of the ganglia, the receptor and the channel.

Both facilitatory and inhibitory actions are exerted by antiChEs at the presynaptic neuromyal sites, both via their cummulating effect on ACh and their direct action on the nerve terminal (cf. Karczmar, 1967, and Zaimis, 1976). Another phenomenon is the repetitive neuromyal discharge (Riker et al., 1957; Karczmar, 1967b; Hobbiger, 1976). This discharge may result from direct action of antiChEs or the action of ACh following its accumulation by antiChEs at the terminal or at the Ranvier node.

Central Cholinergic Synapses: It must be stressed that the muscarinic responses predominate in the CNS, and that many behavioral and mental correlates of cholinergic function are muscarinic in nature (cf.Karczmar, 1967c, 1976). Many central cholinergic neurons exhibit the slow muscarinic potential, some -- show the fast nicotinic potential, while many display a mixed response; still some other central cholinergic synapses exhibit the inhibitory potential, whether elicited mono- or di-synaptically by ACh. It is a moot question at this time whether the synaptic responses responsible for memory and learning are nicotinic, muscarinic, or both. It is important to stress that while it was thought originally that muscarinic rather than nicotinic receptors are downregulated in SDAT, more recently it appeared that only M2 receptors are thus affected, and that nicotinic receptors may also decrease in number (cf. Sims and Bowen, 1987).

Via their control of transmitter release, the presynaptic sites at the central cholinergic nerve terminals are most important in the feedback modulation of the central function (Eccles, 1964). The presence of a negative muscarinic feedback at the central cholinergic synapses is known since the sixties (Polak, 1965; Szerb, 1977); it appears to exist as well at non-cholinergic synapses.

It is clear that antiChEs exert both facilitatory and inhibitory postsynaptic actions at the central cholinergic synapses, the former occurring essentially with smaller doses of antiChEs compared to the inhibitory effects. What must be particularly emphasized that these effects engender actions at non-cholinergic synapses and changes in other than ACh transmitters (Karczmar, 1967c). The inhibitory effects occur both via depolarizing and desensitizing effects of antiChE--induced ACh accumulation and via direct action of OP and reversible antiChEs (VanMeter et al., 1978).

Acetylcholine Kinetics and AntiChEs: Predictions can be made as to the effect of antiChEs on ACh levels and turnover on the basis of their synaptic actions that were discussed above. It was already pointed out that facilitatory as well as blocking synaptic actions of antiChEs depend on accumulation of ACh induced by these agents. Indeed, such accumulation was demonstrated biochemically for even short acting, reversible antiChEs such as physostigmine; it is particularly long lived and conspicuous with OP antiChEs.

Similarly, antiChE-induced decreased release of ACh illustrated electrophysiologically as transmission block, was demonstrated in animals via appropriate chemical measurements (Nilsson et al., 1987, and Nordberg, this volume).

The final effect of antiChEs on ACh dynamics is concerned with ACh turnover. AntiChEs and cholinomimetics decrease ACh turnover in animals, whether in situ or in brain slices (see for example, Hanin and Costa, 1976). This was substantiated at this Symposium by Enz (see Enz, this Book). We may deal here with a negative feedback resulting from antiChE-induced ACh accumulation in the synaptic cleft; or, antiChEs may reduce the amount of choline available for the nerve terminal uptake. Changes in ACh release may also affect ACh turnover.

Comments on Actions of AntiChEs on Synpases With Particular Reference to THA and Metrifonate (MFT): It must be stressed first of all that successful SDAT therapy with antiChEs depends on only one phase of the biphasic action of antiChEs, namely the facilitatory phase. To the contrary, blocking actions of antiChEs, whether direct or indirect, on either the presynaptic or postsynaptic sites, are counterproductive for the treatment of SDAT. It must be added that facilitatory effects of antiChEs are subject to tolerance (Russell, 1977).

It was emphasized that the blocking actions occur with larger doses of antiChEs; and, particularly the direct actions of antiChEs,i.e. those that are not due to ACh accumulation, may require such doses (Mo et al., 1985; Dun et al., 1987; Dun and Karczmar, 1988). Yet, it could be that the blocking effects are present quite early, although initially masked by the facilitatory effects. Furthermore, some of these inhibitory effects -- the diminution of ACh release and turnover -- should be expected to be present as soon as there is any accumulation of ACh.

Unfortunately, there are relatively few data concerning the pertinent actions of THA and MTF. On the whole, however, these compounds do not appear to be exceptional. For example, at this Meeting it was shown (see Harvey,this Book, and Bradley et al., this Book) that similarly to other reversible antiChEs, THA facilitates in animals at low doses neuromyal transmission, affects some ionic currents including K+ current, and blocks the transmission at concentrations many times those exerting facilitation. Similarly, we found in preliminary studies that THA exerts at the ganglia biphasic actions which include both pre- and post-synaptic blocking effects; furthermore, at least some of the blocking actions were not due to ChE inhibition and ACh accumulation (Karczmar and Dun, unpublished). Also, THA exerted facilitatory effects on hippocampal neurons which appeared to be independent of its antiChE action (Stevens and Cotman, 1987). Furthermore, just like other antiChEs THA blocks in normal tissues, ACh release, as well as synthesis, (Hallak and Giacobini, 1988; Nordberg et al., this volume). It is most interesting that, to the contrary, THA facilitates the release of ACh in brain slices obtained post mortum from SDAT patients (Nordberg et al., this Volume). It may be that the results obtained from post mortum brains cannot be readily compared with those obtained with fresh tissues. It may be also speculated that the morphopathological changes that occur in the cholinergic neurons of the SDAT patients alter the neuron and/or its terminal so as to change the effect of antiChEs on ACh release.

Relatively few in depth pharmacological studies of MTF appear to be available, which is surprising in view of its widespread clinical uese in schistosomiasis (cf. Aldridge and

Holmstedt, 1981). MTF, is of course, an OP antiChE (o.c.); at the neuromyal junction it caused in a few human cases a repetitive discharge, this effect being similar to that of classical OP antiChEs (see above; LeQuesne and Maxwell, 1981). As expected, its antiChE action is prolonged, and, at least in one laboratory, its negative effect on ACh release and synthesis was established (Hallak and Giacobini, 1988).

<u>Metabolic and Second and Third Messenger Phenomena</u>: Choline, ACh and phospholipids participate in a joined metabolic process which leads to the generation of membrane phospholipids -- whether in cholinergic or non-cholinergic neurons, as well as non-neuronal cells --and of ACh -- primarily but not only in cholinergic neurons (see Wurtman, this volume; Blusztajn and Wurtman, 1983; and Saito et al., 1986). It should be emphasized that the extent to which either ACh or phospholipids may be generated in this metabolism will depend on the state of synaptic activity and the pharmacological aspects of the situation; when antiChEs change the dynamics of cholinergic transmission, the outcome of this metabolism must be changed (Karczmar, 1986). In fact, antiChEs of the OP type may affect this outcome not only via their effect on ACh dynamics but also via their capacity to phosphorylate serine moieties and thus interfere with phosphorylations that abund in the phospholipid metabolism (Wurtman and Blusztajn, o.c.; Karczmar, o.c.; Saito et al., o.c.; Reinhardt and Wecker, 1987). When antiChEs, particularly of the OP type are used chronically, as they are in the case of their employment in SDAT, these effects may become important; unfortunately, pertinent evidence is scanty with antiChEs generally, and not available in the case of THA and MTF, specifically.

Muscarinic and nicotinic effects of endogenous ACh are generated via their actions on the second -- cyclic nucleotides and phosphatidyl inositol system -- and third --protein substrates for protein kinases -- messengers (Greengard, 1978,1988). Again, antiChE-induced ACh accumulation as well as phosphorylating capacity of OP antiChEs may affect the messenger systems; it must be added in this context that these systems operate both pre- and post-synaptically (Greengard,o.c.). Data are lacking at present as to the actions of antiChEs on the systems in question.

<u>Cellular, Genetic and Morphopathological Effects of AntiChEs</u>: AntiChEs induce a number of morphopathologic and cellular effects in animals and in man. They may vary from reversible and/or relatively readily antagonizable actions at the neuromyal junctions (Dettbarn, 1984) to irreversible actions whether again at the neuromyal junctions or in the brain (see, for example, Sikora-VanMeter et al., 1985; for further references,see Karczmar, 1963,1984,1985). The general consensus is that these actions are not related to antiChE-induced ACh accumulation,

although facilitation or excess of cholinergic activity may be also involved (Dettbarn,o.c.). While pertinent data seem not to be available in the case of THA, congeners or precursors of MTF cause neuromyal and brain pathologies in animals (Kuznetsov et al., 1985; Berge et al., 1986) and in man, as established via post mortum examination (Akimov and Kolesnichenko, 1985),

A particular morphopathology that can be induced by antiChEs, particularly of the OP type, is referred to as axonal neuropathy or delayed neurotoxicity (Johnson, 1980; Abou-Donia et al.,1979). Johnson (1981) studied this effect with respect to dichlorphos and decided that it could occur only with "massive" doses of this compound. MTF was not evaluated in this respect; as this phenomenon is tailored individually for even closely related OP agents, further thought must be given to this matter.

Genetic and gene transcription mechanisms may be involved in some of these morphopathologies. Pertinent findings were considered sporadic in the early eighties (NRC, 1981), however, more recently pertinent information begun to emerge. It is particularly important in the present context that it was described for MTF congeners or its parent compounds (Dedek et al., 1984; Nehez et al., 1987).

A related subject is that of morphogenetic and teratologic actions of antiChEs generally (Karczmar, 1963 ; Buznikov, 1984) and MTF or its congeners, specifically (Fulton and Chambers, 1985; Francis et al., 1985). This phenomenon constitutes the illustration of the concept posited early by Karczmar (cf. Karczmar et al., 1973) that action on or interference with, the developing cholinergic system may lead to teratologies. Of course, in the case of antiChE treatment of SDAT the danger of teratogenesis or spermicidal action (Karczmar, 1963; McGrady and Nelson, 1976) may be not pertinent.

It must be added in this context that there is a close relationship between the cholinergic system and the trophic phenomena (see, for example, Thoenen and Barde, 1980). Data are not available at this time as to whether or not antiChEs and/or ACh accumulation may interfere with these phenomena and, hence, induce target organ atrophy.

<u>Behavioral and Mental Effects of AntiChEs</u>: Cholinergic system is involved in "organic" and mental behavioral phenomena; in fact, there is not a single behavior that is not affected by cholinergic manipulation (see, for example, Karczmar,1976 and 1984). Of course, one feature of this phenomenology, effects on learning and memory, is stressed at this Meeting.

I wish to emphasize a particular aspect of this matter at this time. Both in animals and in man cholinergic agonists produce a syndrome that I referred to as Cholinergic Alert Non-Mobile Behavior (CANMB; Karczmar, 1979); in man this syndrome may include personality changes and affective disorders including depression (Janowsky et al., 1986; Leong and Brown, 1987). These symptoms occur with both reversible and irreversible (OP) antiChEs. A particularly important features of this matter is that with the OP drugs, these effects may persist for years,as shown both for animals and man (Duffy and Burchfiel, 1980; cf. Karczmar, 1984). Pertinent data with respect to THA and MTF are not available at this time, nor is it known whether or not SDAT patients are particularly vulnerable to this phenomenology.

D. FINAL COMMENTS AND CONCLUSIONS

As already stated, the desirable action of antiChEs in SDAT is their facilitatory effect on synaptic cholinergic transmission. As described above, antiChEs exert many additional effects, either counterproductive in nature or manifested as sideactions, or both. Finally, some of the effects in question were described for MTF and/or THA; some require further analysis.

In all fairness, it appears that some of the effects in question -- such as the synaptic block -- may occur with antiChEs at high doses only; on the other hand, it is not clear whether or not the same is true with respect to morphopathological and mental effects of antiChEs. Finally, it should be stressed that many of the effects in question may not be due to ChE inhibition and ACh accumulation, and thus may not be antagonizable by oximes or atropine.

One more final comment is needed. A sufficient number of cholinergic neurons must be left, sufficient ACh release must continue taking place , and the state of cholinergic receptors - whether muscarinic or nicotinic -- must be adequate in SDAT patients treated with these componds, for the latter to be effective . This is then anologous to Parkinsonism, as in this case also dopamine precursors and/or dopaminergic agonists may cease to be effective under certain circumstances.

ACKNOWLEDGEMENT

The research from this laboratory described in this paper was supported in part by NIH Grant No. 18710 and by a Senior Fulbright Fellowship (AGK).

REFERENCES

Abou-Donia, M.D., Graham, D.G., and Komcil, A.A. 1979. Delayed neurotoxicity of O-(2,4-dichlorophenyl)-O-ethyl phosphonothiate. Toxicol. Appl. Pharmacol. 49: 203.

Adams, P.R., Brown, D.A. and Constanti, A. 1982. M-currents and other potassium currents in bullfrog sympathetic neurons. J. Physiol. (Lond.) 330: 537-572.

Adler, M., Albuquerque, E.X., and Lebeda, F.J. 1978. Kinetic analysis of end plate currents altered by atropine and dopamine. Mol. Pharmacol. 14: 514-529.

Akimov, G.A., and Kolesnichenko, I.P. 1985. Morphological changes in the nervous system in acute peroral chlorophos poisoning. Arkh. Patol. 47: 44-51.

Albuquerque, E.X., Aracara, Y., Idriss, M., Shonenberger, B., Bressi, A., and Deshpande, S.S. 1987. Activation and blockade of the nicotinic and glutamatergic synapses by reversible and irreversible cholinesterase inhibitors. In Neurobiology of Acetylcholine: A Symposium in Honor of A.G. Karczmar. N.J. Dun and R.L. Perlman, eds., pp. 301-328, Plenum Press, N.Y.

Aldridge, W.N., and Holmstedt, B. 1981. History and scope of the conference. Acta Pharmacol. Toxicol. 49, Suppl. 5:3-6.

Berge, G.N., Nafstad, I., and Fonnum, F. 1986. Prenatal effects of trichlorfon on the guinea pig. Arch. Toxicol. 59: 30-35.

Blusztajn, J.K., and Wurstman, R.J. 1983. Choline and cholinergic neurons. Science 221: 614-618.

Buznikov, G.A. 1984. The action of neurotransmitters and related substances on early embryogenesis. Phar. Ther. 25: 23-59.

Davis, K.L., Mohs, R.C., and Tinklenberg, J.R. 1979. Enhancement of memory of physostigmine. N.E.J. Med. 301: 946.

Davis, K.L., Mohs, R.C., Davis, B.M., Rosenberg, G.S., Horvath, T.H., and DeNigris, Y. 1981. In Cholinergic Mechanisms. Cholinomimetic agents and human memory: Preliminary observations in Alzheimer's disease. G. Pepeu and H. Ladinsky, Eds. pp. 929-936, Plenum Press, N.Y.

Dedek, W., Grahl, R., and Schmidt, R. 1984. A comparative study of guanine N7-alkylation in mice in vivo by the organophosphorus insecticides trichlorphon, dimethoate, phosmet and bromophos. Acta Pharmacol. Toxicol. (Kopenh.) 55: 104-109.

Dettbarn, W.D. 1984. Pesticide induced muscle necrosis: Mechanisms and prevention. Fund. Appl. Toxicol. 4: S18-S26.

Drachman, D.A., and Leavitt, J.L. 1974. Human memory and the cholinergic system. A relationship to aging. Arch. Neurol. 30: 113-121.

Duffy, F.H., and Burchfiel, J.L. 1980. Long term effects of the organophosphate sarin on EEG in monkeys and humans. Neurotoxicol. 1: 667-689

Dun, N.J., Karczmar, A.G., Lin, C.H., and Mo, N. 1987. Anticholinesterase actions on mammalian sympathetic ganglia. In Cellular and Molecular Basis of Cholinergic Function. M.J. Dowdall, and J.W. Hawthorne, eds. pp. 569-581, Ellis Horwood Ltd., Chichester.

Eccles, J.C. 1964. The Physiology of Synapses. Springer, N.Y.

Fulton, M.H., and Chambers, J.E. 1985. The toxic and teratogenic effects of selected organophosphorus compounds on the embryos of three species of amphibians. Toxicol. Lett. 26: 175-180.

Francis, B.M., Metcalf, R.L., and Hansen, L.G. 1985. Toxicity of organophosphorus esters to laying hens after oral and dermal administration. J. Environ. Sci. Health 20: 73-95.

Greengard. P. 1978. Cyclic nucleotides, phosphorylated proteins and neuronal function. Raven Press, N.Y.

Greengard, P. 1988. Neuronal phosphoproteins and their physiological significance. In Neurochemical Pharmacology: A Tribute to B.B. Brodie E. Costa, ed. Raven Press, N.Y. (in press).

Grob, D. 1963. Anticholinesterase intoxication in man and its treatment. In Cholinesterases and anticholinesterase Agents. G.B. Koelle, ed., Handbch. d. exper. Pharmakol. Erganzungswk. 15: 989-1027.

Hallak, M., and Giacobini, E. 1988. Physostigimine, tacrine and metrifonate: The effect of multiple dosage on acetycholine metabolism in rat brain. Neuropharmacol. (in press).

Hanin, I., and Costa, E. 1976. Approaches used to estimate brain acetylcholine turnover rate in vivo: effects of drugs on brain acetylchoine turnover rate. In Biology of cholinergic function. A.M. Goldberg and I. Hanin, eds. pp. 355-378, Raven Press, N.Y.

Hobbiger, F. 1976. Pharmacology of anticholinesterase drugs. In Neuromuscular Junction, E. Zainais, ed. pp. 487-581, Springer-Verlag, Berlin.

Hodgkins, J.N., and Less, G.M. 1986. Transmission in enteric ganglia. In Autonomic and Enteric Ganglia A.G. Karczmar, K. Koketin, and S. Nishi, eds., pp. 369-408.

Janovsky, D.S., Risch, S.C., Kennedy, B., Ziggles, M., and Huey, L.Y. 1986. Central muscaranic effects of physostigmine on mood, cardiovascular function, pituitary and adreanl neuroendocrine release. Psychopharmacology 89: 150-154.

Johnson, M.K. 1980. Delayed neurotoxicity induced by organophosphorus compounds - areas of understanding and ignorance. Dev. Toxicol. Environ. Sci. 8: 27-38.

Johnson, M.K. 1981. Delayed neurotoxicity - do Trichlorphon and/or Dichlorvos cause delayed neuropathy in man or in test animals? Acta Pharmacol. Toxicol. 49: Suppl. V.: 87-98.

Karczmar, A.G. 1963. Ontogenetic effects of anticholinesterase agents. In *Cholinesterase and Anticholinesterase Agents*. G.B. Koelle, ed., Handbch. d. exper. Pharmakol. Erganzungswk. 15: 779-832, Springer-Verlag, Berlin.

Karczmar, A.G. 1967a. Multiple action of drugs at the neuromyal function as studied in the light of the phenomenon of reversal. *Laval Medical 38*: 465-480.

Karczmar, A.G. 1967b. Neuromuscular pharmacology. *Ann. Rev. Pharmacol. 7:* 241-276.

Karczmar, A.G. 1967c. Pharmacologic, toxicologic, and therapeutic properties of anticholinesterase agents. In *Physiological Pharmacology*. W.S. Root and F.G. Hofman, eds., 3: 163-322.

Karczmar, A.G., Srinivasan, R., and Bernsohn, J. 1973. Cholinergic function in the developing fetus. In *Fetal Pharmacology*. L.O. Boreus, ed., pp. 122-176, Raven Press, N.Y.

Karczmar, A.G. 1969. Quelques aspects de la pharmacologie des synapses cholinergiques et de sa signification centrale. Actualites Pharmacologigues 22: 293-338.

Karczmar, A.G. 1976. Central actions of acetylcholine, cholinomimetics and related drugs. In *Biology of Cholinergic Function*. A.M. Goldberg, and I. Hanin, eds., pp. 395-449, Raven Press, N.Y.

Karczmar, A.G. 1979. Overview: Cholinergic drugs and behavior -- what effects may be expected from a "cholinergic diet"? In *Nutrition and the Brain,* A. Barbeau, J.H. Growdon and R.J. Wurtman, eds., 5: 141-175, Raven Press, N.Y.

Karczmar, A.G. 1984. Acute and long lasting central actions of organophosphorus agents. *Fund. Appl. Toxicol. 4:* S1-S17.

Karczmar, A.G. 1985. Present and future of the development of anti-OP drugs. *Fund. Appl. Toxicol. 5:* 5270-5279.

Karczmar, A.G. 1986. Conference on dynamics of cholinergic function: Overview and comments. In *Dynamics of Cholinergic Function*. I. Hanin, ed., pp. 1215-1259, Plenum Press, N.Y.

Karczmar, A.G., and Dun, N.J. 1986. Pharmacology of synaptic ganglionic transmission and second messengers. In *Autonomic and Enteric Ganglia -- Transmission and its Pharmacology* A.G. Karczmar, K. Koketsu and S. Nishi, eds., pp. 297-337, Plenum Press, N.Y.

Karczmar, A.G., and Ohta, Y. 1981. Neuromyopharmacology as related to anticholinesterase action. *Fund. Appl. Toxicol. 1:* S135-S142.

Katz, B., and Miledi, R. 1973. The characteristics of "end-plate noise" produced by different depolarizing drugs. *J. Physiol. (Lond) 230:* 707-717.

Krnjevic, K. 1987. Role of acetylcholine in the cerebral cortex. In *Neurobiology of Acetylcholine. A Symposium in Honor of A.G. Karczmar*. N.J. Dun and R.L. Perlman, eds., pp. 271-282.

Kuznetsov, V.G., Tomilin. N.V., Cherniak, T.F., and Masharskii, V.F. 1985. Initial ultrastructural changes in muscle fibers of the diaphragm in rats exposed to chlorophos. Tsitologia 27: 411-414.

Leong, S.S., and Brown, W.A. 1987. Acetylcholine and affective disorder. Neural Transm. 70: 295-312.

LeQuesne, P.M. and Maxwell, I.C. 1981. Effect of metrifonate on neuromuscular transmission. Acta Pharmacol. Toxicol. 49, Suppl. 5: 99-104.

McGeer, P.L., McGeer, E.G., Suzuki. J., Dolman, C.E. and Nagai, T. 1984. Aging, Alzheimer disease and the cholinergic system of the basal forebrain. Neurol. 34: 741-745.

McGrady, A.V., and Nelson, L. 1976. Cholinergic effects on bull and chimpanzee sperm motility. Biol. Reprod. 15: 248-253.

Mo, N., Dun, N.J., and Karczmar, A.G. 1985. Facilitation and inhibition of nicotinic transmission by eserine in the sympathetic ganglia of the rabbit. Neuropharmacol. 24: 1093-1101.

National Research Council. 1981. Panel, Anticholinesterase and Anticholinergic Chemicals. F. Marzulli, Convener, Bd.on Toxicol. and Environ. Health Hazards, Washington, D.C.

Nehez, M., Huszta, E., Mazzag, E., Scheuflen, H., Schneider, P., and Fischer, F.W. 1987. Cytogenetic, genetic and embryotoxicity studies with dimethyl 2,2,2-trichloro-1-(2,2,2-trichloro-1-hydroxyethoxy)-ethyl phosphonate, a hypothetical impurity in technical grade trichlorfon. Ecotoxicol. Environ. Safety 13: 216-224.

Nilsson, L., Adem, A., Hardy, J., Winblad, B., and Nordberg, A 1987. Do tetrahydroaminoacridine (THA) and physostigmine restore acetylcholine release in Alzheimer brains via nicotinic receptors? J. Neural Transm. 70: 357-368.

Nishi, S. 1970. Cholinergic and adrenergic receptors at sympathetic preganglionic nerve terminals. Fed. Proc. 29: 1957-1965.

Perry, E.K., Tomlinson, B.E., Blessed, G., Bergman, K., Gibson, P.H. and Perry. R.H. 1978. Correlation of cholinergic abnormalities with senile plaques and mental test scores in senile dementia. Brit. Med. J. 2: 1457-1459.

Polak, R.L. 1965. Effect of hyoscine on the output of acetylcholine into perfused cerebral vessels of cats. J. Physiol. (Lond) 181: 317-323.

Reinhardt, R.R., and Wecker, L. 1987. Regulation of choline phosphorylation in rat striatum. In Neurobiology of Acetylcholine. N.J. Dun and R.L. Perlman, eds., pp. 145-158, Plenum Press, N.Y.

Riker, W.F., Jr., Roberts, J., Standaert, F.G., and Fujimori H. 1957. The motor nerve terminals as the primary focus for drug-induced facilitation of neuromuscular transmission. J. Pharmacol. Expitherap. 121: 286-312.

Rossor, M., and Iversen, L.L. 1986. Non-cholinergic neurotransmitter abnormalities in Alzheimer's disease. Brit. Med. Bul. 42: 70-74.

Russell, R.W. 1977. Cholinergic substrates of behavior. In *Cholinergic Mechanisms and Psychopharmacology*. P.J. Jenden, ed. pp. 709-731, Plenum Press, N.Y.

Saito, M., Kindel, G., Karczmar, A.G., and Rosenberg, A. 1986. Metabolism of choline in brain of the aged CBF-1 mouse. *J. Neurosci. Res.* 15: 197-204.

Shinnick-Gallagher, P., Hirai, K., and Gallagher, J.P. 1987. Muscarinic receptor activation underlying the slow inhibtory postsynaptic potential (S-I.P.S.P.) and the slow excitatory postsynaptic potential (S-E.P.S.P.). In *Neurology of Acetylcholine. Symposium in Honor of A.G. Karczmar*. N. J. Dun and R.L. Perlman, eds., pp. 245-253, Plenum Press, N.Y.

Sikora-VanMeter, K., Ellenberger, T., and Van Meter, W.G. 1985. Neurotoxic changes in cat neurohypophysics after single and multiple exposure to DFP and Soman. *Fund. Appl. Toxicol.* 5: 1087-1096.

Sims, N.R., and Bowen, D.M. 1987. Recent studies of cholinergic and other neurochemical changes in early-onset Alzheimer's disease. In *Cellular and Molecular Basis of Cholinergic Function* M.J. Dowdall and J.N. Hawthorne, eds., pp. 643-857, Ellis Horwood, Ltd., Chichester.

Skok, V.I. 1987. The blockade of open channel of acetylcholine receptor is responsible for selective blockade of nicotinic transmission. In *Neurobiology of Acetylcholine. Symposium in Honor of A.G. Karczmar*. N. J. Dun and R.L. Perlman, eds., pp. 195-210.

Stevens, D.R., and Cotman, C.W. 1987. Excitatory actions of tetrahydro-9-amino-acridine (THA) on hippocampal pyramidal neurons. *Neurosci. Lett.* 79: 301-305.

Szerb, J.C. 1977. Characterization of presynaptic muscarinic receptors in central cholinergic neurons. In *Cholinergic Mechanisms and Psychopharmacology*, D.J. Jenden, ed., pp. 49-60, Plenum Press, N.Y.

Thesleff, S. 1955. The mode of neuromuscular block caused by acetylcholine, nicotine, decamethonium and succinylcholine. *Acta Physiol. Scand.* 34: 218-231.

Thoenen, H.and Barde, Y.-A. 1980. Physiology of nerve growth factor. *Physiol. Rev.* 60: 1284-1335.

VanMeter, W.G., Karczmar, A.G., and Fiscus, R.R. 1978. CNS effects of anticholinesterases in the presence of inhibited cholinesterases. *Arch. Int. Pharmacol. Ther.* 23: 249-260.

Wecker, L. 1969. The utilization of supplemental choline by brain. In *Dynamics of Cholinergic Function*. I. Hanin, ed., pp. 851-858, Plenum Press, N.Y.

Whitehouse, P.J. Price, D.L., Struble, R.G., Clark, A.W. Coyle, J.T., and DeLong, M.R. 1982. Alzheimer's disease and senile dementia: loss of neurons in the basal forebrain. *Science* 215: 1237-1239.

Zaimis, E. (Editor) 1976. Neuromuscular Junction. Springer-Verlag, Berlin.

USE OF LONGACTING ORGANOPHOSPHATES IN ALZHEIMER'S DISEASE

Peter G. Waser and Catherine Streichenberg
Inst. of Pharmacology, Univ. of Zürich, Zürich, Switzerland

The idea of using "irreversible" anticholinesterases in Alzheimer therapy may not be new, but to our knowledge, it has never been tested in man. The reason for this may be the high general toxicity of all organophosphates and especially of sarin used as a nerve gas for chemical warfare. Only few compounds as ecothiopate (Phospholine^R), paraoxone (Mintacol^R) are used therapeutically as local cholinergic agonists against glaucoma. The big advantage of this type of anticholinesterase is their prolonged action on the acetylcholinesterase, which ends only either when the enzyme is reactivated by hydrolysis of the phosphoryl binding or resynthesis of the damaged enzyme. Most organophosphates pass easily through the blood-brain barrier. This is essential for the treatment of Alzheimer's disease. The danger lies in their peripheral action, which gives rise to motor effects of all kinds such as fibrillations, fasciculations, seizures, cramps, paralysis, and to diverse activation of autonomous functions in different organs. These side effects must be prevented for a successful treatment of the diseased central nervous system.

After investigations on distribution and kinetics of labeled oximes (obidoxime, pralidoxime) we believe that these drugs, by their unusual mode of action, may be used for this prevention, perhaps in combination with muscle relaxants and peripherally acting parasympatholytics.

Techniques

The labeled compounds, except diisopropyl-fluorophosphonate (DFP), were synthesized in our laboratories (Chang et al. 1988, Figure 1) and used for distribution studies in mice and rats with the techniques of whole body autoradiography of 20 µm thick sections from deeply frozen animals (Waser et al. 1986). Radioactivity measurements were performed by liquid scintillation counting of organ specimens. In other artificially respirated animals blood pressure, electrocardiogram and elimination of compound derived radioactivity in urine und bile were continuously measured. In vitro interactions of sarin and obidoxime were recorded by TLC, UV-spectra and NMR.

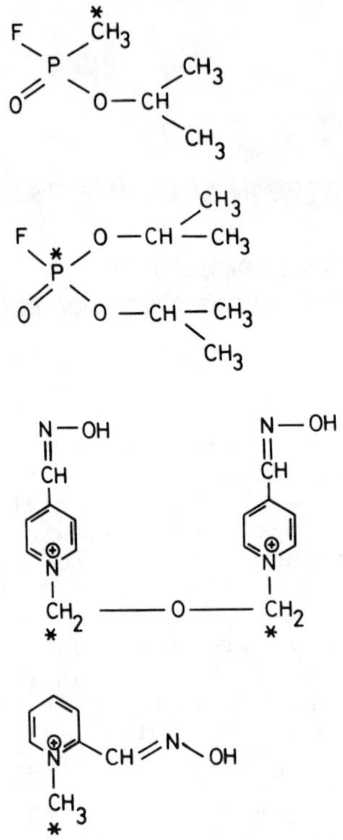

Figure 1. Structural formulas of synthetized compounds.
Asterisk: radioactive label

- ^{14}C-Sarin
  4,48-4,65 mCi/mMol

- ^{32}P-Di isopropyl fluoro phosphate (DFP)
  9,3 mCi/mMol

- ^{14}C-Obidoxime (Toxogonin®)
  4,4 mCi/mMol

- ^{14}C-Pralidoxime (2-PAM)
  1 mCi/mMol

Abbreviations in figures 3, 4, 5a, 5b, 6:

A	aorta	L	liver
AG	adrenal gland	Mu	muscle
Bm	bone marrow	Sc	spinal gland
Br	brain	Sg	salivary gland
E	eye	Sp	spleen
Fd	intervertebral discs	St	sternum
In	intestines	Ub	urinary bladder
K	kidney	Vc	vena cava

Distribution and kinetics of ^{14}C-sarin and ^{32}P-diisopropylfluorophosphonate (DFP) in mice.

When ^{14}C-sarin is injected (1,5 mg/kg i.v. = 20 x LD 50) into mice it enters the central nervous system within 20 seconds. Brain and spinal cord show then the same radioactivity as heart-blood and liver (Figure 2 and 3), which decreases after 30 minutes. The blood-brain barrier is easily permeated by sarin. The radioactivity in the adrenal cortex attains 3 times and lungs, salivary glands and kidney cortex twice this level, until elimination through the kidney pelvis and liver-bile system becomes important. Many other organs (spleen, adrenal medulla, cardiac muscle, intestines, stomach etc.) show an intermediate rise in radioactivity and afterwards a slow decrease within 60 minutes. Plasma half life time of sarin must be well over one hour, because most of its radioactive metabolite is firmly bound in the tissue. Only a small part of free sarin is hydrolized in the plasma at pH 7-8 within 0,5 to 6 hours (Sammet 1983, Waser et al. 1986).

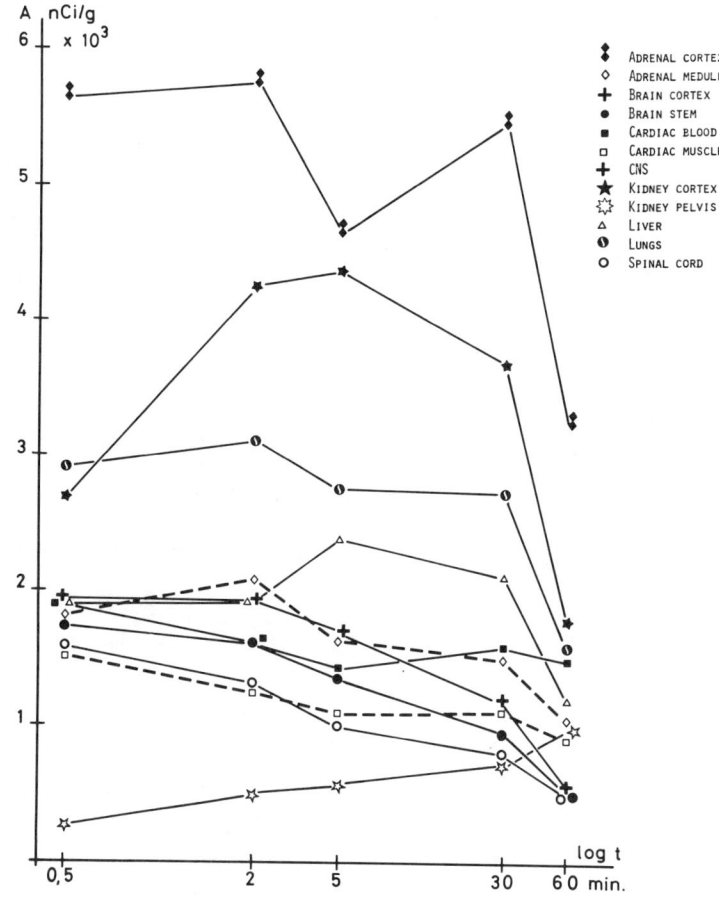

Figure 2. Total radioactivity (μCi x 10^3) in different mouse organs at different times after injection of 1,5 mg/kg of ^{14}C-sarin.

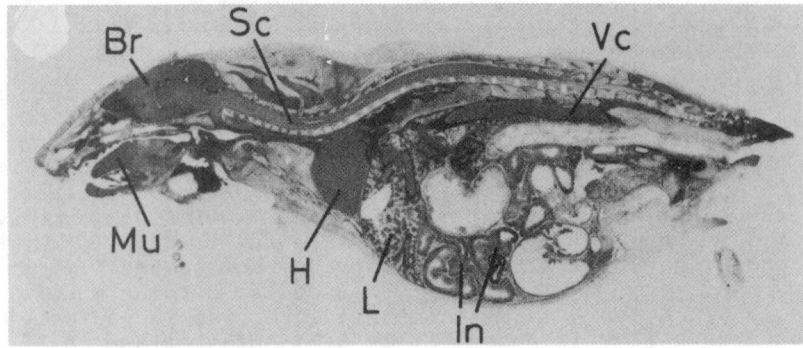

Figure 3. Wholebody autoradiography of mouse section (20 μm) 2 minutes after i.v. injection of 1.4 mg/kg ^{14}C-sarin. Strong radioactivity in brain, spinal cord, heart-blood, liver, spleen, intestines and urinary tract.

The distribution of ^{32}P-DFP is different from sarin, as much less radioactivity enters the central nervous system (Figure 4) 5 minutes after the i.v. injection of 3,4 mg/kg ^{32}P-DFP (= LD 50) most of the radioactivity is found in the liver, the kidney pelvis and the bladder. Twice the amount is found in the blood, whereas brain and spinal cord contain comparably little amounts. After 10-60 minutes small changes in the distribution are observed. Skeletal muscle and especially the bones, accumulate slowly the radioactive phosphate. The bile and the intestines are not much involved in elimination of DFP in contrast to the kidney and the urine. Although quantitative measurements were not made, the half life time in these organs must be several hours (Gotheil 1978, Waser et al. 1986, 1987).

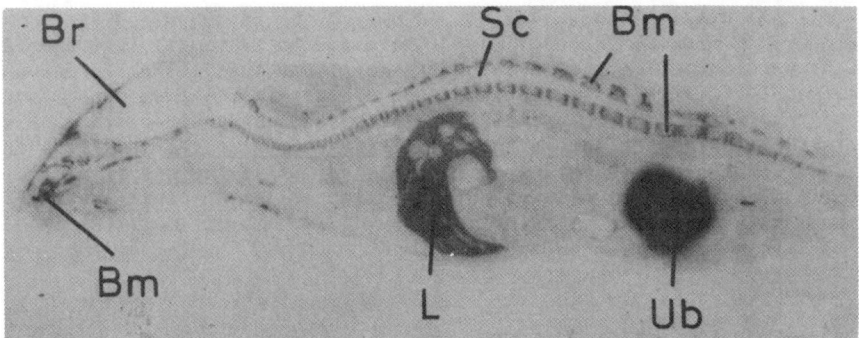

Figure 4. Wholebody autoradiography of mouse section (20 μm) 180 minutes after i.v. injection of 3.4 mg/kg ^{32}P-DFP. Little radioactivity in brain and spinal cord, plenty in liver, urinary bladder and same in bone marrow of spine and skull.

The kinetics and distribution of ^{14}C-pralidoxime (2-PAM) and ^{14}C-obidoxime.

Both compounds are quaternary amines, obidoxime even bisquaternary with a double ionic charge. This makes both polar and easily water soluble. The distribution in the animal body and elimination are entirely different from the lipophilic organophosphates as sarin and DFP.

After intravenous injection most of the radioactivity is found in the elimination organs, kidney and liver, and less in the salivary glands, the suprarenals, the lung, but very little or nothing in the nervous tissue. The maximal values are noticed 5 minutes after intravenous injection. After 10-20 minutes the excretion of obidoxime in the urine starts and much is concentrated in the kidney pelvis and the bladder. Small amounts are found in the liver-bile system with a peak after 20 minutes. The following decrease is very slow, and total excretion takes hours; the half life time is ca. 30 minutes (Figure 5b) (Waser et al. 1986).

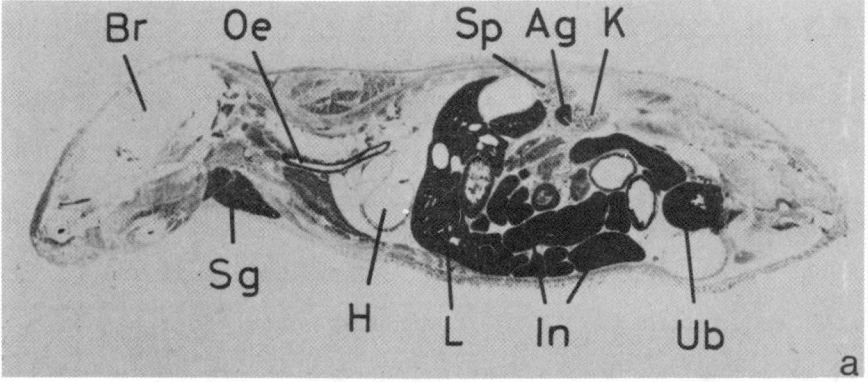

Figure 5a: Wholebody autoradiography of mouse section (20 μm) 30 minutes after i.v. injection of 75 mg/kg ^{14}C-PAM. Brain and spinal cord are free of radioactivity. Radioactivity is mainly in liver, intestines (bile), kidney and urinary-bladder; small amounts are in the salivary gland and traces in heart-blood and lungs.

PAM is readily concentrated to comparable amounts in the kidneys and liver. The half life time is less than 20 minutes (Gotheil 1978, Figure 5a). It appears in the bile and immediately in the lumen and mucosa of the gastrointestinal tract. It is almost completely excreted in 3 hours. Like several typical quaternary molecules as curare drugs, obidoxime and PAM are rapidly concentrated in the cartilage of the intervertebral discs, ribs, joints and other mucopolysaccharide containing tissues.

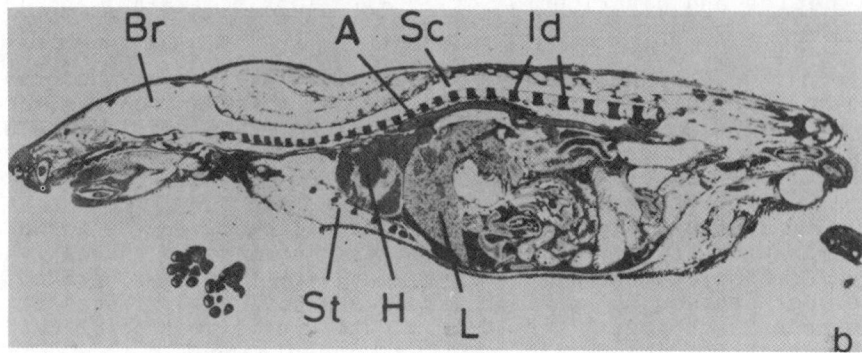

Figure 5b: Wholebody autoradiography of rat section (20 μm) 30 minutes after i.v. injection of 100 mg/kg ^{14}C-obidoxime. Brain and spinal cord are free of radioactivity. Radioactivity is mainly concentrated in heart-blood and cartilage of spinal cord and sternum.

Interaction of organophosphates with obidoxime or pralidoxime (2-PAM)

Pralidoxime and obidoxime were invented as antidotes with the idea of reactivating the phosphorylated AChE. This interesting chemical interaction at the active center of the enzyme has been demonstrated in vitro as a relatively slow process only with elevated concentrations of oximes. A direct action on the phosphorylated enzyme bound to the nervous tissue is more difficult to prove. The interaction of these compounds with the radioactively marked esterases in the organs was then our first interest (Sammet 1983).

Evidently an interaction in the brain and spinal cord cannot take place because the polar "reactivators" do not pass the blood brain barrier, except small traces. Sarin injected before obidoxime does not increase the permeability and change the distribution in favour of the nervous organs. Its main influence must be in the periphery and concerns the distribution of radioactivity in the internal organs.

Cold obidoxime injected 20 minutes i.p. before ^{14}C-sarin i.v. diminishes the cortex radioactivity by half; when obidoxime is injected i.v. 1 or 5 mintues after ^{14}C-sarin, a smaller reduction of radioactivity in the brain follows after 5 or 20-30 minutes respectively. The protective action of obidoxime takes place in the peripheral organs of the body and becomes ineffective when sarin is already fixed in the brain (Figure 6 and 7).

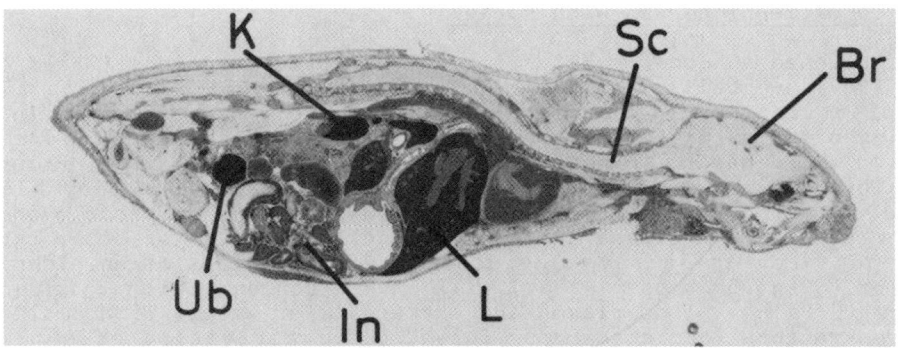

Figure 6. Interaction of obidoxime 80 mg/kg i.p. 20 minutes before ^{14}C-sarin i.v. 1.5 mg/kg, 30 secondes after ^{14}C-sarin. Diminished radioactivity in brain and spinal cord, but increased in liver, intestines, kidneys and bladder.

^{14}C-sarin radioactivity decreases in the cardiac blood, but increases in the liver and the kidney pelvis, and is diminished in the adrenals. Concentration in the kidney cortex and the lungs does not change. ^{14}C-obidoxime and ^{14}C-PAM, being injected following the same experimental protocol, 20 minutes before or 5 minutes after sarin, show no great change of distribution of radioactivity. The amount of radioactivity in blood and in the organs, especially the liver, increases and elimination with the urine is delayed. The protective injection of obidoxime 20 minutes before sarin has a pronounced effect compared to the therapeutic application 1-30 minutes after sarin.

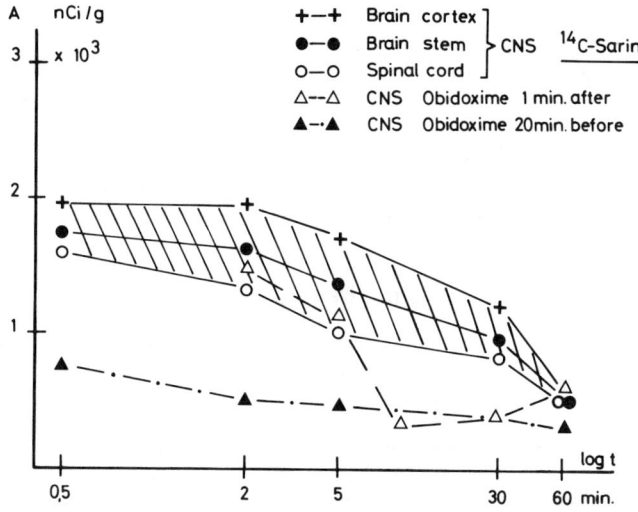

Figure 7. CNS-radioactivity of ^{14}C-sarin and metabolites alone (hatched area) and with different interactions of obidoxime.

How do oximes interact with sarin?

As mentioned in other publication and at former meetings (Waser et al. 1987), we believe the reactivating effect of oximes on the phosphorylated acetylcholinesterase not to be as effective in a living organism, as with pure and fresh enzymes isolated from the electric organs of Torpedo marmorata (Hopff et al. 1984, Figure 8). This process can be performed in vitro only with a very high concentration. Only a toxic dose would produce a comparable concentration in the plasma and the extracellular fluid of mice, rats or man. With an intravenous dose of 250 mg obidoxime an elderly man of 70 kg (50 % body water) will have a concentration of about 20 uMol in his blood plasma and extracellular water. With smaller concentrations (or intravenous doses) the reactivating effect is negligible and below 2 % within 10 minutes.

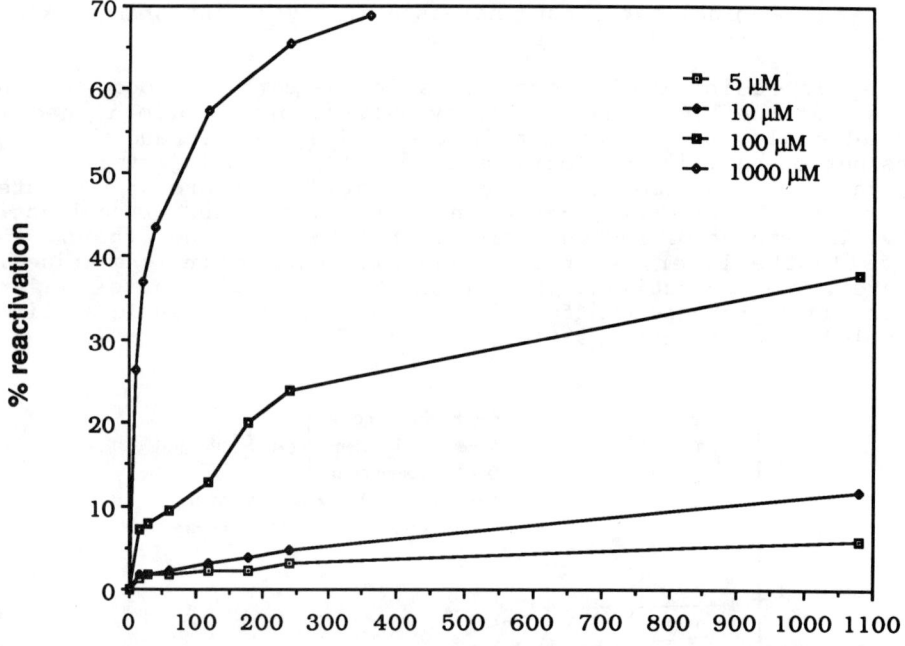

Figure 8. Obidoxime reactivation of AChE inhibited by sarin in stochiometric proportion. AChE from electric organs of Torpedo marmorata, purified by affinity chromatography to cristallizability, was blocked by sarin in small concentration to equimolar porportion, and run through a chromatography column to eliminate the surplus of non-enzyme-bound sarin. The enzyme activity was calculated in % of the non-inhibited starting value.

On frog neuromuscular junctions obidoxime, when applied iontophoretically, has a weak depolarizing effect and furthermore a strong potentiating effect on the acetylcholine induced depolarization (Caratsch, Waser 1984). When sarin (10^{-6}M) is added to the preparation, the peak of the ACh-induced depolarization is potentiated, but after giving a preceding pulse of obidoxime, it is reduced (Figure 9). Obidoxime therefore has a double action as partial agonist on the cholinergic receptor and as an antagonist of the AChE. This action is effective for all nicotinic and muscarinic cholinergic synapses (Amitai et al. 1980; Kloog et al. 1986). Part of the therapeutic effect in sarin poisoning may be due to this curare and atropine-like inhibition and protection of the peripheral cholinergic receptors.

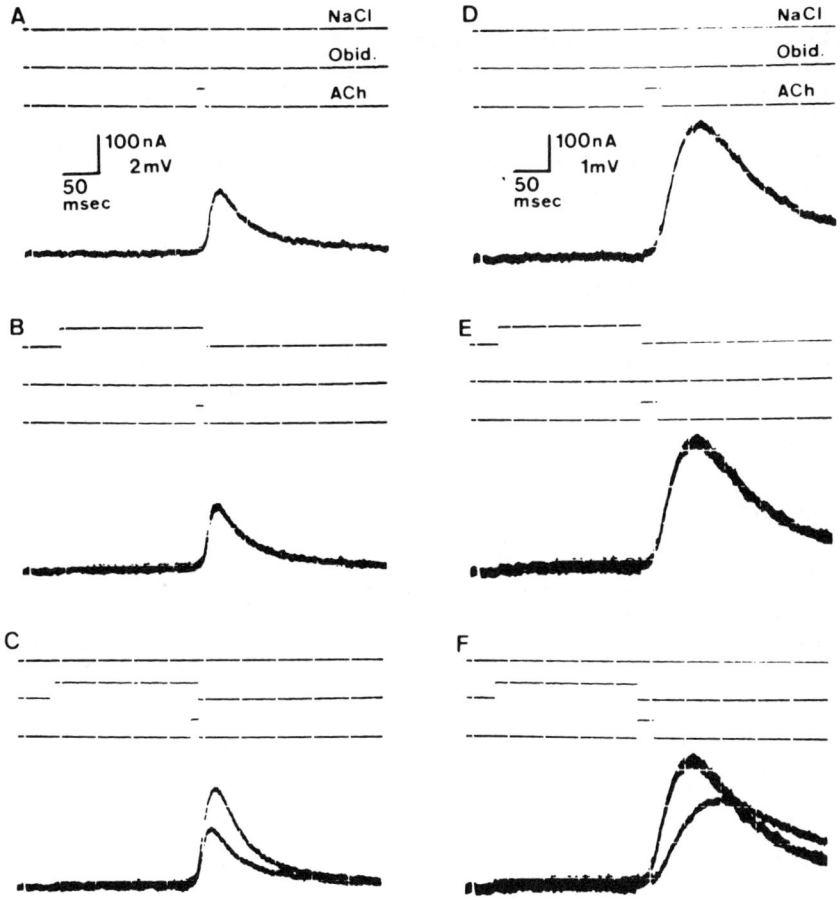

Figure 9. Effect of ACh, NaCl and obidoxime applied iontophoretically to a single endplate. A to C before adding sarin to the bath. A: ACh pulse alone (see top traces), B: preceding NaCl-pulse and ACh pulse with same response, C: preceding obidoxime pulse with potentiation of ACh-response; D to F with sarin (10^{-6}M), D and E similar to A and B but potentiated by blocked AChE, F: with preceding obidoxime reduction of AChE response by partial antagonism on receptor.

Another possibility of interaction concerns the cardiovascular system and the distribution of the blood mass in the body. Sarin (LD50) produces an immediate rise in blood pressure by central stimulation of muscarinic receptors followed by a long depression after 30 minutes which is typical for a state of shock. Obidoxime reduces shortly the blood pressure, probably by blocking ganglionic synapses. Injected before sarin it cannot prevent the rise in blood pressure, but given afterwards it shows a rather short anti-shock activity. These effects are naturally dose dependent and concern only the autonomous regulation of the cardiovascular system, but do not explain the diminished sarin radioactivity in the brain when obidoxime is injected i.p. before. Obidoxime might influence the blood-brain barrier, and less sarin or lipid soluble compounds would enter the brain.

Finally another explanation, easier to accept, concerns the free sarin concentration in the blood plasma. If it is diminished, a smaller amount will reach the brain. A direct reaction between sarin and obidoxime might form a molecular complex with a less lipophilic character than sarin, a higher molecular weight and more difficulty to pass through the barrier. Then only the remaining free sarin molecules would enter the brain.

We have good indications with ultraviolett spectral analysis and TL-chromatography of the existence of such a labile complex, which then has a quite different distribution pattern and elimination kinetics in the animal.

We are now identifying this complex with other physicochemical methods. This detoxification mechanism can probably act only on free sarin in solution but will be of little influence, when sarin is already bound to the cholinesterases in the tissue.

Treatment of Alzheimer's disease with sarin

The use of sarin as therapeutic in man will be a great risk. But this will not be bigger than with other toxic compounds and drugs, as botulinum toxin in treatment of strabism, N-lost (mustargen) in leukemia, high doses of strychnin in non-ketotic hyperglycinemia, curare drugs as muscle relaxants etc. The risk depends on the dose and on the preventive care against intoxication, which must be available. We have experienced a few intoxications in the laboratory probably by inhalation of traces of sarin, from a HPLC and GC apparatus or direct contact, which caused some typical symptoms as asthmatic respiration, dizziness, red face for a short time and miosis for 10-14 days, but no serious side effects even without treatment with atropine. Sarin itself is not the direct toxic principle as death is the result of the accumulation and persistence of endogenous acetylcholine.

Sarin penetrating so easily into the brain would be of great value, although we do not see a selective localization but a general distribution in the brain. DFP enters less easily. Its LD50 is 40 times the LD50 of sarin and therefore much less specific. Soman might be even more suitable as it forms rapidly an irreversible attachement by the "ageing process" of the AChE, which prolongs the action on the enzyme. We are now labeling soman with ^{14}C to investigate its kinetics. Some other organophosphate might have more favourable kinetics than these nerve gases.

The protection of the peripheral cholinergic synapses against sarin or other organophosphates might be achieved, as already mentioned, by pretreatment with obidoxime, either as repeated injection or as infusion. Pralidoxime is perhaps less indicated, being more lipophilic and having a shorter elimination time. The central muscarinic receptors are not influenced by obidoxime, as only traces pass the blood-brain barrier. The cerebrospinal fluid is nearly without radioactivity. In homogenates of whole brains of mice injected intravenously with obidoxime, the acetylcholine concentration is slightly but not significantly lower than in control-animals, whereas atropine produces a decrease of 20 %, probably by inducing an increased turnover after blocking the muscarinic receptors (Vannotti 1979, Table 1).

atropine mg/kg i.v.	obidoxime mg/kg i.v.	acetylcholine $\mu g/g$	n
-	-	2.28 ± 0.22	6
8	-	1.80 ± 0.15	6
-	48	1.96 ± 0.09	6
8	48	1.68 ± 0.04	6

Table 1. The effect of i.v. injection of atropine and obidoxime in mice on the ACh concentration in the brain. Obidoxime alone has little effect, whereas atropine diminishes the ACh concentration to a greater extend. Pyrolysis GC-method.

Another interesting protector might be edrophonium. We have investigated this old anticholinesterase of the carbaminoyl-type in small animals and with electrophysiological technique on the endplate (Caratsch 1981). Confirming earlier reports (Katz, Thesleff 1957), we find beside the enzyme block a direct inhibition of the cholinergic nicotinic receptors. When the AChE is completely blocked, edrophonium inhibits the depolarization of endplates by acetylholine down to 70 % (10^{-5} M sarin) or 40 % (10^{-4} M sarin).

Atropine or scopolamine as very active muscarinic antagonists in the brain, are contraindicated,except in case of sarin intoxication. Quaternary amines as atropinemethonitrate, scopolamine-butylbromide (Buscopan[R]) propantheline bromide (Pro-Banthin[R]) and peripheral ganglionic blockers (mecamylamine, pempidine, trimetaphan) or curare-like compounds might be too selective. Diazepam and other benzodiazepines would be ideal for seizures, cramps etc. but act in the CNS and cannot prevent the direct influence of acetylcholine on the postsynaptical cholinergic receptors. Dantrolene inhibiting muscle contraction by blocking the liberation of sarcoplasmatic calcium might be of use. Other old or new drugs have to be screened for this task in practice.

Conclusions

We propose to use small doses of sarin or other suited organophosphates in the treatment of M. Alzheimer, because of their central action of long duration on the AChE. As protective or prophylactic agents against peripheral sarin induced cholinergic reactions, obidoxime and edrophonium are suggested. This provocative proposal for activating central cholinergic synapses, without overstimulating the damaged presynaptic neurons, might improve their function without accelerating their destruction by over-activation.

References

Amitai, G., Y. Kloog, D. Balderman and M. Sokolovsky. 1980. The interaction of bis-pyridinium oximes with mouse brain muscarinic receptor. Biochem. Pharmacol. 29: 483-488.

Caratsch, C.G. and P.G. Waser. 1984. Effects of obidoxime chloride on native and sarin-poisoned frog neuromuscular junctions. Pflugers Arch. 401: 84-90.

Caratsch, C.G. 1981. Not published investigation.

Chang Sin-Ren, A., G. Riggio, W.H. Hopff and P.G. Waser. 1988. Synthesis of specific labelled ^{14}C-methyl-sarin. J. Labeled Compounds and Radiopharmaceuticals, In Press.

Gotheil, A.M. 1978. Ganztierautoradiographische Untersuchungen uber die Verteilung von Pyridin-2-Aldoxim-^{14}C-Methiodid in Mausen, allein und unter Einwirkung von Mehtylisopropylfluorophosphonat. Doctoral thesis ETH Zurich, No. 6220.

Hopff, W.H., G. Riggio and P.G. Waser. 1984. Sarin poisoning in guinea pigs compared to reactivation of acetylcholinesterase in vitro as a basis for therapy. Acta Pharmacol. et Toxicol. 55: 1-5.

Katz, B. and S. Thesleff. 1957. The interaction between edrophonium (Tensilan^R) and ACh at the Motor Endplate. Brit. J. Pharmacol. 12: 260-264.

Kloog, Y., R. Galron and M. Sokolovsky. 1986. Bisquaternary pyridonium oximes as presynaptic against and postsynaptic antagonists of muscarinic receptors. J. Neurochem. 46(3): 767-772.

Sammet, R. 1983. Kinetic on ^{14}C-Sarin und deren Beeinflussung durch Obidoxim - eine Ganztierautoradiographische Untersuchung an der Maus. Doctoral Thesis ETH Zurich, No. 7288.

Vanotti, R.L. 1979. Determination de l'acetylcholine et de la choline a l'aide de chromatographie en phase gazeuse. Applications dans les cerveaux de souris, de rats et de fourmis. Doctoral Thesis ETH Zurich, No. 6408.

Waser, P.G., C.G. Caratsch, A. Chang Sin-Ren, W.H. Hopff, A. Goteil, E. Kaiser, R. Sammet and G. Streichenberg. 1987. Organophosphate poisoning and its treatment by oximes. In Neurobiology of Acetylcholine, ed. N.J. Dun and R.L. Perlman, 459-471. New York: Plenum Press.

Waser, P.G., R. Sammet, E. Schonenberger and A. Chang Sin-Ren. 1986. Pharmacokinetcs of ^{14}C-sarin and its changes by obidoxime and pralidoxime. In Dynamics of Cholinergic Function, ed. I. Hanin, 743-755. New York: Plenum Press.

ACCUMULATION AND TURNOVER OF ACETYLCHOLINE AFTER ADMINISTRATION OF ACETYLCHOLINESTERASE INHIBITORS IN RAT BRAIN

Albert Enz
Preclinical Research, SANDOZ Ltd., CH-4002 Basle, Switzerland

INTRODUCTION

Alzheimer's disease (AD) is a degenerative disease of the CNS, with severe mental deterioration. Currently, its aetiology is unknown and no rational, effective therapy exists. However, the recent finding that the postmortem number of cholinergic neurons is greatly reduced in AD patients relative to age-matched controls raises the possibility that at least some of the impairments observed in AD might result from the loss of cholinergic activity.

The likely importance of a cholinergic deficit in AD is also indicated by several psychopharmacological observations. Anticholinergic drugs, for example, are known to impair memory in persons not suffering from AD (Drachman, 1977; Mohs et al, 1981), causing symptoms which are similar to those found in early stages of AD (Drachman and Leavitt, 1974). Therefore, drugs known to enhance cholinergic activity have been proposed for the treatment of AD. One possible approach is the inhibition acetylcholinesterase (AChE), the enzyme that inactivates acetylcholine (ACh) and terminates cholinergic activity, both at muscarinic and nicotinic receptors.

Physostigmine (PHYSO), a well known AChE inhibitor (AChE-I), in fact has been shown to lead to some memory improvement in AD patients (Davis et al,1982). The extent of improvement closely correlated with the inhibition of AChE measured in CSF, suggesting that PHYSO penetrates into the CNS in therapeutically significant amounts (Thal et al. 1983). However, the short duration of action (minutes), the low, unpredictable bioavailability after oral administration (Peters and Levin, 1979) and also the pronounced peripheral side effects observed make PHYSO a drug far from ideal for the treatment of AD.

Tacrine (tetrahydroaminoacridine, THA), another AChE-I, also seems to have some beneficial effects in AD patients when given for a relatively short (Summers et al, 1981) or long (average 12 months) period of time (Summers et al, 1986).

RA7 (N-ethyl-3-[(1-dimethylamino)ethyl]-N-methyl-phenylcarbamate) is a novel AChE-I with interesting properties. It reaches the CNS rapidly after s.c., i.p. or p.o. administration, causing less peripheral side effects than, for example, PHYSO (Weinstock et al). In addition, RA7 has a longer duration of action, and is chemically more stable than PHYSO.

In the present experiments we have characterized the properties of RA7 in more detail and compared them to PHYSO and THA. We have measured the inhibition of AChE in vitro and ex vivo (drug application in vivo) and determined the ACh turnover in various regions of the rat brain following administration of the three AChE-I.

Materials and methods.

Chemicals and drugs.
Tacrine (THA, 9-amino-1,2,3,4-tetrahydroacridine HCl) was obtained from SERVA (Heidelberg, FRG) and physostigmine hemisulfate salt (PHYSO) from SIGMA (St.Louis, USA). RA7 (N-Ethyl-3-[(1-dimethyl-amino)ethyl]-N-methyl-phenylcarbamate HCl) was synthesised by R. Amstutz (Preclinical Research SANDOZ). Deuterated compounds used as internal standards were purchased from MSDIsotopes, Montreal, Canada). All other chemicals were of analytical grade.

Animals.
Male OFA rats (200-250 g, KFM, Fuellinsdorf, Switzerland) were either decapitated (if only AChE activities were measured) or sacrificed by microwave irradiation (if ACh levels were measured in addition) (2450 MHz, 6 kW, exposure 1.7 s, focused on the head, Pueschner Mikrowellen-Energietechnik, Bremen, FRG) at different time points after after drug application. The brains were immediately dissected on ice according to Glowinski and Iversen (1966), frozen on dry ice and stored at -70 °C until analysed.

Acetylcholine and choline levels.
The brain tissue was homogenized in 0.1 M perchloric acid containing deuterated internal standards of ACh-d_9 and Ch-d_9. After centrifugation, ACh and Ch were extracted in dichlormethane with dipicrylamine (2,2',4,4',6,6'-hexanitrodiphenylamine) as ion pair. Ch was esterified with propionyl chloride and the resulting mixture of propionylcholine and ACh was analysed, following demethylation with sodium benzenetoluate, by GC-massfragmentography according to Jenden et al. (1973).

ACh turnover.
ACh turnover was determined in various brain regions as described by Jenden et al. (1974). Briefly, the rate of 2H-incorporation into brain ACh pools (1-15 min) following a pulse injection of 2H-choline (20 µmol/kg i.v.) was measured. The turnover rate was calculated using the finite difference method described by Neff et al. (1971).

AChE activity.
AChE activity was measured according to the method described by Ellman (1959). Briefly, the frozen tissue was homogenized in ice cold 0.25 mM phosphate buffer, pH 7.3 containing 0.1 % Triton X-100. After centrifugation the enzyme activity was measured photometrically in the clear supernantant with acetylthiocholine-iodide (0.5 mM) as substrate. When AChI's were added in vitro, the mixture was preincubated at 37°C for 15 minutes before addition of substrate.

RESULTS

Different IC_{50} values were obtained with the three AChE-I's tested when added to the enzyme *in vitro*. As shown in Table 1, PHYSO was about 100 times more potent than RA7, which in turn was 3-4 times more potent than THA. No regional differences were found with either inhibitor when added *in vitro*.

Since both, PHYS and RA7 are, in contrast to THA, pseudoirreversible, active-site directed AChE-I's, it is possible to measure the enzyme inhibition *ex vivo*, i.e. post-mortem in vitro following *in vivo* drug application. The ED_{50} values, obtained from such measurements, are shown in Table 1. The difference in potency between PHYS and RA7 is now reduced by a factor of 5 to 10. In addition, while PHYS showed no regional selctivity, RA7 inhibited AChE preferentially in the cortex and hippocampus.

Table 1. AChE Inhibition in Discrete Rat Brain Regions

ENZYME SOURCE	PHYSOSTIGMINE		RA7		TACRINE
	in vitro IC_{50}	ex vivo ED_{50}	in vitro IC_{50}	ex vivo ED_{50}	in vitro IC_{50}
CORTEX	0.033	0.80	3.2	6.6	13
HIPPOCAMPUS	0.029	0.96	3.9	13.4	15
STRIATUM	0.033	1.02	3.2	17.6	10
PONS/MEDULLA	0.031	0.98	3.4	20.1	12

The IC_{50} values [µM] and ED_{50} values [µmol/kg] were calculated from corresponding enzyme activities, measured at 5 different concentrations resp. doses of inhibitor, as described under Materials and Methods. Physostigmine was injected s.c., RA7 administered p.o.. MEANS from 4 independent experiments, SEM 5-10%.

The ACh levels in all brain regions, with the exception of pons/medulla (data not shown), were increased dose-dependently by the three AChE-I's tested. As shown in Figure 1, PHYS (s.c. injection) was again the most active compound, followed by RA7 and THA (p.o. administration). Interestingly, a regional selectivity was again found for RA7, in that it preferentially increased the ACh levels in the cortex.

A significant increase in brain ACh levels elicited by oral administration of PHYSO (0.3 to 3 µmol/kg) peaked 15 minutes after drug administration and rapidly declined thereafter (data not shown). In contrast, RA7, at an oral dose of 3 to 25 µmol/kg, significantly increased brain ACh levels already at 15 min, peaking at 30 min. The duration of action of PHYSO after s.c. injection and of RA7 and THA after p.o. application, at doses producing about maximal effects, are shown in detail in Figure 2. It is evident that, while the duration of action is about the same for RA7 and THA, it is much shorter for PHYSO.

Effects of Acetylcholinesterase Inhibitors on Acetylcholine Levels in different Rat Brain Regions

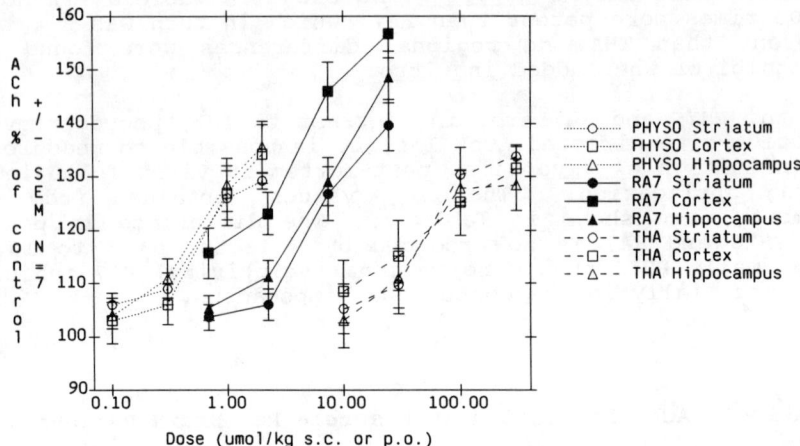

Figure 1. Physostigmine (PHYSO) was administered s.c., RA7 and Tacrine (THA) p.o., 60 minutes before sacrification by microwave irradiation. MEANS +/- SEM in % of control levels, number of animals = 7. Control values in pmole/mg fresh tissue ± SD : Striatum 74.8 ± 5.4; Cortex 13.3 ± 1.55; Hippocampus 19.5 ± 1.05.

Acetylcholine Levels in Rat Cortex at various Times after Administration of Acetylcholinestaerase Inhibitors

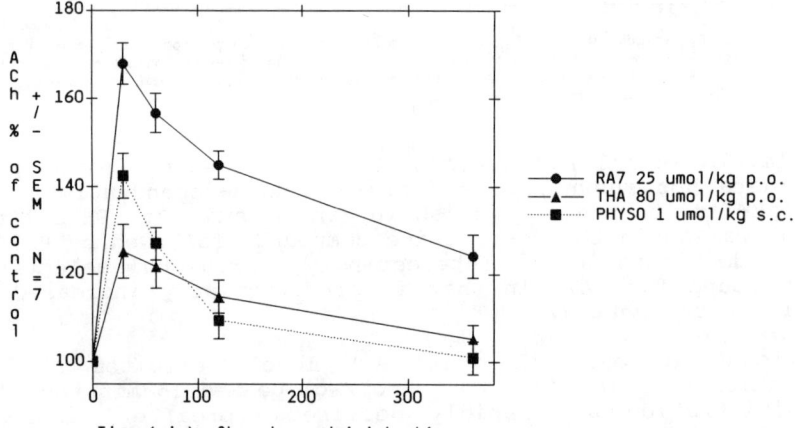

Figure 2. Physostigmine (PHYSO) was administered s.c., RA7 and Tacrine (THA) p.o., 60 minutes before sacrification by microwave irradiation. MEANS +/- SEM in % of control levels, number of animals = 7. Control value, see legend to Figure 1.

Table 2. Influence of Physostigmine and RA7 on ACh Turnover in Different Rat Brain Regions

Brain area/ Treatment	[ACh] (pmol x mg^{-1})	kACh (min^{-1})	Turnover (pmol x mg^{-1} x min^{-1})	Change % of control
STRIATUM				
Control	62.7	0.274	17.2 ± 3.1	100
PHYSO				
0.1 µmol/kg s.c.	71.3	0.193	13.8 ± 2.8*	80
0.3 µmol/kg s.c.	80.9	0.153	12.4 ± 2.4*	72
1.0 µmol/kg s.c.	97.9	0.088	8.6 ± 3.5	50
CORTEX				
Control	12.1	0.199	2.41 ± 0.47	100
PHYSO				
0.1 µmol/kg s.c.	13.5	0.143	1.93 ± 0.51*	80
0.3 µmol/kg s.c.	16.3	0.096	1.57 ± 0.43*	65
1.0 µmol/kg s.c.	18.6	0.058	1.08 ± 0.51	45
HIPPOCAMPUS				
Control	19.3	0.201	3.88 ± 0.70	100
PHYSO				
0.1 µmol/kg s.c.	21.0	0.153	3.21 ± 0.59*	83
0.3 µmol/kg s.c.	23.9	0.100	2.39 ± 0.61*	62
1.0 µmol/kg s.c.	27.7	0.062	1.71 ± 0.53	44
STRIATUM				
Control	70.4	0.232	16.4 ± 2.8	100
RA7				
3 µmol/kg p.o.	81.4	0.162	13.2 ± 2.4*	80
8 µmol/kg p.o.	94.8	0.119	11.3 ± 2.8*	69
25 µmol/kg p.o.	111.3	0.091	10.1 ± 2.3	62
CORTEX				
Control	13.8	0.207	2.85 ± 0.49	100
RA7				
3 µmol/kg p.o.	16.3	0.099	1.62 ± 0.42*	57
8 µmol/kg p.o.	20.2	0.063	1.23 ± 0.32*	43
25 µmol/kg p.o.	21.6	0.044	0.94 ± 0.30	33
HIPPOCAMPUS				
Control	20.2	0.211	4.28 ± 0.78	100
RA7				
3 µmol/kg p.o.	21.4	0.150	3.21 ± 0.63*	75
8 µmol/kg p.o.	25.8	0.111	2.87 ± 0.54*	67
25 µmol/kg p.o.	28.5	0.075	2.14 ± 0.49	50

Experiments were carried out as described in Material and Methods. Physostigmine (PHYSO) was injected s.c., RA7 administered orally 30 min before i.v. pulse injection of 2H-choline. The animals were sacrificed 30 min after drug injection and 0.5 to 15 minutes after 2H-choline chase by micowave irradiation.
The fractional rate constant for ACh efflux (kACh), was calculated according to Neff et al. (1971).The ACh turnover rate was obtained by multiplying the kACh value by the ACh content.Turnover values :MEANS ± SD, N=7 animals each of the 5-10 time points. * 2p< 0.05, t-test.

Since measurement of ACh levels does not reflect subtle changes in ACh utilization, we determined the influence of AChE-I's on ACh turnover in different rat brain regions. As shown in Table 2, PHYSO, at s.c. doses ranging from 0.1 to 1 µmol/kg, dose-dependently reduced the ACh turnover in all brain regions to a similar extent, with the exception of the pons/medulla area, where no effect was seen. The ED_{50} values for reduction of ACh turnover were calculated to be 1.1, 0.7 and 0.9 µmol/kg s.c. for striatum, cortex and hippocampus respectively. The influence of RA7 after oral application on ACh turnover is shown in Table 3. In the dose range tested (3 to 25 µmol/kg p.o.), RA7, similar to PHYSO, reduced dose-dependently the ACh turnover in all brain regions, with the exception of the pons/medulla area. However, in contrast to PHYSO, the anticipated regional selectivity was again observed with RA7. The corresponding ED_{50} values were calculated to be 5.1, 27 and 71 µmol/kg p.o. for cortex, hippocampus and striatum respectively. The reduction in ACh turnover induced by RA7, in contrast to PHYSO, correlated closely with the extent of AChE inhibition observed in all brain regions, with exception of pons/medulla. Thus, the <u>in vivo</u> potency of RA7 for reducing ACh turnover in rat cortex differs from that of PHYSO by no more than a factor of 7, whereas in other areas (e.g. peripheral organs) the potency (i.e. toxicity) of PHYSO is 40 to 100 times that of RA7.

THA, the weakest AChE inhibitor investigated, reduced the ACh turnover as expected, without regional selectivity, by only about 20%, at a p.o. dose of 100 µmol/kg (see Table 3). No attempts were made to further decrease ACh turnover with THA, since the dose used seems, in comparison to PHYSO and RA7, already quite high.

Table 3. Influence of THA on ACh Turnover in Different Rat Brain Regions

Brain area/ Treatment	[ACh] (pmol × mg^{-1})	kACh (min^{-1})	Turnover (pmol × mg^{-1} × min^{-1})	Change % of control
STRIATUM				
Control	68.5	0.247	16.9 ± 2.1	100
THA 100 µmol/kg p.o.	87.9	0.153	13.4 ± 2.5*	79
CORTEX				
Control	13.5	0.185	2.50 ± 0.31	100
THA 100 µmol/kg p.o.	16.9	0.117	1.97 ± 0.28*	79
HIPPOCAMPUS				
Control	18.5	0.214	3.97 ± 0.41	100
THA 100 µmol/kg p.o.	24.0	0.129	3.10 ± 0.28*	78

Experiments were carried out as described in Material and Methods. Tacrine (THA) was injected p.o. 30 min before i.v. pulse injection of 2H-choline. The animals were sacrificed 30 min after drug injection and 0.5 to 15 minutes after 2H-choline chase by micowave irradiation. The fractional rate constant for ACh efflux (kACh), was calculated according to Neff et al. (1971). The ACh turnover rate was obtained by multiplying the kACh value by the ACh content. Turnover values :MEANS ± SD, N=7 animals each of the 5-10 time points.
* $2p< 0.05$, t-test.

DISCUSSION

RA7, a miotine derivative, is a novel AChE-I which is about 100 times less potent than PHYSO, a classical carbamate derivative, when the inhibitors are added <u>in vitro</u> to enzyme preparations from various rat brain regions.

The pseudoirreversible mechanism common to RA7 and PHYSO enabled us to measure <u>ex vivo</u> the inhibitory effects of these drugs after p.o. or s.c. applications. PHYSO, following subcutaneous administration, inhibited the enzyme in all brain regions with the same potency. However, RA7, in contrast to PHYSO, displays a regional selectivity by preferentially inhibiting <u>ex vivo</u> AChE extracted from cortex and hippocampus. The rank order of inhibition was cortex > hippocampus > striatum = pons/medulla. A possible mechanism for this brain selectivity of RA7 might be the recent findings by Weinstock et al (1987) who reported distinct differences between PHYSO and RA7 regarding enzyme kinetics <u>in vitro</u>: both drugs act according to mechanisms of active-site-activated inhibition, which is typical for carbamates (Main 1974). However, while the enzyme with PHYSO was carbamylated in all brain regions at an equal rate, RA7 carbamylated the enzyme more rapidly in those brain regions (cortex and hippocampus), in which the enzyme was also more potently inhibited. Since THA, another clinically investigated AChE-I, follows simple mixed competitive inhibition kinetics, its IC_{50} values cannot be directly compared with those of PHYSO and RA7 and no effects <u>ex vivo</u> can be measured.

This regional selectivity offered by RA7 was also reflected by the extent to which ACh levels were increased in the various brain regions: the effect of RA7 was more pronounced in cortex than in other brain areas, while PHYSO showed no regional preference.

Measurements <u>in vivo</u> of the incorporation of deuterated choline into the ACh pools resulted in a dose-dependent reduction of de-novo ACh synthesis by PHYSO and RA7. With RA7, this reduction of ACh turnover again was more pronounced in the cortex and hippocampus than in the striatum. The ED_{50} values for the reduction of ACh turnover obtained with RA7 closely correlated with the ED_{50} values for AChE inhibition. No such correlation was found for PHYSO.

It is interesting to note that all AChE-Is tested showed no effects on ACh levels and turnover in the pons/medulla region, in spite of the fact that they clearly inhibited AChE activity measured <u>ex vivo</u>. Although the exact reason for this phenomenon is currently unknown, it might reflect a regional difference in the way ACh turnover is regulated in this brain area (Enz, 1987).

Several properties of RA7 suggest that this AChE-I might elicit fewer side effects than e.g. PHYSO when used for the treatment of AD. The relatively weak inhibition of AChE observed in the pons/medulla region as compared to cortical areas could be essential, since cholinergic stimulation in this region can cause respiratory arrest (Machne and Unna, 1963). RA7 is about 100 times less potent that PHYSO in inhibiting AChE <u>in vitro</u>. In <u>in vivo</u> experiments, this factor is becomes less than 10, depending on the anatomical location. In cortex e.g., RA7 is only 7 times less potent than PHYSO. These findings strongly suggests, that RA7 will probably cause less side-effects, since it is more potent in brain areas which are presumably most important for the treatment of AD.

Studies with PHYSO in AD patients have led to the assumption that 20 to 40 % inhibition of AChE in the CNS is required for optimal therapeutic effects (Levy et al., 1986). When brain AChE is inhibited by 40% with RA7 in the rat, almost no peripheral side effects, such as salivation or diarrhoea, are seen.

The results with RA7 indicate that the disadvantages of the clinically used AChE inhibitors may be overcome by improving CNS selectivity, decreasing peripheral side effects and toxicity, as well as by prolonging their duration of action after administration of a single dose. Drugs like RA7, acting preferentially in brain areas known to be more affected in Alzheimer's disease, may be useful in the treatment of senile dementia and of senile mental decline.

LITERATURE

Davis,K.L. and Mohs,R.C. 1982. Enhancement of memory processes in Alzheimer's disease with multiple-dose intravenous physostigmine. Am.J.Psychiatry 139: 1421-1424.

Drachman, D.A. 1977. Memory and cognitive function in man: Does the cholinergic system have a specific role? Neurology 27: 783-790.

Drachman, D.A. and Leavitt, J. 1974. Human memory and the cholinergic system. A relationship to aging? Archs.Neurol. 30: 113-121.

Ellman,G.L., Courtney, K.D., Andres, V. and Featherstone, R.M. 1961. A new and rapid colorimetric determination of acetylcholinesterase activity. Biochem. Pharmacol. 7: 88-95.

Enz,A. 1987. Influence of acetylcholinesterase inhibitors (AChE-I) on ACh levels and turnover in rat brain. Experientia 43: 702.

Glowinski, J. and Iversen,L.L. 1966. Regional studies of catecholamines in the rat brain. I. The disposition of ^{3}H-noradrenaline, ^{3}H-dopamine and ^{3}H-DOPA in vatious regions of the brain. J.Neurochem. 13: 655-665

Jenden, D.J., Roch, M. and Booth, R.A. 1973. Simultaneous measurement of endogenous and deuterium-labelled tracer variants of choline and actylcholine in sub-picomole quantities by gaschromatography/mass spectrometry. Anal.Biochem.22: 438-448.

Jenden, D.J., Choi, L., Silverman, R.W., Steinborn, J.A., Roch, M. and Booth, R.A. 1974. Acetylcholine turnover estimation in brain by gaschromatography/mass spectrometry. Life.Sci.14: 55-63.

Levy,D.,Glickfeld,P.,Grunfeld,Y.,Grunwald,J.,Kushnir,M.,Levy,A., Meshulam,Y., Spiegelstein,M., Zehavi,D. and Fisher,A. 1986.A novel transdermal therapeutic system as a potential treatment for Alzheimer's disease. In <u>Advances in behavioral biology</u>, Vol. 29, ed. A.Fisher, I.Hanin and C.Lachman, 557-563. New York: Plenum Press.

Main, A.R. 1973. Kinetics of Active-Site-Directed irreversible inhibition.In <u>Essays in Toxicology</u>, Vol.4, ed. Wayland and Hagen, 59-105. New York: Academic Press. , Academic Press.

Machne,X. and Unna,K.W.R. 1963. Actions at the central nervous system. In: <u>Cholinesterases and anticholinesterase agents</u>, ed.G.B. Koelle, Berlin: Springer Verlag.

Neff,N.H., Spano,P.F., Groppetti,A., Wang,C.T. and Costa,E. 1971. A simple procedure for calculating the synthesis rate of norepinepherine, dopamine and serotonin in rat brain. <u>Pharmacol.Exp.Ther. 176:</u>701-710.

Peters, B.H. and Levin, H.S. 1979. Effects of physostigmine and lecithin on memory in Alzheimer's disease. <u>Ann.Neurol. 6:</u> 219-221.

Thal,L.J., Fuld,P.H.,Masur, M.S. and Sharpless,N.S. 1983. Oral physostigmine and lecithin improve memory in Alzheimer's disease. <u>Ann.Neurol. 13:</u> 491-496.

Summers,W.K., Viesselman,J.O., Marsh,G.M. and Candelora,K. 1981. Use of THA in treatment of Alzheimer-like dementia: Pilot study in twelve patients. <u>Biol.Psychiatry 16:</u> 145-53.

Summers,W.K., Majovski,L.V., Mars,G.M., Tachiki,K. and Kling,A. 1986. Oral tetrahydroaminoacridine in long-term treatment of senile dementia, Alzheimer type. <u>N.Engl.J.of Med. 315:</u> 1241-45.

Weinstock,M., Razin,M., Chorev,M., Tashma,Z. 1986. Pharmacological activity of novel anticholinesterase agents of potential use in the treatment of Alzheimer'disease. In <u>Advances in Behavioral Biology</u>, Vol.29, ed. A.Fisher, I.Hanin and C.Lachman, 539-549. New York: Plenum Press.

Weinstock,M., Kay,G., Razin,M. and Enz,A. 1987. Selective inhibition of acetylcholinesterase and acetylcholine turnover in the rat cortex and hippocampus. In <u>International symposium on muscarinic cholinergic mechanisms</u>, ed. S.Cohen and M.Sokolovsky, 362-366. London: Freund Publishing House.

SENESCENCE-LIKE LEARNING/RETENTION DEFICITS IN AUTOIMMUNE MICE: Reversal by Physostigmine

Michael J. Forster, Konrad C. Retz, and Harbans Lal
*Dept. of Pharmacology, Texas College of Osteopathic Medicine,
3516 Camp Bowie Blvd., Fort Worth, TX*

INTRODUCTION

Although progress continues, a great deal is now known about the neurological/neurochemical defects responsible for cognitive decline related to normal aging and Alzheimer's disease (AD). Application of existing knowledge to development of effective therapeutic interventions has occurred (as attested to by this volume) and will continue as new knowledge is accumulated. The study of animal models has also contributed to identification of potentially successful therapeutic strategies (cf. Bartus & Dean, 1985; Bartus et al., 1986; Giacobini, 1987; Moos, 1988) in addition to improving current understanding of those neurological targets responsible for cognitive impairment in AD. It is recognized that the degree to which successful clinical application can be predicted based upon animal studies depends upon the similarities of the model to the target clinical syndrome, in terms of both the nature of the cognitive impairment and the underlying neurological defect. As a result, recent research has focused upon improved models in which behavioral and neurological defects are engineered to reflect current knowledge (cf. Haroutunian et al., this volume; Wenk et al., 1987).

Ultimately, successful approaches to prevention or arrest of cognitive decline related to AD will require knowledge of those biological processes which are involved in the etiology of relevant neurological defects. In addition, the success of treatments targeted at age-associated cognitive decline or the early primary symptoms of AD could be enhanced considerably by such knowledge. While the etiology of AD remains unclear, if AD-like cognitive impairment could be identified in animals exhibiting physiological abnormalities with a possible etiological role, then these animals could serve both as valuable research tools and as novel models for development and testing of therapeutics. Toward development of such models, studies in this laboratory have investigated the possible relationship between autoimmunity and neurobehavioral change in aging mice.

AUTOIMMUNITY AND COGNITIVE IMPAIRMENT IN AGING AND AD

It has been suggested that immunological disturbances may play an important role in aging-associated cognitive decline and/or AD (Fudenberg et al., 1984; Khansari et al., 1985; Lal & Forster, 1986; 1988; McRae-Degueurce et al., 1987; Nandy, 1985). This speculation is based, in part, on evidence that autoimmune activity associated with aging appears to involve components of the central nervous system. It has been known for some time that in both aged humans and animals, there is an increased incidence of brain-reactive antibodies (BRA) in sera (for review see Baldinger & Blumenthal, 1982; Lal & Forster, 1986; 1988). The possible mechanisms of interactions between BRA and the CNS (and the potential outcome of such interactions) have been discussed with respect to cognitive decline in aging (Lal & Forster, 1986; 1988) and AD (cf. Nandy, 1978; 1985). Numerous investigations have suggested that AD patients can be differentiated from healthy aged patients or patients with other diseases, based upon the presence or level of BRA (Fillit et al., 1985; Gaskin et al., 1986; Nandy, 1978; Singh & Fudenberg, 1986a; see Lal & Forster, 1986; 1988 for reviews). Of particular interest are recent studies suggesting antibodies recognizing structural or enzymatic components of cholinergic neurons present in the sera (Fillit et al., 1985) and cerebrospinal fluid (Dahlstrom et al., 1987; McRae-Degueurce et al., 1987; 1988) of subgroups of AD patients.

COGNITIVE DECLINE AND AUTOIMMUNITY IN MOUSE MODELS

The relationship between BRA and cognitive impairment in human studies has led us (and others, e.g. Fudenberg et al., 1984; McRae-Degueurce et al., 1988; Nandy, 1985) to speculate that immunological factors may be involved in aging-associated cognitive decline, and that age-related immunological aberrations may be accelerated in AD. Animal studies in our laboratories have supported a parallel relationship between brain autoimmunity and cognitive impairment. In our initial studies of C57BL/6 mice we compared age-related declines in learning ability and accumulation of BRA in both normal aged mice and young recipients of immunopoietic cells from old mice. It was found that the formation of BRA (as indicated by an indirect immunofluorescence method) paralleled age-related declines in the ability of the mice to learn a simple one-way avoidance learning task (Nandy et al., 1983). BRA and learning decline were evident in young mice following transfer of immunopoietic cells from aged mice (Lal et al., 1986; Nandy & Bennett, 1983), suggesting that factors present in the aged immune system were sufficient for concurrent acceleration of BRA formation and decline of avoidance learning ability.

Behavioral impairments of autoimmune mice

As a further test of the relationship between autoimmunity and age-related cognitive decline, we compared age-related behavioral changes in non-autoimmune (C57BL/6NNia) and genetically autoimmune-prone mice (Forster et al., 1988). The latter strains exhibit an accelerated development of autoimmunity, including the formation of BRA in sera (Hoffman et al., 1983; 1987). If autoimmunity (in particular, the formation of BRA) was related to development of behavioral deficits, we expected that the autoimmune strains would show more rapid age-related cognitive decline when compared with the non-autoimmune mice.

Active avoidance learning. Figure 1 shows age-related declines in the ability of non-autoimmune (C57BL/6NNia) and autoimmune mouse strains to learn a one-way active avoidance task (Forster, et al., 1988). Overall, the strains showed a heterochronic appearance of avoidance acquisition deficits with age which was predictable based upon the formation of BRA. The results with C57BL/6NNia mice were in accordance with previous investigations suggesting formation of BRA and appearance avoidance acquisition deficits after 6 months of age (Nandy et al., 1983; for review see Lal & Forster, 1986; 1988). However, NZB/BlNJ and BXSB/MpJ strains which develop autoimmunity and BRA soon following puberty (Hoffman et al., 1983; 1987; Lal & Forster, 1988; Nandy et al., 1983), showed accelerated declines in avoidance acquisition ability occurring between one and three months of age. BXSB/MpJ mice are extremely short-lived (few survive

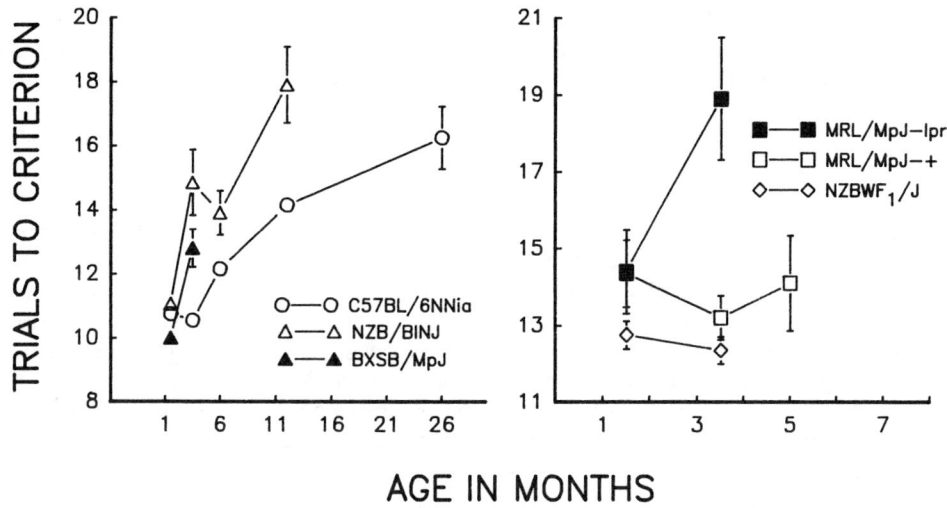

Figure 1. Age-related differences in ability of non-autoimmune C57BL/6NNia and autoimmune mouse strains to acquire a one-way avoidance response (see left and right panels). Twenty male mice of each strain and age received active avoidance training in a single session until an avoidance response had occurred on at least eight of ten consecutive trials. On each trial, the mouse was placed in a test chamber and a tone was presented. If the mouse failed to reach a safe platform within 10 sec of tone initiation, shock was administered through the grid floor of the apparatus until the mouse reached the platform. Latencies of less than 10 sec from tone onset were scored as avoidance responses. Learning ability was considered inversely proportional to the mean number (± SEM) of trials required to reach the acquisition criterion (note that different ordinate scales are used in left and right panels). (Adapted from Forster, et al., 1988).

beyond six months of age) and were not tested beyond three to four months of age. However, avoidance acquisition ability of the somewhat longer-lived NZB/BINJ mice showed further decline between 3 and 12 months of age.

Autoimmune MRL/MpJ-*lpr* mice (see Figure 1, right panel) also showed rapid decline of avoidance learning abilities with age, whereas MRL/MpJ-+ and NZBWF$_1$/J failed to show declines within the same period. These results were also in accordance with the onset of autoimmunity and BRA formation (see Lal & Forster, 1988). The mutant gene *lpr* is responsible for accelerated autoimmunity in the MRL/MpJ-*lpr* strain, whereas mice of the congenic MRL/MpJ-+ strain develop autoimmunity at a slower rate and to a smaller extent (Izui et al., 1984; Murphy & Roths, 1978). The autoimmune disease of NZBWF$_1$ mice is sex-linked (cf. Quimby & Schwartz, 1982), with males (used in this experiment) showing an onset at ages older than those tested.

The results of our studies of avoidance learning suggested a clear parallel between autoimmunity, the formation of BRA, and one behavioral impairment known to occur as a consequence of aging in mice (an avoidance acquisition deficit). While these findings provided support for the hypothesis that immune factors may play some role in age-related cognitive decline, it was not clear that poor performance on the avoidance problem was strictly a consequence of impaired learning or memory ability as opposed to other factors (see Forster et al., 1988; Lal & Forster, 1986; 1988 for discussions). Nevertheless, other studies of NZB/BINJ mice have revealed accelerated age-related learning and memory impairments in other tasks, including step-through passive avoidance (Spencer et al., 1986) and discrimination (Forster et al., in preparation; Schwegler et al., 1988) tasks.

Retention of habituation. To further investigate the nature of age-related learning/memory impairments in autoimmune mice, we recently compared age-related patterns of between-session habituation of locomotor activity in C57BL/6NNia and NZB/BINJ strains (Forster et al., 1987). The between-session habituation of locomotor activity has been employed as a model of memory for simple nonassociative learning, and appears to have validity across numerous parametric and pharmacological manipulations (cf. Hughes, 1982; Platel & Porsolt, 1982). A decline in exploratory activity from one session to the next (between-session habituation) can be considered evidence of retention for the initial exposure, whereas a lack of decline is analogous to forgetting. Of particular interest in our investigations were findings indicating that aged mice exhibit impairments in between-session habituation of locomotor activity and exploratory behavior (Brennan, et al., 1981; 1984; Forster, 1987; Fraley & Springer, 1981a; 1981b). Based upon these and our previous findings with avoidance learning, we expected that autoimmune NZB/BINJ mice would exhibit age-related changes in between-session habituation at earlier ages than non-autoimmune C57BL/6NNia mice.

To evaluate this expectation, separate age groups of C57BL/6NNia and NZB/BINJ mice were tested in an Omnitech Digiscan activity monitoring apparatus, which provided simultaneous measurement of horizontal and vertical components of closed-field locomotor behavior, as well as the time spent in various locations within the field. The mice were administered four, 20-minute locomotor activity sessions under dim illumination. In accordance with a previous investigation (Forster, 1987), the most robust age-related changes were observed in the spatial components of closed-field behavior. Young C57BL/6NNia mice tended to spend considerably less time in the center zone of the apparatus as a function of test sessions, whereas 26-month-olds showed much slower between-session decreases in center time. Because the center of the apparatus is less preferred at all ages, these results were consistent with a deficient between-session habituation of exploratory activity. NZB/BINJ mice also showed a similar age-related change in between-session habituation, albeit at much earlier ages when compared with the C57BL/6NNia mice. These findings suggested that NZB/BINJ mice exhibit accelerated age-related deficits in the retention of habituation (Forster et al., 1987).

Reversal of autoimmune-related retention deficits by physostigmine
In mature mice, the retention of habituation has been shown to be impaired following postsession administration of the anticholinergic, scopolamine, and improved following similar treatment with cholinomimetics (see Platel & Porsolt, 1982). Moreover, impairments in the retention of habituation have been described in a hypocholinergic model of age-related memory dysfunction (Johns et al., 1985). In order to test for the possible involvement of the cholinergic system in the altered habituation of young NZB/BINJ mice, preliminary studies have examined the ability of the cholinesterase inhibitor, physostigmine, to improve habituation retention in this strain. These studies were conducted in

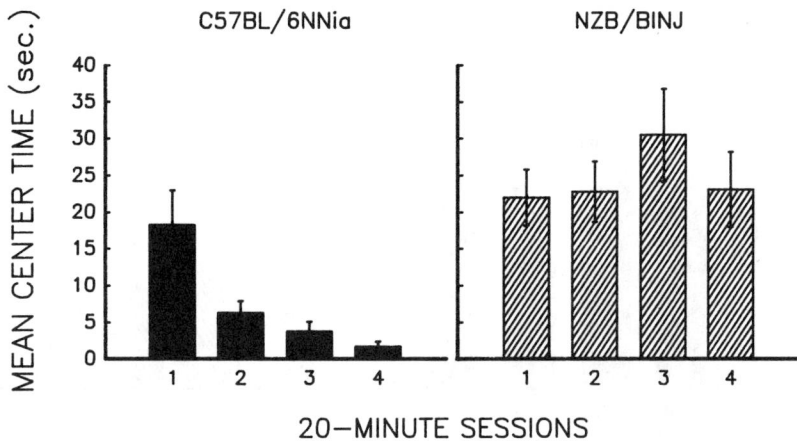

Figure 2. Between-session changes in center zone time (± SEM) of 3-4 month-old C57BL/6NNia (left) and NZB/BINJ (right) mice. Mice of each strain received four, 20-min locomotor activity sessions (spaced at 24-h intervals) in commercial activity monitors (Digiscan system, Omnitech Electronics, Columbus, OH). The apparatus quantified (using photocells) the time spent in each of 9 equal floor zones of the 40 X 40 X 30.5 cm acrylic chamber. Time spent within the center zone showed between-session decline in C57BL/6NNia, but not NZB/BINJ mice. This pattern yielded a significant Strain X Trials interaction [$F(3,81)=4.1$, $p<0.01$] when center time was subjected to a two-way analysis of variance. (Extrapolated from Forster et al., 1987).

NZB/BINJ mice aged 3-4 months, when a robust lack of between-session habituation is observed relative to age-matched C57BL/6NNia mice (see Figure 2).

To avoid confounding by state-dependent effects and the direct effects of physostigmine on locomotor activity, a postsession drug administration paradigm was used (see Platel & Porsolt, 1982) to test the effect of physostigmine upon habituation retention of NZB/BINJ mice. When NZB/BINJ mice received from 0.04 to 0.16 mg/kg physostigmine immediately following an initial 20-min activity session, all drug groups spent less time in the center zone of the apparatus during a second test conducted 24-h later (see Figure 3). As observed previously in untreated mice of this age, mice receiving postsession saline injections showed no between-session reduction in center time, suggesting impaired retention.

Because physostigmine was not present during the initial or retest session, it is likely that modification of between-session habituation by physostigmine involves some process taking place following the initial test session. While that process may involve memory consolidation, such an interpretation must be considered cautiously. Although deficits in between session habituation may be a valid model of aging-related memory deficits (see Fraley & Springer, 1981a; 1981b), these deficits have also been attributed to age-related changes in the arousal-inducing properties of novel stimuli (cf. Brennan et al., 1984). Regardless of the interpretation, the findings do suggest that a cholinergic dys-

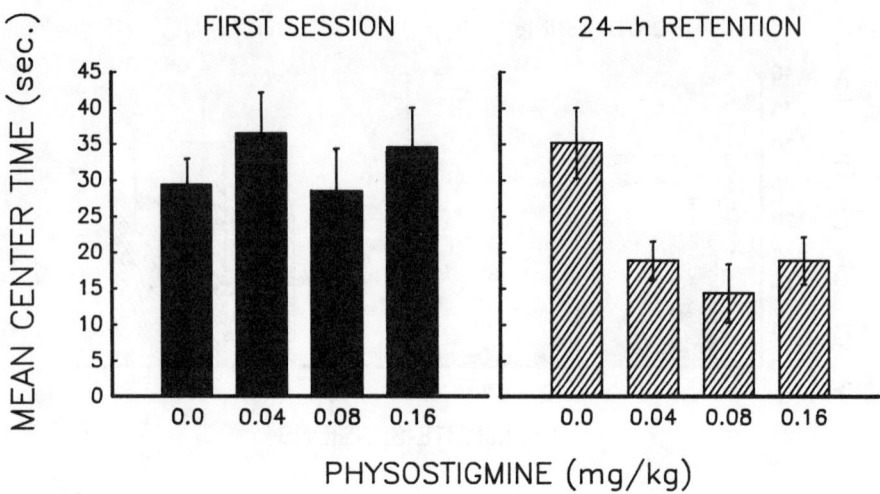

Figure 3. Effects of post-session administration of physostigmine upon habituation retention of 3-4 month old NZB/BINJ mice. Saline, 0.04, 0.08, or 0.16 mg/kg physostigmine (by free drug) was injected i.p. to separate groups of 10-15 NZB/BINJ mice immediately following a 20-min activity session in a Digiscan automated activity-monitoring apparatus. Each mouse was retested in the same activity monitor 24 h following the first activity session. The figure shows mean center time ± SEM during the first session (left) and 24-h retention test (right). All doses of physostigmine resulted in a reduction of center time from the first session to the 24-h retention test. A two-way analysis of variance indicated a Dose X Session interaction [$F(3,43)=3.5$, $p<0.025$], and individual comparisons revealed significant between-session decrements in center time for each of the groups receiving physostigmine (all ps<0.05).

function could underlie the senescence-like deficits in between-session habituation exhibited by the NZB/BINJ mice. This finding is consistent with other observations in our laboratory indicating that mature NZB/BINJ mice exhibit an senescence-like abnormality in locomotor responses following pre-session administration of cholinomimetics (Retz et al., 1987a; 1987b).

PRECLINICAL APPLICATIONS OF AUTOIMMUNE MICE

Although the nature of learning/retention deficits associated with autoimmune disorders has yet to be fully investigated, findings to date suggest that autoimmune mice could prove useful as model systems for preclinical evaluation of drug therapy targeted at age-related cognitive dysfunctions. The validity of autoimmune models in that capacity will depend upon the extent to which autoimmune-related learning/memory deficits involve neurological defects which are similar to those which occur during normal aging or Alzheimer's disease and, ultimately, by the extent to which the autoimmune models predict clinical efficacy of a particular therapy. At the present time autoimmune NZB/BINJ mice appear to meet some of these requirements, as these mice exhibit senescence-like deficits in several learning and memory tasks. The appearance of those deficits with age is accelerated relative to normal mice, and at least one memory deficit can be reversed by physostigmine, an agent known to yield some limited improvement in AD patients.

The most important potential advantage of the autoimmune mouse models relates to the possibility that autoimmunity may play some role in the etiology of age-related disorders of cognition. If such is the case, then effective therapy for prevention or reversal of disease processes and their symptoms may be targeted at the immune system (cf. Singh and Fudenberg, 1986b). Autoimmune mice could provide a valid and efficient means of predicting the clinical efficacy of such therapy.

ACKNOWLEDGMENTS

This work was supported by NIH grants AG06182 (M.J.F.), RR05879 (T.C.O.M.) and a research grant from the Bristol Myers Company (K.C.R).

REFERENCES

Baldinger, A. and H.T. Blumenthal. 1982. Neuroimmunology of the aging brain. In *Geriatrics*, ed. D. Platt, 283-299. Berlin: Springer Verlag.

Bartus, R.T., and R.L. Dean. 1985. Developing and utilizing animal models in the search for an effective treatment for age-related memory disturbance. *In Physiological Aging and Dementia*, ed. C. Gottfries, 231-267. Basle: S. Karger.

Bartus, R.T., R.L. Dean, and S.K. Fisher. 1986. Cholinergic treatment for age-related memory disturbances: Dead or barely coming of age? *In Treatment Development Strategies for Alzheimer's Disease*, eds. T. Crook, R.T. Bartus, S. Ferris and S. Gershon, 421-450. Madison, Connecticut: Mark Powley Associates, Inc.

Brennan, M.J., A. Dallob, and E. Friedman. 1981. Involvement of hippocampal serotonergic activity in age-related changes in exploratory behavior. *Neurobiol. Aging 2*: 199-203.

Brennan, M.J., D. Allen, D. Aleman, E.C. Azmitia, and D. Quartermain. 1984. Age differences in within session habituation of exploratory behavior: Effects of stimulus complexity. *Behav. Neural Biol. 42*: 61-72.

Dahlstrom, A., S. Booj, K. Haglid, L. Rosengren, A. Wallin, J. Karlsson, L. Svennerholm, C.G. Gottfries, A. McRae-Degueurce. 1987. CSF from subgroups of Alzheimer's patients contain antibodies recognizing cholinergic neurons in the rodent central nervous system. *Soc. Neurosci. Abstr. 13*: 1462.

Fillit, H., V.N. Luine, B. Reisberg, R. Amador, B. McEwen, and J.B. Zabriskie. 1985. Studies of the specificity of antibrain antibodies in Alzheimer's disease. In *Senile Dementia of the Alzheimer Type*, eds. J.T. Hutton and A.D. Kenny, 307-318. New York: Alan R. Liss.

Forster, M.J. 1987. Age differences in locomotor behavior of C57BL/6Nnia mice: A data management approach. *Pharmacol. Biochem. Behav. 27*: 545-551.

Forster, M.J., K.C. Retz, T.L. Johnson, M.D. Popper, and H. Lal. 1987. Habituation retention: Age differences in young NZB/BINJ mice parallel senescence-related changes in C57BL/6NNia mice. *Soc. Neurosci. Abstr. 13*: 441.

Forster, M.J., Popper, M.D., Retz, K.C. and Lal, H. 1988. Age differences in acquisition and retention of one-way avoidance learning in C57BL/6NNia and autoimmune mice. *Behav. Neural Biol. 49*: 139-151.

Fraley, S.M. and A.D. Springer. 1981a. Memory of simple learning in young, middle aged, and aged C57BL/6 mice. *Behav. Neural. Biol. 31*: 1-7.

Fraley, S.M. and A.D. Springer. 1981b. Duration of exposure to a novel environment affects retention in aging mice. *Behav. Neural Biol.* 31:1-7.

Fudenberg, H.H., H.D. Whitten, P. Arnaud, and N. Khansari. 1984. Is Alzheimer's disease an immunological disorder? Clin. Immunol. Immunopathol. 32: 127-131.

Gaskin, F., B.S. Kingsley, and S.M. Fu. 1986. Autoantibodies in Alzheimer's disease and age-matched normal controls. *Soc. Neurosci. Abstr. 12*: 36.

Giacobini, E. 1987. Models and strategies of cholinergic therapy of Alzheimer disease. In *Cholinergic Transmission in Pathophysiological Conditions*, ed. M.Dowdall, 882-901. West Sussex England: Ellis Horwood Ltd.

Hoffman, S.A., D.N. Arbogast, P.N. Ford, D.W. Shucard, and R.J. Harbeck. 1983. Anticerebellar antibody levels in the sera of autoimmune mice. *Soc. Neurosci. Abstr.* 9: 269.

Hoffman, S.A., D.N. Arbogast, P.M. Ford, D.W. Shucard, and R.J. Harbeck. 1987. Brain-reactive autoantibody levels in the sera of aging autoimmune mice. *J. Clin. Exper. Immunol. 70*: 74-83.

Hughes, R.N. 1982. A review of atropinic drug effects on exploratory choice behavior in laboratory rodents. *Behav. Neural. Biol. 34*: 4-41.

Izui, S., V.E. Kelley, M. Kazushige, H. Yoshida, J.B. Roths, and E.D. Murphy. 1984. Induction of various autoantibodies by mutant gene *lpr* in several strains of mice. *J. Immunol. 133*: 227-233.

Johns, C.E., V. Haroutunian, B.S. Greenwald, R.C. Mohs, B.M. Davis, P. Kanof, T.B. Horvath, and K.L. Davis. 1985. Development of cholinergic drugs for the treatment of Alzheimer's disease. *Drug Dev. Res. 5*: 77-96.

Khansari, N., H.D. Whitten, Y.K. Chou, and H.H. Fudenberg. 1985. Immunological dysfunction in Alzheimer's Disease. *J. Neuroimmunol. 7:* 279-285.

Lal, H., and M.J. Forster. 1986. Cognitive disorders related to immune dysfunction: Novel animal models for drug development. *Drug Dev. Res. 7*: 195-208.

Lal, H., and M.J. Forster. 1988. Autoimmunity and age-related cognitive decline. *Neurobiol. Aging*, in press.

Lal, H., M. Bennett, D. Bennett, M.J. Forster, and K. Nandy. 1986. Learning deficits occur in young mice following transfer of immunity from senescent mice. *Life Sci. 39*: 507-512.

McRae-Degueurce, A., S. Booj, K. Haglid, L. Rosengren, J.E. Karlsson, I. Karlsson, A. Wallin, L. Svennerholm, C.G. Gottfries, and A. Dahlstrom. 1987. Antibodies in the CSF of some Alzheimer's patients recognize cholinergic neurons in the rodent central nervous system. *Proc. Natl. Acad. Sci. USA 84*: 9214-9218.

McRae-Degueurce, A. K. Haglid, L. Rosengren, A. Wallin, K. Blennow, C.-G. Gottfries and A. Dahlstrom. 1988. Antibodies recognizing cholinergic neurons and thyroglobuline are found in the CSF of a subgroup of AD/SD patients. *Drug Dev. Res.*, in press.

Moos, W.H., Davis, R.E, Schwartz, R.D. and Gamzu, E.R. 1988. Cognition activators. *Medicinal Res. Rev.*, in press.

Murphy, E.D. and J.B. Roths. 1978. Autoimmunity and lymphoproliferation: Induction by mutant gene *lpr*, and acceleration by a male-associated factor in strain BXSB mice. In *Genetic Control of Autoimmune Disease*, eds. N.R. Rose, P.E. Bigazzi, and N.L. Warner, 207-221. New York: Elsevier/North-Holland.

Nandy, K. 1978. Brain-reactive antibodies in aging and senile dementia. In *Alzheimer's Disease--Senile Dementia and Related Disorders*, eds. R. Katzman, R. Terry, and K. Bick, 503-512. New York: Raven Press.

Nandy, K. 1985. Immunopathology of aging and dementia. In *Senile Dementia of the Alzheimer Type*, eds. J.T. Hutton and A.D. Kenny, 293-305. New York: Alan R. Liss.

Nandy, K. and Bennett, M. 1983. Immune manipulations and brain- reactive antibody formation in aging mice. *Mech. Ageing Devel., 22*: 287-293.

Nandy, K., H. Lal, M. Bennett, and D. Bennett. 1983. Correlation between a learning disorder and elevated brain-reactive antibodies in aged C57BL/6 and young NZB mice. *Life Sci. 33*: 1499-1503.

Platel, A. and R.D. Porsolt. 1982. Habituation of exploratory activity in mice: A screening test for memory enhancing drugs. *Psychopharmacol. 78*: 346-352.

Quimby, F.W. and R.S. Schwartz. 1982. Systemic Lupus Erythematosus in mice and dogs. In *Clinical Aspects of Immunology* Vol. 2, eds. P.J. Lachman and D.K. Peters, 1217-1230. Oxford: Blackwell Scientific.

Retz, K.C., M.J. Forster, N. Frantz, and H. Lal. 1987a. Differences in behavioral responses to oxotremorine and physostigmine in New Zealand Black (NZB/BINJ) and C57BL/6 mice. *Neuropharmacol. 26:* 445-452.

Retz, K.C., C.K. Trimmer, M.J. Forster, and H. Lal. 1987b. Motor responses of autoimmune NZB/BINJ and C57BL/6NNia mice to arecoline and nicotine. *Pharmacol. Biochem. Behav. 28*: 275-282.

Schwegler, H., H.-P. Lipp, and W.E. Crusio. 1988. The NZB mouse: Hippocampal mossy fiber patterns and behavioral profiles of young and older animals. *Drug Dev. Res.*, in press.

Singh, V.K. and H.H. Fudenberg. 1986a. Detection of brain autoantibodies in the serum of patients with Alzheimer's disease but not Down's syndrome. *Immunol. Lett. 12*: 277-280.

Singh, V.K., and H.H. Fudenberg. 1986b. Immunopharmacological approach to the study of chronic brain disorders. *Prog. Drug Res. 30*: 345-363.

Spencer, D.G., K. Humphries, D. Mathis, and H. Lal. 1986. Behavioral impairments related to cognitive dysfunction in the autoimmune New Zealand Black mouse. *Behav. Neurosci. 100*: 353-358.

Wenk, G., D. Hughey, V. Boundy, A. Kim, L. Walker, and D. Olton. 1987. Neurotransmitters and memory: Role of cholinergic, serotonergic, and noradrenergic systems. *Behav. Neurosci. 101*:325-332.

PHARMACOLOGICAL CONSEQUENCES OF CHOLINERGIC PLUS NORADRENERGIC LESIONS

V. Haroutunian, G.K. Tsuboyama, P.D. Kanof, and K.L. Davis
*Dept., of Psychiatry, The Mount Sinai School of Medicine,
One Gustave Levy Place, New York, NY*

INTRODUCTION
Results of studies on autopsy and biopsy materials obtained from Alzheimer's disease (AD) patients have overwhelmingly established the involvement of forebrain cholinergic systems in AD (eg. Whitehouse et al, 1982; Bowen et al, 1983). Cholinergic neurons in the basal forebrain (nucleus basalis of Meynert-nbM) degenerate during the course of this disease and the degeneration of these cholinergic neurons leads to marked decreases in neocortical markers of cholinergic function (eg. Davies and Maloney, 1976). Animal studies indicate that the cholinergic cells of the nbM not only participate in cortical electro-physiological manifestations of associative learning (eg. Rigdon and Pirch, 1986), but that lesions of this cell group result in significant learning and memory impairments (see Haroutunian et al, 1986 for review). These findings have led to the view that the reversal of the cholinergic deficits must be a part of any symptomatic treatment of AD. Despite the overwhelming evidence implicating the nbM cholinergic system in the cognitive deficits of AD, clinical experience with cholinomimetic therapy has led to only partial improvements in the mnemonic function of a subgroup of AD patients (Mohs et al, 1985). The results of these clinical studies are in marked contrast to findings in nbM lesioned animals in which cholinomimetic treatments appear to almost completely ameliorate cognitive deficits (eg Haroutunian et al, 1985, 1986; Murray and Fibiger, 1985, Tilson et al, 1988). This discrepancy and the plethora of evidence showing abnormalities in several other neurotransmitter systems has given impetus to the view that non-cholinergic neurotransmitter and neuroanatomical deficits must also play a role in the cognitive dysfunctions associated with this disease.

It has been consistently established that AD does not only affect forebrain cholinergic systems, but often involves peptidergic and catecholaminergic systems as well. For example, cortical somatostatin-like immunoreactivity and CRF-like immunoreactivity are significantly reduced in postmortem samples from AD patients (eg, Davies et al, 1980, DeSouza et al, 1986) as are cortical somatostatinergic receptors. It has been difficult, however, to ascribe a cognitive role to the depletion of cortical somatostatin- and CRF-like immunoreactivity (Haroutunian et al, 1986, 1988a). Although the role of somatostatin-like immunoreactivity depletion in the cognitive impairments of AD cannot be dismissed, the results of animal and clinical studies suggest that changes in other neural systems must also be involved in AD dementia.

It is now apparent that a significant proportion of AD patients suffer considerable cortical noradrenergic depletion and decreased NE turnover (eg. Adolfsson et al, 1979; Cross et al, 1983; Palmer et al, 1987). These cortical noradrenergic deficits are coupled with an approximately 60% cell loss in the pigmented neurons of the locus coeruleus with virtually no overlap between AD and control samples. In general, it appears that the greatest NE system deficits are observed in cases of AD with the earliest age of onset. These results which have been replicated many times provide a strong basis for the belief that noradrenergic deficits contribute to the pathophysiology of AD in at least a significant subpopulation of AD patients.

The evidence that noradrenergic dysfunction influences the cognitive deficits of AD is less well established, and somewhat contradictory. Significant correlations between brain NE markers and performance on cognitive tests have been reported in AD patients (eg Adolfsson et al, 1979, although some studies, eg Palmer et al, 1987, have failed to find a significant correlation). Animal studies have shown a strong relationship between noradrenergic function and performance on learning and memory tasks (Gold and Zornetzer, 1983; Arnsten and Goldman-Rakic, 1985a,b). Furthermore, NE receptor stimulation by high doses of systemically administered clonidine can attenuate age-related cognitive deficits in monkeys (Arnsten et al, 1985b) and can prevent the amnesia which results from dopamine-B-hydroxylase inhibition. Thus, forebrain noradrenergic systems do participate in the processes which subserve learning and memory, however, the precise function and mechanism of NE influences on cognition remains to be elucidated.

Irrespective of whether it can be demonstrated that NE system dysfunction has a direct influence upon cognition, locus coeruleus cell loss and forebrain noradrenergic deficits could affect cognitive performance indirectly by a) depriving forebrain cholinergic neurons of their noradrenergic influences; or b) altering the action of acetylcholine at a post-synaptic site where NE and ACh systems converge and interact. For example, decreased alpha$_2$ receptor binding sites (which are often located on presynaptic noradrenergic terminals) and decreased NE levels in the nbM/substantia innominata region have been noted in autopsy samples from AD patients (Shimohama et al, 1986). If the noradrenergic system normally modulates the activity of forebrain cholinergic cells, then the NE deficits noted above could further impair the function of the already compromised forebrain cholinergic cells of the AD patient, resulting in additional cholinergic deficits and disregulation. It should be noted, however, that noradrenergic damage is probably not causally related to forebrain cholinergic cell loss since lesions of the forebrain projecting NE system do not affect high affinity choline uptake in the nbM (Decker and Gallagher, 1987). Another source of adrenergic regulation of the activity of nbM cells could be through the perturbation of the noradrenergic cells projecting to the frontal cortex which could influence forebrain cholinergic function by altering the cortical feedback to the nbM, or by affecting the action of acetylcholine upon the target cells (Waterhouse et al, 1981).

Noradrenergic deficits can also influence the responsivity of the CNS to cholinomimetic drugs and may constitute one reason for the inability of cholinomimetics to achieve significant symptomatic relief in a subpopulation of AD patients. One way in which the efficacy of cholinomimetics can be diminished in AD is through the loss of their influence upon the noradrenergic system. As Cross et al (1983) have suggested, noradrenergic deficits, such as those noted above, may severely hamper the efficacy of cholinomimetics in the treatment of AD and may necessitate treatment with combinations of drugs aimed at simultaneous potentiation of NE and ACh activity. One hypothesis resulting from this argument is that adrenergic hypoactivity can attenuate the efficacy of cholinomimetic agents in reversing forebrain cholinergic lesion-induced deficits.

TABLE 1.
Effects of Ibotenic Acid Lesions of the nbM on Cortical Values of Some Neurochemical Markers. Values Expressed as Mean [+/-SEM] (n)

Marker	Sham	nbM	% Depletion
Choline Acetyltransferase (nmol Ach/hr/mg prot)	19.2 [0.3] (12)	14.1 [0.7] (14)	26.3*
Acetylcholinesterase (nmol/hr/mg prot)	1512 [24] (12)	1070 [54] (14)	29.1*
Norepinephrine (ng/100mg tissue)	30.1 [0.8] (12)	29.1 [0.9] (14)	3.2
Dopamine (ng/100mg tissue)	6.7 [0.2] (12)	6.4 [0.3] (14)	5.2
Serotonin (ng/100mg tissue)	32.2 [0.8] (11)	29.0 [0.8] (12)	9.9
Somatostatin-LI (pg/mg tissue)	377.8 [11] (12)	387 [18] (12)	2.6
CRF (pg/mg tissue)	1.9 [0.1] (12)	1.8 [0.1] (12)	3.1

* nbM vs Sham, p<0.01

RESULTS
Over the past few years we and others have conducted experiments on the pharmacological alleviation of cholinergic lesion-induced cognitive deficits (see Haroutunian et al, 1986). Several studies have demonstrated that excitotoxic lesions of the nbM (bilateral injections of 5ug/ul ibotenic acid into the nbM) result in significant deficits in cholinergic markers (Table 1) as well as learning and memory deficits on a number of tasks (Murray and Fibiger, 1985; Haroutunian et al, 1985; Tilson et al, 1988). An example of the results of such a study are shown in Figure 1. In this experiment male Sprague-Dawley rats received either sham operations or ibotenic acid-induced lesions of the

nbM (1ul of 5ug/ul of ibotenic acid injected bilaterally at stereotaxic coordinates Bregma -0.3, +/- 3.0 relative to midline and 7.8 below skull, leading to a 26-32% depletion of cortical choline acetyltransferase and acetylcholinesterase activity). After a two week recovery period, each rat was trained on a one trial passive avoidance task and received one of a several doses of physostigmine. Passive avoidance training consisted of placing each rat into the lit compartment of a light/dark shuttle box and allowing the rat to step into the dark compartment. Upon entry into the dark compartment each rat received a two second long 0.6 mA scrambled foot shock. One minute following entry into the dark compartment each animal was removed and received a sc. injection of either saline or physostigmine (0.03, 0.06 mg/kg). Retention test performance was assessed 72 hours later by placing the rat into the lit compartment of the shuttle box and recording the latency to cross into the dark, previously shocked compartment. The results of this study showed that nbM lesioned rats were significantly impaired on the 72 hour retention of passive avoidance ($p<0.01$), and that the administration of a relatively low dose of physostigmine (0.06 mg/kg) reversed the lesion-induced retention test performance deficit. The performance of the sham operated rats was also improved ($p<0.05$) by physostigmine, however, the effective dose of physostigmine was lower (0.03 mg/kg) in these rats. These results have been replicated many times by us as well as several other groups (eg, Tilson et al, 1988). In addition, we have extended these findings with physostigmine to other classes of cholinomimetics, namely the muscarinic agonist oxotremorine and the acetylcholine releasing agent 4-aminopyridine (Haroutunian et al, 1986). The results of these and other experiment suggest that forebrain cholinergic lesions can cause cognitive impairments on a wide variety of cognitive tasks which can be reversed by cholinomimetic therapy.

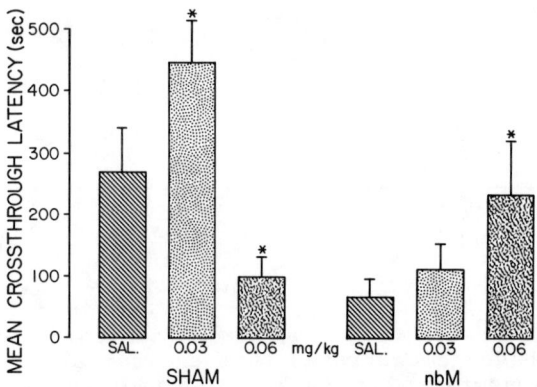

Figure 1. Effect of post-acquisition physostigmine on the 72 hour retention of one trial passive avoidance in nbM lesioned rats. Ns=11, * vs SAL ps<0.05.

The experiments outlined above address the effects of

cholinomimetics on purely cholinergic lesion-induced deficits and ignore the contribution of other neurotransmitter deficits in AD. Recently, several experiments have shown that the functional integrity of the noradrenergic system is of critical importance to the ability of physostigmine to improve retention test performance in nbM lesioned rats. In a single surgical session, rats received ibotenic acid induced lesions of the nbM coordinates as described above) as well as 6-OHDA induced lesions of the ascending noradrenergic bundle (2ul of 8ug/ul 6-OHDA injected bilaterally at coordinates Bregma -6.0, +/- 0.8 relative to midline and 6.5 below the surface of the skull, resulting in 92% mean depletion of cortical NE). After a two week recovery period, each rat was trained on the passive avoidance task described above and then received one of a wide range of doses of physostigmine (0.0-0.24 mg/kg) immediately following training. Significant (ps<0.01) seventy-two hour retention test deficits were found in all lesioned animals, regardless of the dose of physostigmine administered (Figure 2). Concurrently run nbM lesion alone animals, on the other hand, demonstrated the well known physostigmine enhancement of retention test performance effect. Similar results were obtained in another study when oxotremorine was used as the cholinomimetic agent.

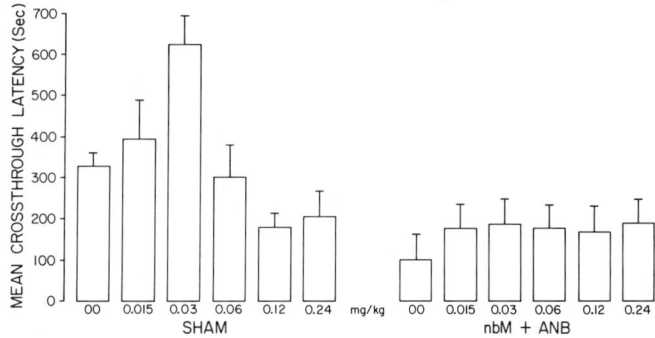

Figure 2. Effects of post-acquisition administration of several doses of physostigmine to rats with combined lesions of the nbM and the ascending noradrenergic bundle (ANB). Ns=9-11.

The inability of physostigmine to enhance retention test performance in nbM+NE lesioned rats is reminiscent of the work of Mason and Fibiger (1979), who in an entirely different paradigm (cholinergic drug-induced catalepsy) also found that noradrenergic lesions blocked cholinomimetic drug effects. Similarly, catecholamine synthesis inhibition by alpha-methyl-p-tyrosine has been shown to block the passive avoidance memory facilitation induced by the systemic administration of oxotremorine (Huygens et al, 1980). In addition, an NE lesion potentiation of anti-cholinergic drug effects has been noted in a learning and memory paradigm where 6-OHDA induced lesions of the ascending NE bundle have no significant effects upon performance in a radial arm maze, but dramatically potentiated the disruptive effects of scopolamine on this task (Decker and Gallagher, 1987). These results combined with the data presented above, confirm our hypothesis and suggest that an intact noradrenergic system is a prerequisite for cholinomimetic enhancement of retention test deficits in cholinergic lesioned animals.

Three studies have recently been initiated to evaluate the pharmacological variables which might reestablish cholinomimetic responsivity in nbM + NE lesioned animals. Studies by Arnsten and Goldman (1985b), provide one possible avenue for pharmacotherapy. In that study, the cognitive performance of aged, forebrain norepinephrine depleted monkeys was reversed by the administration of relatively high doses of the alpha$_2$ adrenergic agonist clonidine. The ability of clonidine, administered alone or in combination with cholinomimetics, to reverse the retention test deficits of nbM + NE lesioned rats was investigated in the following studies. Lesioned rats were prepared in exactly the same manner as those described above. Two weeks following the lesion procedure, sham and NE + nbM lesioned rats were trained on the one trial passive avoidance task and received a sc. post acquisition injection of physostigmine (0.0 or 0.06 mg/kg) alone clonidine (0.0, 0.01 or 0.5 mg/kg) alone or a combination of the two drugs. Retention of passive avoidance was assessed 72 hours later. The results of this study demonstrated that cognitive deficits resulting from combined nbM + NE lesions can be reversed by the simultaneous administration of physostigmine and the alpha$_2$-adrenergic receptor agonist clonidine. Thus, while neither physostigmine, nor a low dose of clonidine alone were able to improve retention test performance in nbM + NE lesioned rats, the combination of a 0.06 mg/kg dose of physostigmine with a 0.01 mg/kg dose of clonidine resulted in significant (ps<0.01) improvements in 72 hour retention (Figure 3). Interestingly, the high dose of clonidine (0.5 mg/kg) also reversed the passive avoidance retention test deficits of the nbM + NE lesioned rats. This dose of clonidine is high enough, however, to preclude realization of pharmacological specificity and application to an elderly population with potential cardiovascular system disease. It is also noteworthy that while physostigmine by itself did not affect the retention test performance of sham operated rats, the combination of physostigmine and clonidine led to significant impairment of retention test performance.

These results have now been replicated several times and additional experiments have shown that even lower doses of clonidine (eg, 0.005 mg/kg) can reverse the retention test deficits when combined with a 0.06 mg/kg dose of physostigmine. The generality of these findings was assessed in the next experiment, where the dose of clonidine was held constant at 0.01 mg/kg, but was combined with different doses of the muscarinic agonist oxotremorine (0.0, 0.01, 0.05 or 0.10 mg/kg). As was the case with physostigmine, oxotremorine alone (0.01 mg/kg) enhanced the retention test performance of sham operated rats (Haroutunian et al, 1986), but failed to affect the 72 hour passive avoidance retention test performance of nbM + NE lesioned rats. Effective enhancement of 72 hour retention test performance was attained in combined lesion animals which received a 0.01 mg/kg dose of clonidine in addition to either a 0.01 or 0.05 mg/kg dose of oxotremorine.

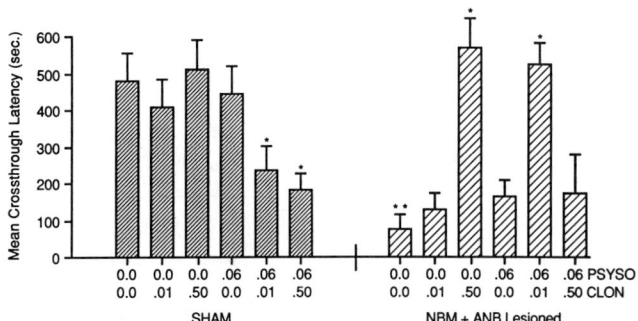

Figure 3. Effects of post-acquisition administration of physostigmine, clonidine or physostigmine and clonidine to rats with combined lesions of the nbM and the ANB. Ns=9-11. * vs 0.0 0.0 dose ps<0.03, ** vs Sham p<0.01.

The studies outlined above have shown that in rat model systems a) cognitive deficits resulting from forebrain cholinergic lesions alone can be reversed by the administration of cholinomimetic drugs such as physostigmine and oxotremorine; b) the efficacy of cholinomimetic therapy is significantly diminished when a forebrain noradrenergic deficit is combined with the cholinergic lesion; and c) the efficacy of cholinomimetic "therapy" can be restored by the combination of low doses of the alpha$_2$ adrenergic agonist clonidine with low doses of cholinomimetics such as physostigmine and oxotremorine. A draw back of the studies outlined must be pointed out however. The 6-OHDA-induced lesions of the ascending noradrenergic system in these experiments was profound and in most cases complete. Neuropathologic and neurochemical studies of AD indicate that some minimal cortical NE activity is preserved in AD (Palmer et al, 1987). It remains to be determined therefore, whether less extensive NE lesions in rats also block cholinomimetic efficacy and to define the limits of the NE lesion blockade of cholinomimetic efficacy.

The animal studies described above indicate that there are not only behavioral and cognitive deficits which can be associated with the various neurotransmitter systems affected in AD, but that deficits in each system can affect the functioning of other neuronal systems. These studies, however, do not imply that every neurochemical deficit observed in AD has an impact on cognitive or therapeutic function. For example, studies similar to those described above have shown that somatostatinergic lesions play a less critical mnemonic role in the present paradigm and fail to alter the responsivity of nbM lesioned rats to cholinomimetic agents (Haroutunian et al, 1986). Thus, although the multitransmitter dysfunction characteristic of AD can have an impact on therapeutic approaches which seek to normalize the functional deficits, the contribution of each lesion to overall responsivity to pharmacological treatment must be evaluated separately. On the other hand, the results of the present experiments show that even when multiple system lesions block the ability of a therapeutic agent such as physostigmine to reverse lesion-induced memory deficits, pharmacotherapeutic approaches can be designed to overcome this blockade.

REFERENCES

Adolfsson, R., C.G. Gottfries, B.E. Roos, and B. Winblad. 1979. Changes in the brain catecholamines in patients with dementia of Alzheimer type. Brit. J. Psychiat. 135: 216-223.

Arnsten, A.F.T and P.S. Goldman-Rakic 1985a. Catecholamines and cognitive decline in aged nonhuman primates. Ann. N.Y. Acad. Sci. 218-234.

Arnsten, A.F.T and P.S. Goldman-Rakic. 1985b. a_2-adrenergic mechanisms in prefrontal cortex associated with cognitive decline in aged nonhuman primates. Science, 230: 1273-1279.

Bowen, D.M., S.J. Allen, J.S. Benton, M.J. Goodhardt, E.A. Haan, A.M. Palmer, N.R. Sims, C.C. Smith, J.E. Spillane, M.M. Esiri, D. Neary, J.S. Snowden, G.K. Wilcock, and A.N. Davison. 1983 Biochemical assessment of serotonergic and cholinergic dysfunction and cerebral atrophy in Alzheimer's disease. J. Neurochem., 41:266-272.

Cross, A.J., T.J. Crow, J.A. Johnson, M.H. Joseph, E.K. Perry, R.H. Perry, G. Blessed, and B.E. Tomlinson. 1983. Monoamine metabolism in senile dementia of Alzheimer's type. J. Neurol. Sci., 60:383-392.

Davies, P., R. Katzman, and R.D. Terry. 1980. Reduced somatostatin-like-immunoreactivity in cerebral cortex from cases of Alzheimer's disease Alzheimer's senile dementia. Nature, 288: 279-280.

Davies, P. and A.J.F. Maloney. 1976. Selective loss of central cholinergic neurons in Alzheimer's disease. Lancet, 2: 1403.

Decker, M.W. and M. Gallagher. 1987. Scopolamine-disruption of radial arm maze performance: modification by noradrenergic depletion. Brain Research, 417:59-69.

DeSouza, E.B., P.J. Whitehouse, M.J. Kuhar, D.L. Price, and W.W. Vale. 1986. Reciprocal changes in corticotropin-releasing factor (CRF)-like immunoreactivity and CRF receptors in cerebral cortex of Alzheimer's disease. Nature, 319:593-595.

Gold, P.E. and S.F. Zornetzer. 1983. The mnemon and its juices: neuromodulation of memory processes. Behavioral and Neural Biology, 38:151-189.

Haroutunian, V., P.D. Kanof, G.K. Tsuboyama, G.A. Campbell, and K.L. Davis. 1986. Animal models of Alzheimer's disease: Behavior, pharmacology, transplants. Can. J. Neurol. Sci., 13:385-393.

Haroutunian, V., P.D. Kanof, and K.L. Davis. 1985. Pharmacological alleviation of cholinergic lesions induced memory deficits in rats. Life Sci., 37:945-952.

Haroutunian, V., P.D. Kanof, K.L. Davis, and W.W. Vale. 1988a. Lesions of the rat nucleus basalis of Meynert do not affect cortical levels of corticotropin-releasing factor.

Submitted.

Haroutunian, V., P.D. Kanof, G. Tsuboyama, and K.L. Davis. 1988b. Restoration of cholinomimetic activity by clonidine in cholinergic / noradrenergic lesioned rats. Submitted.

Huygens, P., C.M. Baratti, J.L. Gardella, and E. Filinger. 1980. Brain catecholamine modifications. The effects on memory facilitation induced by oxotremorine in mice. Psychopharm., 69:291-294.

Mason, S.T. and H.C. Fibiger. 1979. Possible behavioral function for noradrenaline-acetylcholine interaction in brain. Nature, 277:396-397.

Mohs, R.C., B.M. Davis, C.A. Johns, A.A. Mathe, B.S. Greenwald, T.B. Horvath, and K.L. Davis. 1985. Oral physostigmine treatment of patients with AD. Amer. J. Psychiatry, 142:28-33.

Murray, C.L. and H.C. Fibiger. 1985. Learning and memory deficits after lesions of the nucleus basalis magnocellularis: Reversal by physostigmine. Neuroscience, 19:1025-1032.

Palmer, A.M., P.T. Francis, D.M. Bowen, J.S. Benton, D. Neary, D.M.A. Mann, and J.S. Snowden, J.S. Catecholaminergic neurons assessed ante-mortem in Alzheimer's disease. Brain Research, 414: 365-375.

Rigdon, G.C. and J.H. Pirch. 1986. Nucleus basalis involvement in conditioned neuronal responses in the rat frontal cortex. J. Neurosci., 6:2535-2542.

Shimohama, S., T. Taniguchi, M. Fujiwara, and M. Kameyama. 1986. Biochemical characterization of a-adrenergic receptors in human brain and changes in Alzheimer-type dementia. J. Neurochem., 47:1294-1301.

Tilson, H.A., R.L. McLamb, S. Shaw, B.C. Rogers, P. Pediaditakis, and L. Cook. 1988. Radial-arm maze deficits produced by colchicine administered into the area of the nucleus basalis are ameliorated by cholinergic agents. Brain Research, 438:83-94.

Waterhouse, B.D., H.C. Moises, and D.J. Woodward. 1981. Alpha-receptor-mediated facilitation of somatosensory cortical neuronal responses to excitatory synaptic inputs and ionto-phoretically applied acetylcholine. Neuropharm., 20:907-920.

Whitehouse, P.J., D.L. Price, R.G. Struble, A.W. Clark, and J.T Coyle. 1982. Alzheimer's disease and senile dementia: loss of neurons in the basal forebrain. Science, 215:1237-1239.

BLOOD CHOLINESTERASE INHIBITION AS A GUIDE TO THE EFFICACY OF PUTATIVE THERAPIES FOR ALZHEIMER'S DEMENTIA:
Comparison of Tacrine and Physostigmine

Kathleen A. Sherman and Erik Messamore
Dept. of Pharmacology, S.I.U. School of Medicine, Springfield, IL

INTRODUCTION

Alzheimer disease (AD) is a slowly progressing neurodegenerative disorder which is the leading cause of dementia in the elderly. The discovery in 1976 that choline acetyltransferase, the synthetic enzyme for acetylcholine (ACh), is markedly reduced in cortex and hippocampus of patients with AD suggested a possible neurotransmitter basis for the disease (Bowen et al., 1976; Davies, 1979; Perry et al., 1978). Although other neurotransmitters can also be affected in AD, the earliest and most consistent deficit involves brain cholinergic mechanisms (Francis et al., 1985; Neary et al., 1986). Considerable evidence indicates that this loss of cholinergic function contributes to the debilitating memory loss and cognitive decline in AD (reviews: Bartus et al., 1982; Coyle et al., 1983; Sherman et al., 1987). The finding of a relatively specific transmitter deficit prompted hope that pharmacological treatment for AD could then be rationally devised. However, until recently treatments designed to "restore" or "replace" brain cholinergic function have met with limited and/or variable success.

Physostigmine (PHYSO), a "reversible" carbamate inhibitor of the metabolic enzyme acetylcholinesterase (AChE), produced only modest improvement of cognitive performance in AD patients treated acutely (Johns et al., 1983; Schwartz and Kohlstaedt, 1986; Blackwood and Christie, 1986). Significant improvement does occur when AD patients are treated repeatedly with oral PHYSO every 2-3 hr (Mohs et al., 1985; Thal et al., 1986). However, this improvement is restricted to particular tasks and only a subgroup of AD patients show "clinically significant" change. Many factors have been considered as possible explanations of the weak and variable response to cholinomimetics including (1) the severity of Alzheimer dementia and the possibility that inhibitors of ACh metabolism cannot be effective when cholinergic neurons are extensively lost or multiple transmitters are affected in severe illness (Hollander et al., 1987); (2) the possibility that only certain cognitive tests are sensitive indicators of cholinergic response in Alzheimer patients (Brinkman and Gershon, 1983); (3) the possibility that drugs such as PHYSO are too short acting and/or induce limiting side effects, and (4) the possibility that cholinomimetics improve cognition only in a very narrow therapeutic window as is commonly observed in unlesioned animals (Flood et al., 1985; Haroutunian, this volume).

Given that pharmacokinetic factors had not been extensively considered as a basis of the weak or negative results with PHYSO, we examined whether the ChE activity in blood could be used to establish the timing of drug absorption and to determine individual differences in absorption or metabolism after oral PHYSO which might explain the variable clinical response (Sherman et al., 1987). Our results suggest that failure to achieve sufficient inhibition of AChE may explain the lack of response to acute PHYSO treatment in the dose range tolerated by most AD patients. After 60 µg/kg PHYSO, plasma ChE was transiently inhibited a maximum of 25% at 60 min after ingestion (range 12 to 33%, n=10) and red blood cell AChE was inhibited only 11% (4 to 21%). None of the battery of cognitive tests administered every 30 min for 3 hr was significantly improved after acute oral PHYSO. Elevation of plasma neuroendocrine levels consistent with central cholinergic stimulation occurred only in patients with greater than 27% inhibition of plasma ChE, but three of the five patients with this degree of inhibition also experienced adverse side effects (see Kumar et al., this volume). Thus, blood ChE is measurably inhibited after PHYSO administration and the degree of plasma ChE inhibition appears to be related to indices of central cholinergic response; however, the window of central stimulation without side effects is very low and narrow. We have extended these findings to show that plasma ChE inhibition is well correlated with the dose of i.v. PHYSO infusion.

In addition to serving as a quantitative measure of efficacy for a given drug in clinical trials, inhibition of blood ChE may also provide a useful comparison between the various ChE inhibitors used as putative treatments for AD, and a comparison of efficacy across species so that animal models can be made maximally relevant to the clinical situation. We therefore investigated the effects of tacrine in rats.

In contrast to the weak response to PHYSO and other cholinergic replacement strategies, Summers et al. (1981; 1986) reported improvement in 75% of the dementia patients after acute i.v. treatment with tacrine (THA) and substantial improvement in 16 of 17 Alzheimer patients treated chronically with up to 200 mg/d oral THA. Dramatic results were reported in 10 patients including resumption of more normal daily function as well as improved psychometric test scores. The most severely demented patients are less responsive to THA (ibid) and short-term, low dose oral THA (30 mg given over 14 hr) improved performance only in the less impaired patients (Kaye et al., 1982). Summers et al. (1986) report the therapeutic range of THA in plasma to be above 5-70 ng/ml (2-30 x 10^{-8}M) and suggest that the therapeutic index of THA is superior to PHYSO.

THA has been used in several other clinical situations including treatment of myasthenia gravis and tardive dyskinesias (Ingram and Newgreen, 1983). THA is more efficacious for reversal of anticholinergic delirium and memory loss than PHYSO and the reversal is more prolonged (Gershon and Olariu, 1960; Itil and Fink, 1966). In several animal models, relatively low doses are effective in memory paradigms. In old monkeys, 5 mg/day oral THA improved cognitive function (Kling et al., 1985). In mice, optimum improvement of memory retention resulted from 1 to 2.5 mg/kg THA; 5 mg/kg THA had no effect (Flood et al., 1985). Thus, THA does show the typical bell-shaped dose-response curve. Low doses of THA (1-2 mg/kg) blocked the deleterious effect of anticholinergic drugs on retention in avoidance (Flood and Cherkin, 1986) and habituation in mice (Brown, 1971), and reversed the anticholinergic syndrome in dogs (Bell et al., 1964). However, despite the many studies of THA's interaction with other drugs, there is little information on the _in vivo_ neuropharmacology of THA alone and there are few direct comparisons of THA with other ChE inhibitors.

Although numerous studies have shown that _in vitro_ THA inhibits plasma ChE and tissue AChE, and potentiates cholinergic transmission in neuromuscular preparations, few studies have tested the role of this action as the mechanism of THA's effect on brain and behavior. The affinity for BuChE is 6-180 fold greater than for AChE (Bajgar and Patocka, 1977; Barrow and Johnson, 1966; Heilbronn, 1961; Ho and Freeman, 1965; Tonkopii et al., 1976): IC_{50}'s ranged from 1 to 11 x 10^{-8}M for BuChE and from 0.1 to 118 µM for AChE (most report an IC_{50} of about 1 µM, which is well above the plasma concentration reported as therapeutic threshold for THA). These values are roughly comparable to those for carbamate inhibitors of ChE such as PHYSO (Bajgar and Patocka, 1977; Barrow and Johnson, 1966). However, whereas carbamate inhibitors interact with the catalytic site, THA produces allosteric inhibition by binding to a hydrophobic region (the γ-anionic site) on the active surface of AChE and BuChE (Patocka et al., 1976; Tonkopii et al., 1976; Steinberg et al., 1975). Inhibition by THA is predominantly noncompetitive at ACh concentrations below 2.5 mM (ibid; Beneviste et al., 1967; Heilbronn, 1968). In neuromuscular preparations, THA potentiates the response to nerve stimulation or ACh at concentrations ranging from 0.1 to 74 µM (de la Lande and Porter, 1963; Freeman and Turner, 1970; Karis et al., 1966; Porter, 1965). Whereas up to 10 to 30 µM THA may be required for optimal potentiation, further increase of THA concentration produces less effect or actual blockade of neuromuscular transmission (Freeman and Turner, 1970; Karis et al., 1966). Moreover, at high concentrations THA has other actions, such as inhibition of monoamine oxidase (Kaul, 1962) and cyclic AMP-phosphodiesterase (Curley et al., 1984).

The data available on ChE inhibition after in vivo treatment with THA do not clarify whether this is likely to be the mechanism of the drug's effects. Marked and prompt inhibition of serum ChE was observed in humans (Barrow and Johnson, 1966; Beneviste et al., 1967; El-Kammah et al., 1975) and dog (Heilbronn, 1961) after i.v. bolus THA. Peak inhibition (over 80% in man, 60% in dog) occurred 3-5 min after THA and was followed by relatively rapid partial recovery of enzyme activity. However, RBC AChE was inhibited much less than plasma ChE after THA in the dog. In rats, THA was a weak inhibitor of whole blood AChE activity compared to PHYSO: less than 30% inhibition resulted after the LD_{50} dose for THA (34 mg/kg) (Bajgar et al., 1979). AChE inhibition was relatively low in brain compared to other tissues after 10 mg/kg THA in dogs and rats (Fusek et al., 1975). In mice, 4 mg/kg i.p. THA produced 70% inhibition of brain AChE by 30 min, but inhibition was gone by 3 hr (Tonkopii and Padinker, 1975). In rats, peak inhibition of brain AChE at 30 min was dose-dependent ranging from 20% after 3.7 mg/kg to 70% after 30 mg/kg and inhibition was reversed by 6 hr after THA (Rosic and Milosevic, 1967). After 10 mg/kg THA i.p., peak AChE inhibition varied from 5% to 25% in specific regions of the rat pons-medulla (Bajgar et al., 1984). Thus, neither blood nor brain AChE results have consistently shown superior magnitude or duration of efficacy for THA.

By contrast, a much different pharmacokinetic pattern is suggested by measurement of THA levels. In humans and monkeys, plasma THA concentrations peak at 2-3 hr after i.v. or oral administration and decline with a half-life of 5-6 hr in both species (Park et al., 1986). In rats, 20 mg/kg i.p. THA resulted in maximal plasma levels of 2.8 µg/ml (12 µM) from 15 min to 45 min post-drug, after which blood levels fell rapidly (Hsieh et al., 1983). In rat brain, THA concentration rose more gradually and plateaued at 80% of the plasma value by 30 min (ibid). These findings confirm that THA readily penetrates the blood brain barrier and can reach concentrations which inhibit AChE near maximally in vitro.

The relatively weak inhibition of AChE after THA in vivo appears discrepant with the superior efficacy and duration of THA compared to PHYSO or organophosphate ChE inhibitors in certain tests such as antagonism of the anticholinergic syndrome. This may be related to the unique mechanism of AChE inhibition by THA which does not involve alteration of the catalytic site. Alternatively, THA may be distinguished by virtue of other properties such as actions on cholinergic receptors or cation permeability. We therefore reexamined the effect of THA in a systematic comparison to PHYSO and examined the relationship of blood ChE inhibition to behavioral and neurochemical actions in brain after acute and chronic THA.

METHODS

Cholinesterase Activity in plasma, RBC or brain was measured according to a modification of the radioenzymatic method of Johnson and Russell (1975) which utilizes [^{3}H]-ACh as substrate for hydrolysis. Tissue is homogenized by Polytron or diluted in 0.5% Triton X-100 containing 10 mM EDTA, pH 7.4 and stored at -70°C for assay. Incubation is started by addition of 2.5 mM ^{3}H-ACh (final concentration, 0.1 µCi of 100 mCi/mmol ^{3}H-AChI from New England Nuclear, Boston, MA, adjusted with ACh·Cl from Sigma, St. Louis, MO) in reaction media containing 120 mM NaCl and 50 mM Tris, pH 7.4. The interval of incubation at 37°C was adjusted to utilize less than 20% of the substrate at the tissue concentration tested (specified below). At the end of incubation, the reaction is stopped and the product of hydrolysis ([^{3}H]-acetate) is protonated by 1 M chloroacetic acid and extracted into toluene-based scintillation cocktail containing 10% isoamyl alcohol. Unreacted ^{3}H-ACh is retained in the aqueous stopping solution (1 M chloroacetic acid, 0.5 M NaOH and 2 M NaCl). Blanks (buffer without tissue; 3,000-4,000 DPM) are subtracted.

RESULTS AND DISCUSSION

Clinical Studies: Plasma ChE Inhibition as a Function of PHYSO Dose

The effect of the dose of PHYSO infused on the degree of plasma ChE inhibition was determined in ten AD patients. Figure 1 shows a linear increase in plasma ChE inhibition at steady state (from 105 to 135 min) as a function of PHYSO dose. Inhibition of RBC AChE was considerably less than the reduction of plasma ChE activity after 900 µg/m^2, consistent with our findings after oral PHYSO in dementia patients. Variability between subjects appears to be considerably less with infusion when the dosage is based on body surface area rather than body weight, as in the oral study.

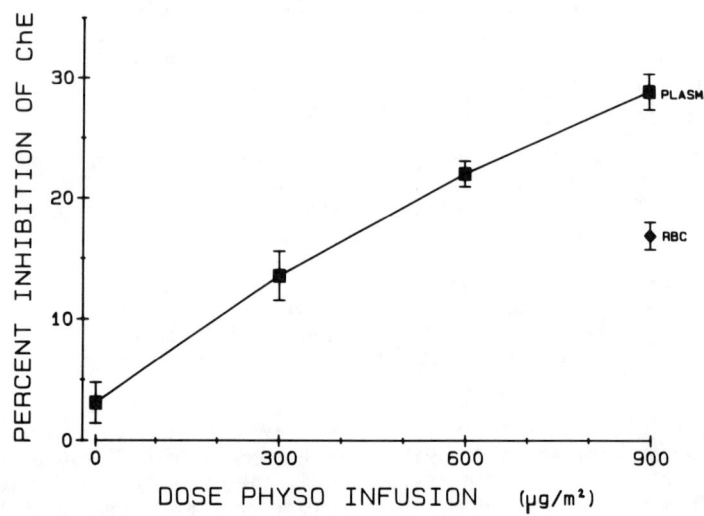

Figure 1. Dose-response of plasma ChE inhibition with i.v. PHYSO infusion. Ten patients with Alzheimer-type dementia received saline or total doses of 300, 600 and 900 µg PHYSO per m^2 body surface area according to the infusion program described in detail by Elble et al., this volume. Plasma and RBC ChE inhibition were determined at 105 and 135 min of infusion compared to pre-drug baseline the same day.

Comparison of tacrine and physostigmine in animal models

First, we compared the inhibitory potency of THA and PHYSO in vitro against human plasma ChE and RBC AChE and rat brain homogenate AChE. Few studies of THA have directly compared the potency of inhibition against brain or RBC AChE and plasma ChE, or the potency of other inhibitors under identical conditions and the available data are conflicting. We found that the IC_{50} for THA inhibition was 0.25 μM for plasma ChE, and 1 μM for RBC and cortical AChE; IC_{90} concentrations were 2.5 and 10 μM for plasma ChE and for AChE, respectively. For PHYSO the IC_{50} was 0.6-0.9 μM for plasma ChE and 0.3 μM for RBC AChE. Thus, in the in vitro affinity of the two inhibitors is fairly comparable.

Next, we compared the behavioral actions of THA and PHYSO after s.c. injection of male Sprague-Dawley rats pretreated with methylatropine. At timed intervals after drug, rats were placed in hindlimb splay posture and then the latencies to resume normal posture and forward locomote in an open field were measured. 10 mg/kg THA·HCl had a longer latency, but much more prolonged effect than PHYSO hemisulfate (0.88 mg/kg). PHYSO produced both behavioral deficits by 15 min and the effects declined from 1 to 2 hr after injection. The effect of THA on walking in an open field had gradual onset, peaked from 2-4 hr and persisted at least 6 1/2 hr post-injection (Figure 2). By contrast, the effect of THA on splay posture did not begin until 2 hr and declined more abruptly after 3 1/2 hr (not shown).

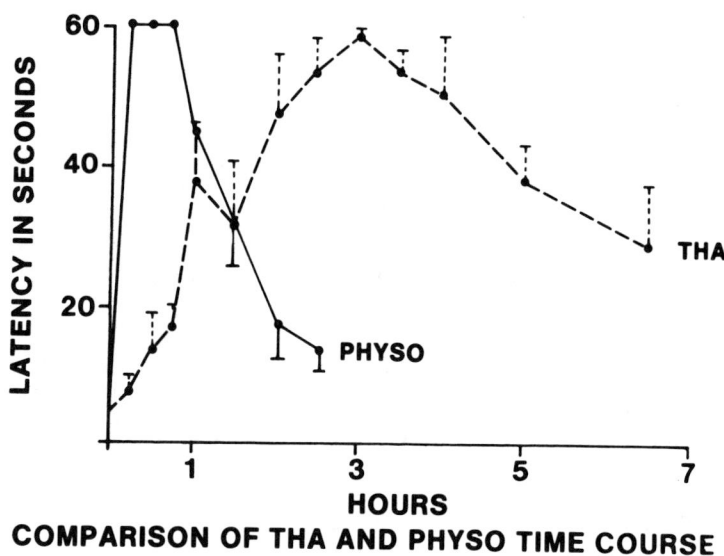

Figure 2. Time course of behavioral effects of tacrine compared to physostigmine. Rats were pretreated with atropine methylnitrate (5 mg/kg base weight, i.p.) 10 min prior to THA·HCl (10 mg/kg, s.c.) or PHYSO hemisulfate (0.88 mg/kg, s.c.) or saline (not shown) (N = 5 per group). A criterion of 60 sec was used for tests of splay and walk latency at 30 min intervals.

The effect of tissue dilution on the measured ChE inhibition was examined in rats treated with THA (10 mg/kg, 2 hr) or PHYSO (0.88 mg/kg, 15 min). Plasma ChE inhibition declined from 70% to 54% in the PHYSO treated rat when final dilution at assay was varied from 2-fold to 100-fold. By contrast, inhibition of plasma activity in the THA treated rat declined from 80% to 12%. In cortex, inhibition after PHYSO ranged from 84% at 6.8 X to 78% at 340 X dilution (Figure 3). After THA, the apparent inhibition in cortex was 66% at 6.8 X but declined with increasing dilution such that by 200 X dilution the inhibition is abolished. Similar effects were attained in striatum, hippocampus and RBC. The decline in THA inhibition with tissue dilution is exponential yielding a linear decrease when plotted against log of dilution. Thus, the dilution curves for THA indicate that the magnitude of in vivo inhibition is underestimated when tissue is diluted for assay. This may explain the apparent discrepancy between the relatively weak inhibition of blood and brain ChE after THA in vivo reported previously and its reported superior efficacy as an anticholinergic antagonist. The plot of inhibition versus log dilution may provide a means of accurately estimating in vivo inhibition for reversible allosteric inhibitors such as THA.

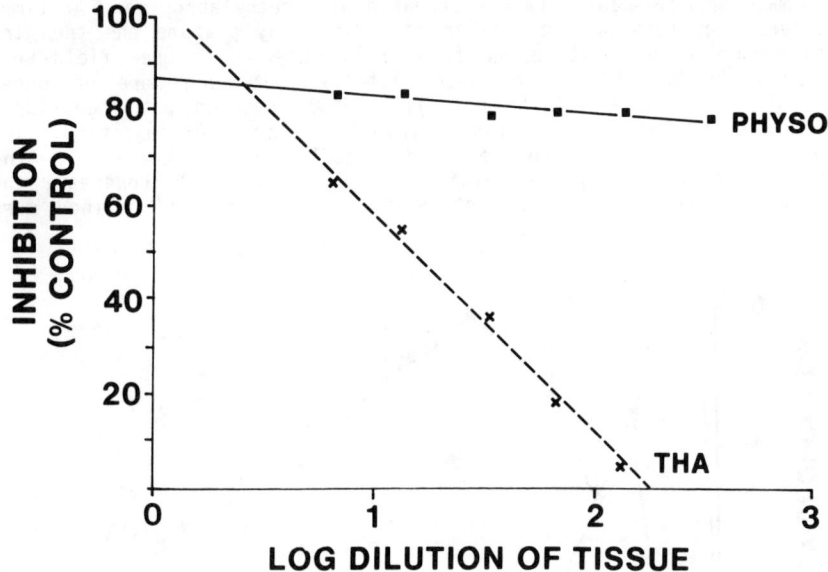

Figure 3. Effect of dilution of tissue for acetylcholinesterase assay after s.c. administration of PHYSO hemisulfate (0.88 mg/kg) or THA (10 mg/kg). Reduction in AChE activity is plotted as a function of final dilution of tissue after in vivo treatment with either PHYSO or THA compared to saline controls at the same dilution. Similar results were obtained in plasma, hippocampus and striatum.

Dose-response and time course of ChE inhibition after s.c. tacrine

Given the dramatic effect of tissue dilution on measured enzyme inhibition after systemic THA administration, we reinvestigated the dose-response and time course functions for ChE inhibition in blood and specific brain regions using the minimal dilution feasible for tissue homogenization and in vitro assay. Enzyme activity was maintained within the range of linearity by reduction of the incubation time to 30 or 60 sec. Behavior was measured as above and then rats were killed 3 hr after doses of THA varied from 1.25 to 20 mg/kg s.c. At the lowest doses (1.25 and 2.5 mg/kg) there was little effect on forward walking and differences between tissues in the magnitude of enzyme inhibition were most pronounced (Figure 4). After 1.25 mg/kg THA, inhibition ranged from 10% in RBC to 30% in plasma and brain AChE inhibition was 14-21%. This dose range is of particular interest because it corresponds to that which optimally improved memory retention in mice (Flood et al., 1985). Maximal inhibition after 20 mg/kg THA was 80% in plasma and RBC, and 72-80% in the brain regions. There was a very close correspondence between the dose-related inhibition of forward walking, cortical AChE inhibition and RBC AChE inhibition from 5 to 20 mg/kg. At 5 mg/kg and below, plasma ChE consistently showed greater effect of THA than AChE in RBC or brain at 3 hr. When this experiment was repeated at 2 hr after THA, we observed that plasma ChE inhibition was greater than in brain after 1.25 mg/kg but less than in hippocampus and cortex after 0.625 mg/kg (not shown).

Rats were killed at various times after an intermediate dose of THA (5 mg/kg, s.c.). Again, inhibition was consistently greater in plasma than cortex, hippocampus or cerebellum (Figure 5). Moreover, near maximal inhibition of plasma ChE occurred by 15 min after drug, whereas in all brain areas, AChE inhibition increased gradually to a maximum maintained from 1 to 3 hr post-THA. Although inhibition in both plasma and cortex declined by 5 hr after THA, it is noteworthy that at least half of the maximal effect remained at this time in cortex. In a second experiment on the time course of THA-induced inhibition, we observed that the inhibition of plasma ChE is already near maximal by 5 min after s.c. injection, whereas RBC and cortical AChE activity were not maximally inhibited until 30 to 60 min after injection (Figure 6). The magnitude of peak

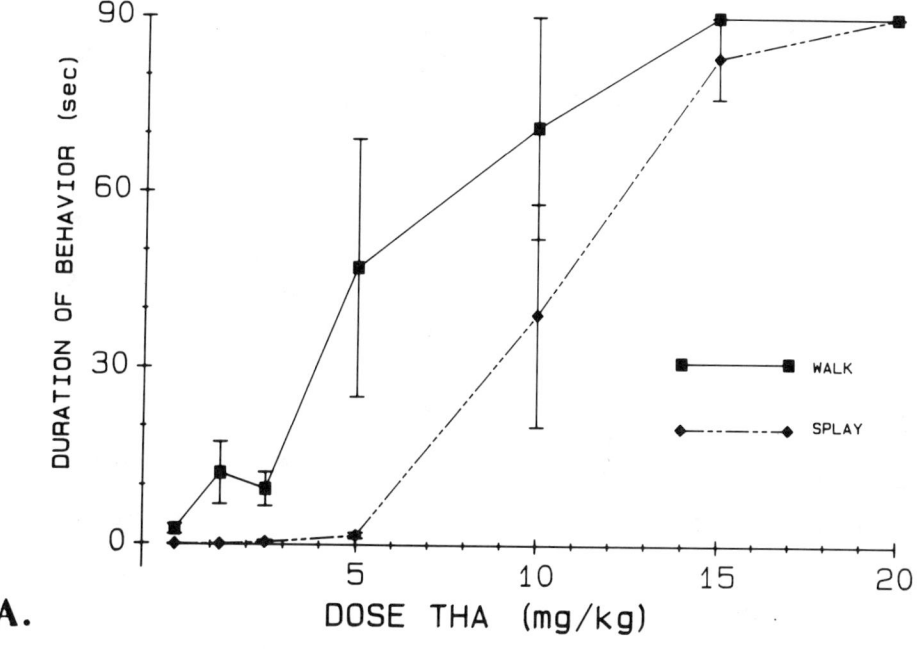

Figure 4. Dose-response curves for behavioral and cholinesterase inhibition following THA administration. A. The latency to forward walk and duration of splay posture were measured 3 hrs after various doses of THA by s.c. injection, 1.25 - 20 mg/kg (n = 3-4 per dose). The same rats were then killed for ChE inhibition studies. B. Trunk blood was collected in 2 ml tubes containing 7.15 U of heparin. Plasma and red blood cell (RBC) fractions were separated by repeated centrifugation at 15,400 Xg. Plasma was assayed at final dilution of 2.5 fold (shown) and 1.25 fold (not shown) with similar results. RBC were homogenized in 1:1 dilution with Triton-X-EDTA buffer pH 7.4 and assayed at final dilution of 4 X. RBC AChE activity was standardized to the hemoglobin content of parallel aliquots. Results are expressed as percent ChE inhibition relative to saline controls and shown as group mean ± S.E. (continued on next page)

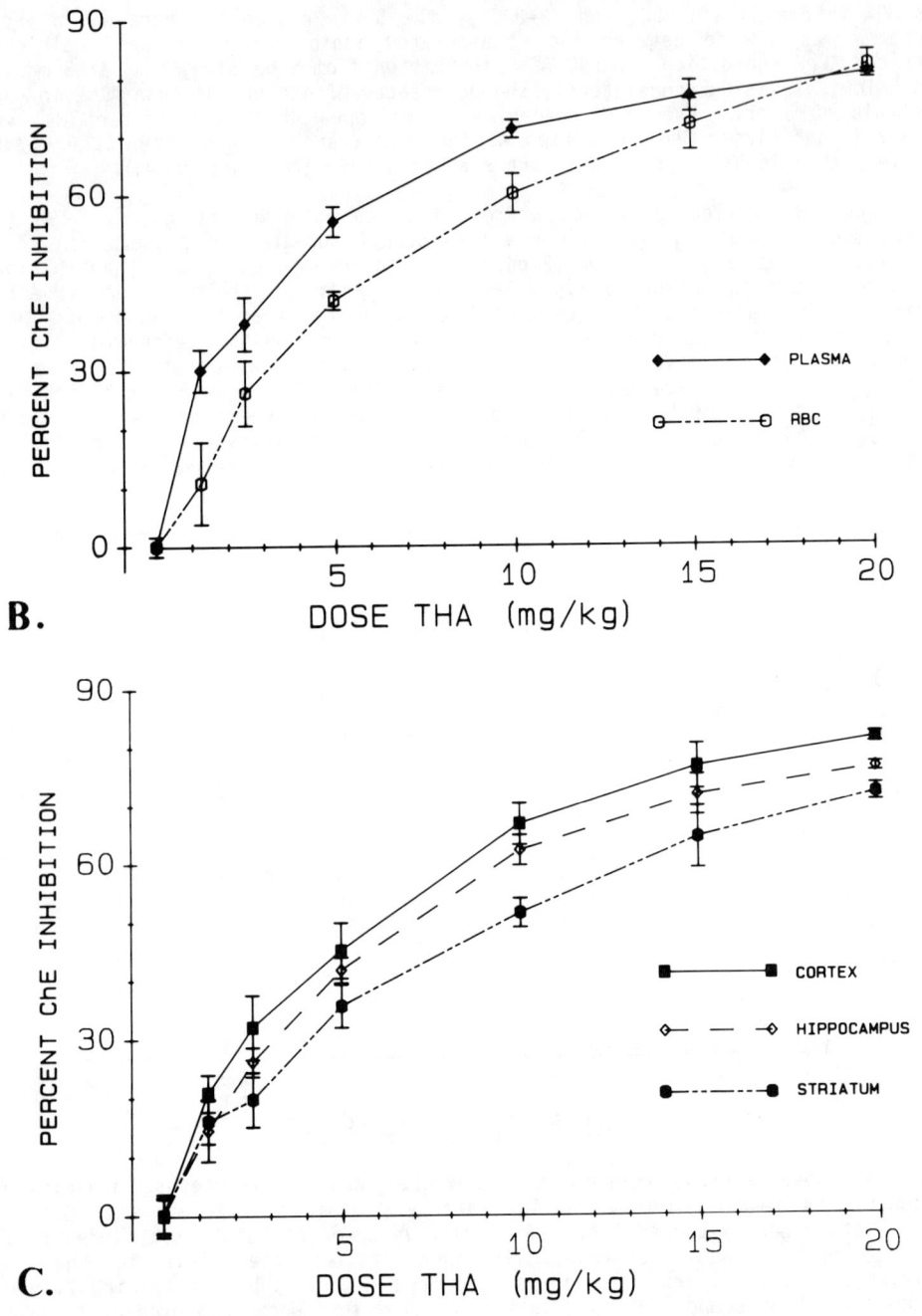

(Figure 4-con't.) C. Dissected brain regions were homogenized by Polytron in Triton-X-EDTA buffer; cortex was assayed at final dilution of 8 X, hippocampus at 10 X and striatum at 21 X. Dose-response functions for AChE inhibition in brain and RBC were parallel; however, after lower doses of THA, plasma ChE inhibition was greater than AChE inhibition in RBC or brain.

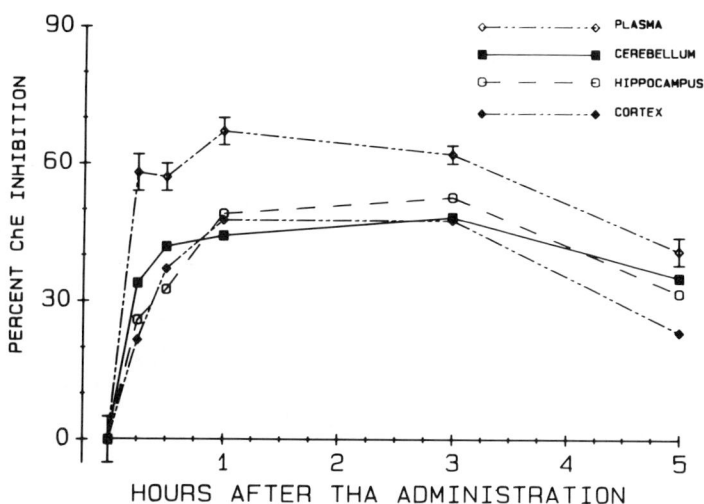

Figure 5. Time course of cholinesterase inhibition after THA administration: Experiment I. Rats were killed from 15 min to 5 hr after 5 mg/kg s.c. injection of THA (N=4-7 per time point) and ChE activity was determined. Inhibition is expressed as percent reduction compared to controls pretreated with saline at various times (N = 10 pooled). Dilutions were as specified in Figure 4 and 8-fold for cerebellum.

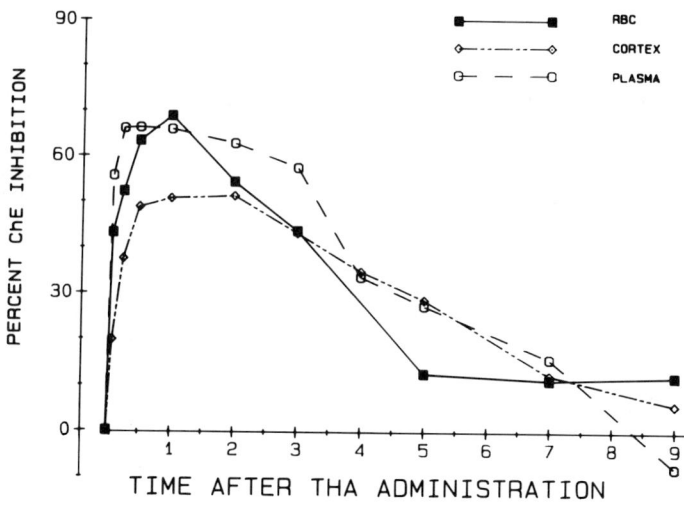

Figure 6. Time course of cholinesterase inhibition: Experiment II. The preceding experiment was replicated to further characterize the onset and duration of ChE inhibition after THA (5 mg/kg, s.c.). Rats were killed from 5 min to 9 hr after drug (N=4-7 per time point) and ChE activity was determined in cortex, plasma and RBC as described in Figure 4 compared to saline controls (N = 17).

inhibition was greater in both plasma and RBC compared to cortex. Inhibition of RBC AChE activity was less sustained than plasma or cortex, and the decline in inhibition in plasma and cortex was parallel; both returned to control activity by 9 hr after drug.

Thus, our results show a greater magnitude and duration of ChE inhibition after THA than was anticipated on the basis of literature reports. The long-lasting behavioral and biochemical effects we observed in THA treated rats agree with the pharmacokinetic studies in monkey and man indicating a long half-life of THA (Park et al., 1986), and slower accumulation of THA in brain compared to plasma found after i.p. injection in rats (Hsieh et al., 1983).

Chronic PHYSO and tacrine: Behavioral and presynaptic tolerance

The possible development of tolerance to effects of reversible ChE inhibitors has not been systematically investigated in man or animal models. During the first few days of oral PHYSO, increasing cognitive improvement was found in AD patients (Thal et al., 1986). However, patients who initially improved after PHYSO or THA may become unresponsive after months of continued treatment (ibid, Summers et al., 1986; Thal et al., 1986). This may reflect the progression of neuronal degeneration such that ChE inhibitors are no longer able to overcome the cholinergic deficiency. Alternatively, the loss of efficacy may reflect the development of tolerance to the actions of these drugs on cholinergic transmission. Whereas tolerance to the therapeutic action may be a limitation of chronic cholinomimetic treatment, tolerance to the adverse actions of these drugs, e.g. gastrointestinal distress, may broaden the therapeutic window. For this reason, gradual increase of dosage may be an advantage (Summers et al., 1986). Studies in experimental animals have shown the development of tolerance to several physiologic and behavioral actions of irreversible ChE inhibitors paralleled by the down-regulation of nicotinic and muscarinic receptors (e.g. Costa and Murphy, 1983; McKinney and Coyle, 1982). However, the long-lasting effects of these irreversible drugs on AChE (up to 14 days) complicate true tests of tolerance and withdrawal phenomena.

To evaluate the potential of PHYSO to produce tolerance, we injected a high dose (0.88 mg/kg s.c.) of PHYSO hemisulfate twice daily for up to one month. Rats were pretreated with methylatropine to reduce peripheral muscarinic effects. Behavioral observations throughout the course of this treatment indicated a reduction in the intensity and duration of tremors and splay posture in 7 of 8 rats. After 2-3 weeks of PHYSO treatment, the drug began to induce stereotyped movements (repetitive rearing, foraging movements and grooming of the forepaws). The inhibition of brain regional AChE activity 35 min after drug was equivalent after acute and 30 days chronic PHYSO. Effects on high affinity choline uptake (HACU) were examined as a measure of the impact of PHYSO on presynaptic cholinergic activity. Neither the inhibition of hippocampal HACU (44%) nor the lack of effect in striatum were altered by the chronic drug pretreatment. However, cortical HACU was inhibited after chronic PHYSO (-20%), whereas it was slightly increased (+20%) after acute treatment ($p < .05$). Moreover, muscarinic receptor binding was slightly, but significantly, reduced in the cortex (-10%; not shown). A similar trend occurred in striatum, but hippocampal ligand binding was unaffected. Thus, chronic PHYSO pretreatment produced clear alterations in the behavioral consequences of the drug which might be related to alterations in both pre- and postsynaptic mechanisms. Nonetheless, the changes were less pronounced than those after irreversible ChE inhibitors (Yamada et al., 1983) and were only significant in cortex.

Our behavioral results (above) and neurochemical studies in rat brain (Hallak and Giacobini, 1986) indicate that even at the near-maximal dose of PHYSO used for our chronic studies, PHYSO is effective for less than 2 hr. This short-action may explain the minimal effect on muscarinic receptors and

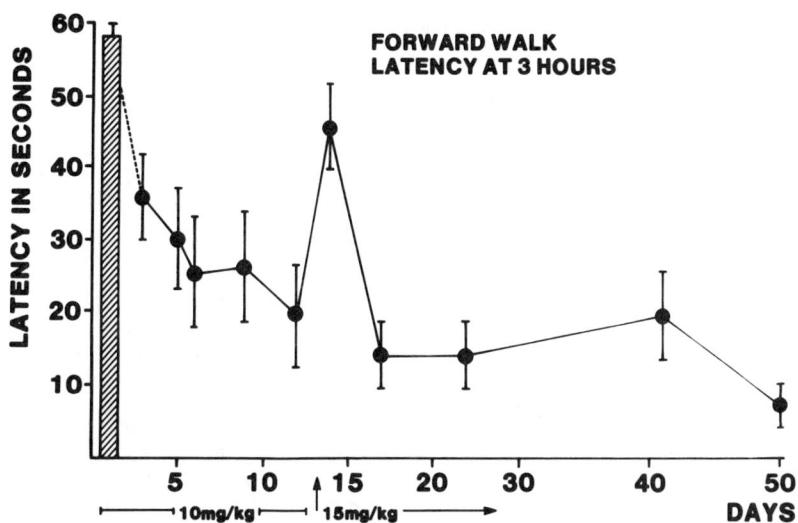

Figure 7. **Behavioral inhibition during chronic THA administration.** Rats received 10 mg/kg THA s.c. b.i.d. from day 1 to day 14. The dose was increased to 15 mg/kg, b.i.d. for one week and then lowered back to 10 mg/kg. Forward walk latency and duration of splay were tested after the A.M. dose. The effect of 10 mg/kg THA on walk latency at 3 hr after drug decreased during the first two weeks. The deficit was reinstated by 15 mg/kg, but tolerance developed rapidly to this dose.

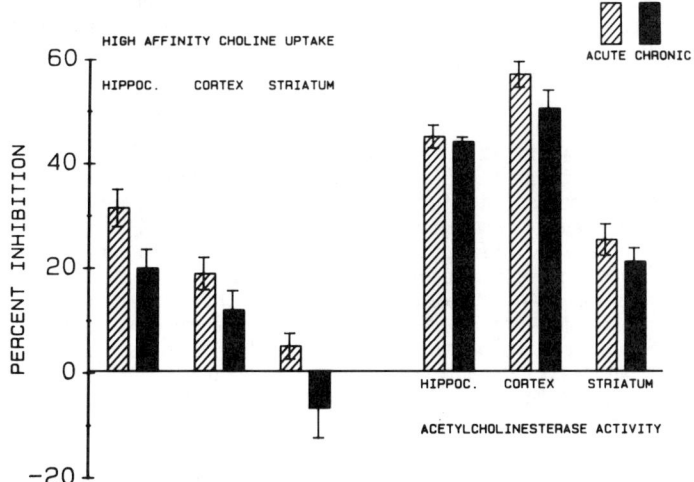

Figure 8. **Effect of acute and chronic THA on high affinity choline uptake and brain acetylcholinesterase activity.** Rats were pretreated for 48-54 days with 10-15 mg/kg b.i.d. THA or vehicle s.c. (details in Figure 7). On the day of experiment, rats (N= 7-9 per group) were killed 2 hrs after 10 mg/kg THA and compared to saline controls. HACU was reduced after acute THA in hippocampus and to a lesser extent in cortex, but not in striatum. Effects on HACU were less after chronic THA. Plasma and brain ChE activity was comparably reduced after acute and chronic THA. The tissue dilution at assay was 29 X for cortex, 24-26 X for hippocampus and 52-65 X for striatum. Plasma ChE was assayed at 2 X.

incomplete tolerance after one month of b.i.d. injection. The relatively long action of THA is ideal for testing the possible development of tolerance.

We used the behavioral tests described above to quantify the development of tolerance to THA. For the first 14 days, rats were injected b.i.d. with 10 mg/kg THA after methylatropine pretreatment. The latency of forward walk at 3 hr after THA declined progressively during the first week of injection (Figure 7). On day 14, the rats received a challenge dose of 15 mg/kg and were compared to drug-naive controls receiving the same dose (acute). The inhibition of forward walking was reinstated by the increased dose but tolerance rapidly developed as the higher dose was continued. By contrast, the effect of THA on splay showed marked tolerance and was not reinstated by the challenge dose. The THA injections were continued at 15 mg/kg b.i.d. for one week, but the sporadic development of seizures at this dose prompted a return to the 10 mg/kg dose.

After 48-54 days of b.i.d. treatment, marked tolerance to effects on both locomotion and splay was observed in comparison to rats receiving an acute injection of 10 mg/kg THA. The rats were sacrificed by decapitation 2 hr after drug for analysis of plasma and brain ChE inhibition and HACU in specific brain regions (Figure 8). There was no significant change in the inhibition of plasma ChE or AChE activity in cortex, hippocampus or striatum after chronic THA pretreatment. HACU was measured in synaptosomal P_2 fractions prepared from each region as described by Sherman et al. (1978), except that 10 µM hemicholinium-3 (rather than sodium-free medium) was used to define the HACU in the presence of 0.75 µM ^{3}H-Ch. HACU in hippocampal synaptosomes was reduced 31 ± 4% from control after acute THA and cortical HACU was reduced 20 ± 4%. After 48-54 days of administration, THA had less effect on choline uptake. The reduction was only 19 ± 3% in hippocampus and 12 ± 3% in cortex. Striatal HACU was unaffected by either acute or chronic THA. These results contrast with PHYSO in that the effect of acute THA treatment is less marked in hippocampus and less regionally specific; and tolerance to HACU inhibition was not observed after the 30 day PHYSO.

CONCLUSIONS

Our results from clinical trials of oral and i.v. PHYSO indicate that blood ChE inhibition provides a sensitive and quantitative measure of biological efficacy. In future clinical studies, this measure would offer a means of defining the therapeutic window associated with cognitive improvement and help titrate dosage to avoid inter-patient variability due to pharmacokinetic differences. Our results in experimental animals indicate that blood ChE inhibition may also serve as a means to compare putative therapies for AD and to compare species so that the doses tested in animals can be more relevant to the clinical situation. However, we found that the methodology for estimation of ChE inhibition after THA administration must take into account the reversibility of this drug's action. The dilution of tissue for ChE assay results in underestimation of the degree of inhibition after THA, but not PHYSO which carbamylates the catalytic site of ChE. Using minimum tissue dilutions we have shown that THA is less potent but equally efficacious compared to PHYSO. Our behavioral results and analysis of brain AChE inhibition both indicate that THA is slower in onset, but much longer in duration of effect than PHYSO. These properties are very important considerations for comparisons of the behavioral (especially cognitive) effects of the drugs. The results with THA indicate that plasma ChE inhibition overestimates the degree of inhibition in brain under some circumstances, and further studies are needed to define the relationship in the low dose range which results in enhanced cognitive performance. We examined high affinity choline uptake as a measure of the inhibitory action of these drugs on presynaptic cholinergic mechanisms and show quantitative differences between THA and PHYSO. THA produced less marked inhibition of HACU in hippocampus than PHYSO, but cortical HACU was also inhibited. The effect of THA but not PHYSO on HACU showed tolerance with repeated administration. Behavioral tolerance was also more marked with THA. Further studies will be necessary to determine whether tolerance occurs to lower doses of THA and whether this limits the long-term efficacy of the inhibitor for improvement of cognitive performance. The inhibition of brain AChE does not show tolerance after either chronic THA or chronic PHYSO.

ACKNOWLEDGEMENTS

This work was supported by SIU Central Res. Funds, ADRDA, and IL Dept. of Pub. Health Alzheimer Research Funds. Technical assistance by Angela Guerrettaz, and manuscript preparation by Roberta Melton are gratefully acknowledged.

REFERENCES

Bajgar, J., Fusek, J., Patocka, J. and Hrdina, V. (1979) Physiol. Bohemoslov 28:31-34.
Bajgar, J. and Patocka, J. (1977) Collect. Czech. Chem. Commun. 42:2723-2727.
Bajgar, J., Patocka, J., Fusek, J., Hrdina, V. (1984) Sb. Ved. Pr. Lek. Fak. Univ. Karlovy 27:425-435.
Barrow, M.E.H. and Johnson, J.K. (1966) Brit. J. Anaesth. 38:420-431.
Bartus, R.T., Dean, R.L., III, Beer, B., Lippa, A.S. (1982) Science 217:408-417.
Bell, C., Gershon, S., Carroll, B. and Holan, G. (1964) Arch. Int. Pharmacodyn. 147:8-25.
Beneveniste, D., Hemmingsen, L., Juul, P. (1967) Acta Anaesth. Scand. 11: 297-309.
Blackwood, D.H.R. and Christie, J.E. (1986) Biol. Psychiat. 21:557-560.
Bowen, D.A., Smith, C.B., White, P., and Davison, A.M. (1976) Brain 99:459-496.
Brinkman, S.D. and Gershon, S. (1983) Neurobiol. Aging 4:139-145.
Brown, H. (1971) Pharmacologia 21:294-301.
Costa, L.G. and Murphy, S.D. (1983) J. Pharmacol. Exp. Ther. 226:392-397.
Coyle, J.T., Price, D.L. and DeLong, M.R. (1983) Science 219:1184-1190.
Curley, W.H., Standaert, F.G. and Dretchen, K.L. (1984) J. Pharmacol. Exp. Ther. 228:656-661.
Davies, P. (1979) Brain Res. 171:319-327.
de la Lande, I.S. and Porter, R.B. (1963) Austral. J. Exp. Biol. 41:149-162.
Ehlert, F.J., Kokka, N. and Fairhurst, A.S. (1980) Molec. Pharmacol. 17:24-30.
El-Kammah, B.M., El-Gafi, S.H., El-Sherbiny, A.M. and Abdel-Kader, M.M. (1975) J. Egypt. Med. Assoc. 58:559-567.
Flood, J.F., Smith, G.E. and Cherkin, A. (1985) Psychopharmacology 86:61-67.
Flood, J.R. and Cherkin, A. (1986) Behav. Neural. Biol. 45:169-184.
Francis, P.T., Palmer, A.M., Sims, N.R. et al. (1985) N. Engl. J. Med. 313: 7-11.
Freeman, S.E. and Turner, R.J. (1970) J. Pharmacol. Exp. Ther. 174:550-559.
Fusek, J., Patocka, J., Bajgar, J., Urban, R., Kolesar, J., Herink, J., Hrdina, V. (1975) Activ. Nerv. Super. 17:252.
Gershon, S., Olariu, J. (1960) J. Neuropsychiat. 1:283-292.
Hallak, M. and Giacobini, E. (1986) Neurochem. Res. 11:1037-1048.
Heilbronn, E. (1961) Acta Chem. Scand. 15:1386-1390.
Ho, A.K.S. and Freeman, S.E. (1965) Nature 205:1118-1119.
Hsieh, J.Y.K., Yang, R.K. and Davis, K.L. (1983) J. Chromatogr. 274:388-392.
Hollander, E., Davidson, M., Mohs, R.C., Horvath, T.B., Davis, B.M., Zemishlany, Z. and Davis, K.L. (1987) Biol. Psychiat. 22:1067-1078.
Ingram, N.A.W. and Newgreen, D.B. (1983) Am. J. Psychiat. 140:1629-1631.
Itil, T. and Fink, M. (1966) J. Nerv. Ment. Dis. 143:492-507.
Johns, C.A., Levy, M.I., Greenwald, B.S., Rosen, W.G., Horvath, T.B., Davis, B.M., Mohs, R.C. and Davis, K.L. (1983) Banbury Report 15:435-449.
Johnson, C.D. and Russell, R.S. (1975) Analyt. Biochem. 64:229-238.

Karis, J.H., Nastuk, W.L. and Katz, R.L. (1966) Brit. J. Anaesth. 38:762-774.
Kaul, P.N. (1962) J. Pharm. Pharmacol. 14:243-248.
Kaye, W.H., Sitaram, N., Weingartner, H., Ebert, M.H., Smallberg, S. and Gillin, J.C. (1982) Biol. Psychiat. 17:275-280.
Kling, A., Fitten, L.J., Perryman, K., Tachiki, K. (1985) Soc. Neurosci. Abstr. 11:381.
McKinney, M. and Coyle, J.T. (1982) J. Neurosci. 2:97-105.
Mohs, R.C., Davis, B.M., Johns, C.A., Mathe, A.A., Greenwald, B.S., Horvath, T.B. and Davis, K.L. (1985) Am. J. Psychiat. 142:28-33.
Neary, D., Snowden, J.S., Mann, D.M.A., Bowen, D.M., Sims, N.R., Northern, B., Yates, P.O., Davison, A.N. (1986) J. Neurol. Neurosurg. Psychiat. 49:229-237.
Park, T.H., Tachiki, K.H., Summers, W.K., Kling, D., Fitten, J., Perryman, K., Spidell, K. and Kling, A.S. (1986) Analyt. Biochem. 159:358-362.
Patocka, J., Bajgar, J., Bielavsky, J. and Fusek, J. (1976) Coll. Czechoslov. Chem. Commun. 41:816-824.
Perry, E.K., Tomlinson, B.E., Blessed, G., Bergmann, K., Gibson, P.H., and Perry, R.H. (1978) Brit. J. Med. 42:1457-1459.
Porter, R.B. (1965) Brit. J. Pharmac. Chemother. 25:179-186.
Rosic, N. and Milosevic, M.P. (1967) Yugosl. Physiol. Pharmacol. Acta 3:43-47.
Schwartz, A.S. and Kohlstaedt, E.V. (1986) Life Sci. 38:1021-1028.
Sherman, K.A., Kumar, V., Ashford, J.W., Murphy, J.M., Elble, R.J. and Giacobini, E. (1988) Aging, Vol. 33 (R. Strong, W.G. Wood and W.J. Burke, Eds.), Raven Press, NY, pp. 71-90.
Sherman, K.A., Zigmond, M.J. and Hanin, I. (1978) Life. Sci. 23:1863-1870.
Steinberg, G.M., Mednick, M.L., Maddox, J., Rice, R., Cramer, J. (1975) J. Med. Chem. 18:1056-1061.
Summers, W.K., Majorski, L.V., Marsh, G.M., Tachiki, K., Kling, A. (1986) N. Eng. J. Med. 315:1241-1245.
Summers, W.K., Viesselman, J.O., Marsh, G.M., Candelora, K. (1981) Biol. Psychiat. 16:145-153.
Thal, L.J., Masur, D.M., Sharpless, N.S., Fuld, P.A. and Davies, P. (1986) Prog. Neuro-Psychopharmacol Biol. Psychiat. 10:627-636.
Tonkopii, V.D., Padinker, E.P. (1975) Biull. Eksp. Biol. Med. 79:51-52.
Tonkopii, V.D., Prozorovskii, V.P. and Suslova, I.M. (1976) Biull. Eksp. Biol. Med. 82:947-950.
Yamada, S., Isogai, M., Okudaira, H. and Hayashi, E. (1983) J. Pharmacol. Exp. Ther. 226:519-525.

CHOICE REACTION TIME IN AGED RHESUS MONKEYS AS A MODEL OF AD

J.W. Ashford and J.L. Bice
Dept. of Psychiatry, S.I.U. School of Medicine, Springfield, IL

INTRODUCTION

Alzheimer's disease (AD) has been a recognized clinical entity for 80 years, yet has still not been reproduced in the laboratory. The only factor clearly associated with the general incidence of this disease is increasing age (Horner, 1987; Shoenberg et al., 1987). (Note: Down's syndrome, history of head trauma and autosomal dominant inheritance apply clearly only to small portions of the population). Therefore, an important line for developing a model of AD is the investigation of aged animals.

Several animal models show neuropathological changes similar to those of AD, but none are identical (Wisniewski et al., 1970). However, the aged monkey is susceptible to the development of senile plaques (Wisniewski et al., 1973) and these plaques show a distribution similar to that of the aged human (Struble et al., 1985). The senile plaques seen in aged animals of many higher mammalian species are immunologically similar to those of aged humans and Alzheimer patients (Selkoe et al., 1987). The consistent pathological distinction is that the neurofibrillary tangles, which are characteristic of AD and are commonly seen in elderly humans (Ulrich, 1985), are not seen in the monkey; understanding this distinction may be a key to elucidating the etiology of AD.

While the aged monkey does not have the complete picture of Alzheimer neuropathology, this model also shows memory deficits which are comparable to impairments seen in aging humans and AD patients (Bartus et al., 1978; Bartus and Dean, 1987; Davis et al., 1982; Presty et al., 1987). Short-term memory in the young adult monkey has been further shown to be modified by cholinergic (Bartus, 1979) and anticholinergic manipulations (Bartus and Johnson, 1976; Penatar and McDonough, 1983; Pontecorvo and Evans, 1985; Levin and Bowman, 1986) in a manner similar to that seen in humans (Aigner and Mishkin, 1986). However, short-term memory in monkeys shows only a small impairment after damage to the basal forebrain cholinergic system (Aigner et al., 1987).

Underlying the memory deficit of AD, there is likely to be a basic deficit of neuroplasticity (Ashford and Jarvik, 1985). This link suggests that all information processing functions which access memory will be impaired by AD, while those processes associated strictly with perception will be preserved. In support of this concept, choice reaction time is slowed by AD while simple reaction time is not (Pirozzolo et al., 1981; 1985). Therefore, these two types of information processing pathways should be differentially affected in an Alzheimer model. We chose to study the relation between simple and choice reaction time in the aged monkey model, adding the challenge of cholinergic and anticholinergic agents to more extensively test for a similarity to the Alzheimer syndrome (Bartus et al., 1982).

Cholinergic neurons function in several different brain systems including the nucleus basalis of Meynert and the basal ganglia. Those neurons in the nucleus basalis of Meynert are highly affected in AD (Whitehouse et al., 1982) though acetylcholine neurons in other areas, such as the basal ganglia, are also affected (Rossor et al., 1982). The existence of neurons using the same neurotransmitter but anatomically localized in vastly different situations has made it difficult to explore the pharmacologic properties of cholinergic neurons. Yet, the possible role of cholinergic drugs is of importance in the treatment of AD (Ashford et al., 1981; Brinkman and Gershon, 1983; Jorm, 1986), and the effect of antimuscarinic agents on memory has been advocated as a model of AD (Drachman, 1977).

METHODS

Subjects

This study used three male rhesus monkeys, Macaca mulatta, ages over 22 years. The animals, retired breeders, were experimentally naive prior to behavioral training. Animals were maintained on a diet of biscuits (Purina monkey chow) and fresh fruit daily. To enhance performance motivation, fluid was restricted for up to 23 hours prior to testing. This schedule of fluid deprivation has been used with adequate motivational effects and without harmful health effects. Fluid intake was closely monitored daily. The animals were housed individually in a room with automatically controlled lighting (light:dark, 12:12).

Behavioral paradigm

The paradigm assesses the speed of simple and complex (choice) information processing and the accuracy of binary memory during task performance in the context of a simultaneous simple and choice reaction time task.

The first step in training the monkeys was to adapt them to the restraining chair and testing booth. As the animals learned to sit quietly, facing the video screen in the testing booth, they were introduced to the reward delivery system. The monkeys quickly learned to manipulate the joystick lever to obtain a squirt of sweetened grape juice for reward. The next step was to pair the correct motor response (a lever press) to its corresponding screen color. A red screen required a downward deflection, while green demanded an upward deflection. The task was entirely computer automated (Tandy Radio Shack Color Computer). Once all three monkeys had mastered the choice reaction time task, they were shown a blue screen as a cue stimulus. Correct response to the blue stimulus was a downward deflection, a simple reaction. Once the second task was well learned, the two tasks were combined. Criterion for training completion was performance for 5 consecutive days above 90%

correct. (Training these animals took about 8 months.)

In the combination task used for drug testing, the monkey was first presented the blue video screen. Correct response was required for the trial to continue. After a 2-3 second delay, the red or green screen was presented. Correct response was then rewarded (1 cc of juice). The response to each stimulus was recorded by the computer (correctness and reaction time).

Anticholinergic agent study

Each monkey was entered into a counterbalanced order of drug (scopolamine hydrobromide, .4 mg/cc, Elkins-Sinn, Cherry Hill, NJ), active placebo (glycopyrrolate, .2 mg/cc, Elkins-Sinn, Cherry Hill, NJ), and inactive placebo (saline). The drug and the placebo sessions were separated by sessions in which no agent was administered. Drug doses were determined in an earlier dose-finding phase. The maximum tolerable dose (dose 5) was used as a high dose in the experiment. In these monkeys, the maximum tolerable dose at which some performance occurred was approximately 50 micrograms per 2/3 root of body weight in kilograms (surface area). In the experiment, 1/5, 2/5, 3/5, 4/5, and 5/5 of the maximum dose was given. Each alternating week for 9 weeks, monkeys received the drug and complementary placebos in a specific order, unique to that monkey. The doses were also given in an order unique to each monkey. The dose of scopolamine and glycopyrrolate was the same for a given monkey in a given week. For drug/placebo sessions, each monkey received an intramuscular injection of either scopolamine, glycopyrrolate, or saline while in his home cage. All injections were brought to a constant volume of 3 cc with sterile saline. Two hours following the injection, the monkey was placed in the testing booth and performance tested for 1 hour (160 trials). Therefore, the measured reaction times occurred 2-3 hours following drug administration, giving the scopolamine sufficient time to develop its behavioral effects (Safer and Allen, 1971).

Performance of the animals during scopolamine administration was analyzed and compared to both control conditions (glycopyrrolate and saline) for responsiveness (was the animal working), accuracy (generally the performance should exceed 90% correct for each aspect of the task) and reaction time: simple and choice. These aspects of behavioral performance test for effects of general motor function and allow for different aspects of behavior to be distinguished.

Cholinergic Agent Study

During the weeks between scopolamine, glycopyrrolate and saline testing, the animals were studied following administration of a cholinergic drug (physostigmine salicylate, 1 mg/cc, Forest Pharmaceutical, St. Louis, MO), active placebo (pyridostigmine bromide, 12 mg/cc, Roche Laboratories, Nutley, NJ) and inactive placebo (saline). The cholinergic drug testing was carried out in the same manner as the anticholinergic study. In the dose-finding phase, the maximum tolerable dose of physostigmine was determined to be 160 micrograms per 2/3 root of body weight in kilograms. The maximum dose of pyridostigmine was determined to be 5 milligrams per 2/3 root of body weight in kilograms. As in the anticholinergic study, doses 1-5 and placebo agents were given in a specific, unique order to each monkey. The cholinergic agents and saline were administered orally one hour prior to testing. Each dose was brought to a constant volume of 5 cc with sweetened grape juice. The animals were tested for 1 hour, 160 trials. Performance for each monkey during cholinergic trials was analyzed in the same manner as the anticholinergic data.

RESULTS

Glycopyrrolate, a polar drug, which is not supposed to enter the brain, may have had some effect on the animals' performance, but this was not consistent. One strictly peripheral effect would be a dryness of the mouth, which should increase the animals' thirst and enhance performance. Other peripheral side-effects could distract the animals (e.g.: dry eyes, blurred vision). Glycopyrrolate is thought to have twice the peripheral potency of scopolamine, but some of the compound may enter the brain as does methscopolamine (Bohdanecky et al., 1967).

Scopolamine showed obvious effects, even at the lower doses. The 2/5 dose of scopolamine caused a substantial decrease in responsiveness. This effect clearly contrasted with the inactive and active placebo data. In fact, across the three elderly monkeys, the major effect of scopolamine was to reduce responsiveness. Little fluctuation of responsiveness was noted for the placebo conditions, levels varying for two animals between 85-95%, and for the third, 65-75 %. The two more responsive animals were more sensitive to low doses of scopolamine, dropping below 20% responsiveness at the 3/5 dose. The responsiveness problem was mainly manifest as a failure to respond to the blue stimulus. If the animal made that initiating step, even in the presence of higher doses of scopolamine, responses to the blue stimulus may actually have been more rapid than usual. When the animal made that initial response, the response correctness and the choice response latency were minimally affected. Because of the decrease of responsiveness, the effect of scopolamine on reaction time relative to the placebos could not be assessed.

Pyridostigmine, the active control, had no consistent effect on either simple or choice reaction time. Similarly, in all but the highest dose (dose 5), no consistent effects were seen on correctness or responsiveness relative to the saline control. At the highest dose, pyridostigmine had an adverse effect on the performance of two of the monkeys, perhaps due to cholinergic side-effects.

Physostigmine generally had no consistent effect on either reaction time measure compared to saline and pyridostigmine controls. No major effects of the drug were detected in responsiveness or correctness at the doses administered. Following the initial study, a higher dose of physostigmine (192 ug/2/3 root body weight) was administered to each monkey in two separate weeks. Again, no effect on behavior or reaction time was observed.

CONCLUSIONS

Scopolamine toxicity has been advocated as a model for

Alzheimer's disease because of its severe effect on memory function in normal humans. Further, scopolamine has been shown to slow choice reaction in normal adults (Caine et al., 1981; Nissen et al., 1987). Choice reaction time has been shown to be an index of the severity of Alzheimer's disease. However, in this experiment, scopolamine made its major impact on responsiveness, masking an effect on either simple or choice reaction time (in some cases a more rapid reaction time was noted). Failure to respond might be due to effects on any of the numerous central cholinergic pathways, and might best be related to the delirium which occurs with toxic anticholinergic serum levels in humans (Miller et al., 1988). The monkeys showed erratic behavior, more aggressiveness and less coordination at the higher doses. Though failure to recall the rules of the task and psychosis may be compared to deficits seen in severe Alzheimer's patients, we were not able to document a scopolamine-induced prolongation of choice reaction time in aged monkeys as is seen in mild AD.

There has been considerable difficulty in the literature with documenting a beneficial effect from physostigmine (Ashford and Jarvik, 1981; Jorm, 1986). However, some primate studies have shown positive effects in spatial attentive memory (Bartus and Johnson, 1976), in the delayed match-to-sample task (Penatar and McDonough, 1983) and in a short-term memory task (Aigner and Mishkin, 1986; Aigner et al., 1987). Moreover, bethanechol can improve reaction time in AD patients (Davous and Lamour, 1985). The findings in these positive studies have generally been small and have represented extremely diligent work. AD affects many brain systems other than the cholinergic system, and some of the behavioral deficits of AD may be related more to those other deficits than the cholinergic deficits.

The possibility that refinements of cholinergic enhancement may lead to more substantial benefits for AD patients has served as an impetus to a host of scientific studies. To more clearly understand the positive effect of cholinergic agents, more simple models must be designed to demonstrate the value of these drugs. In the course of these studies, a better animal model for AD is likely to emerge. The approach of analyzing information processing utilizes easily measured biological phenomena related to AD (Ashford and Fuster, 1985) and is one approach which warrants further study.

SUMMARY

Choice reactions are slowed relative to simple reactions in Alzheimer's disease (AD) patients (Pirozzolo et al., 1981; 1985). To further elucidate this behavioral component of information processing as a model of AD, we administered anticholinergic and cholinergic agents to three elderly monkeys (>22 years) performing a concurrent simple and choice reaction time task. Scopolamine was administered (doses of 10, 20, 30, 40, 50 ug/2/3 root of body weight in kg) once a week. Inactive placebo (saline) and active placebo (glycopyrrolate) were given on the other days (drug administered 2 hours before testing). At low doses, scopolamine slightly impaired responsiveness but not correctness, while reaction time was slightly improved. However, at higher doses, scopolamine caused a major decrease of animal responsiveness and correctness. Only at the highest dose was reaction speed slowed below the saline mean. Low doses of scopolamine may improve motor function by affecting the basal ganglia as such drugs benefit Parkinsonian patients. However, the behavioral disruption at the higher doses is similar to that seen with AD patients performing related tasks. Glycopyrrolate showed no effect on performance, though delayed effects were suspected. Oral physostigmine (5 graded doses up to a maximum of 160 ug/2/3 root of body weight) and an active control (pyridostigmine) were tested using the same paradigm in another study using the same elderly animals. No significant effects of either drug were observed in the monkeys at the doses administered. Two additional trials with a 20% higher dose still showed no effects on behavior. Thus, we were unable to demonstrate selective effects of cholinergic drugs on choice response latencies, even though this measure is impaired in AD.

ACKNOWLEDGEMENTS

The authors would like to acknowledge the technical assistance of Harvey Edwards in training and testing the monkeys.

REFERENCES

Aigner, T.G. and M. Mishkin. 1986. The effects of physostigmine and scopolamine on recognition memory in monkeys. Behav. and Neural Biology 45:81-87.

Aigner, T.G., S.J. Mitchell, J.P. Aggleton, M.R. DeLong, R.G. Struble, D.L. Price, G.L. Wenk, and M. Mishkin. 1987. Effects of scopolamine and physostigmine on recognition memory in monkeys with ibotenic-acid lesions of the nucleus basalis of Meynert. Psychopharm. 92:292-300.

Ashford, J.W. and L.F. Jarvik. 1981. Alzheimer's disease: Does neuron plasticity predispose to axonal neurofibrillary degeneration? New Engl. J. Med. 313(6):388-389.

Ashford, J.W., S. Soldinger, J. Schaeffer, L. Cochran and L.F. Jarvik. 1981. Physostigmine and its effect on six patients with dementia. Am. J. Psychiatry 138:829-830.

Ashford, J.W. and J.M. Fuster. 1985. Occipital and inferotemporal responses to visual signals in the monkey. Exp. Neurol. 90:444-466.

Bartus, R.T. and H.R. Johnson. 1976. Short-term memory in the rhesus monkey:Disruption from the anticholinergic scopolamine. Pharm. Biochem. & Behav. 5:39-46.

Bartus, R.T., D. Fleming and H.R. Johnson. 1978. Aging in the rhesus monkey:Debilitating effects on short-term memory. J. Geront. 33(6):858-871.

Bartus, R.T. 1979. Physostigmine and recent memory: Effects in young and aged nonhuman primates. Science 206:1087-1089.

Bartus, R.T., R.L. Dean III, B. Beer and A.S. Lippa. 1982. The cholinergic hypothesis of geriatric memory dysfunction Science 217:408-417.

Bartus, R.T. and R.L. Dean. 1987. Animal Models for age-related memory disturbances. In Liss, A.R. (ed.) Animals Models of Dementia 69-79.

Bohdanecky, Z., M.E. Jarvik and J.L. Carley. 1967. Differential impairment of delayed matching in monkeys by scopolamine and scopolamine methylbromide. Psychopharmacologia (Berl.) 11:293-299.

Brinkman, S.D. and Gershon, S. 1983. Measurement of cholinergic drug side effects on memory in Alzheimer's disease.

Neurobiol. of Aging 4:139-145.

Caine, E.D., H. Weingartner, C.L. Ludlow, E.A. Cudahy and S. Wehry. 1981. Psychopharm. 74:74-80.

Davis, R.T., C.L. Bennett, W.P. Weisenburger. 1982. Repeated measurements of forgetting by rhesus monkeys (macaca mulatta). Perceptual and Motor Skills 55:703-709.

Davous, P. and Y. Lamour. 1985. Bethanecol decreases reaction time senile dementia of the Alzheimer type. J. Neurol. Neurosurg. Psychiat. 48:1297-1299.

Drachman, D.A. 1977. Memory and cognitive function in man:Does the cholinergic system have a specific role? Neurol. 27:783-790.

Horner, RD. 1987. Age at onset of Alzheimer's disease:Clue to the relative importance of etiologic factors? Am. J. Epidemiol 126(3):409-414.

Jorm, A.F. 1986. Effects of cholinergic enhancement therapies on memory function in Alzheimer's disease:A meta-analysis of the literature. Australian New Zealand J. Psychiat. 20:237-240.

Levin, E.D. and R.E. Bowman. 1986. Scopolamine effects on Hamilton search task performance in monkeys. Pharmacol. Biochem. & Behav. 24:819-821.

Miller, P.S., J.S. Richardson, C.A. Jyu, J.S. Lemay, M. Hiscock, D.L. Keegan. 1988. Association of low serum anticholinergic levels and cognitive impairment in elderly presurgical patients. Am. J. Psychiat. 145:342-345.

Nissen, M.J., D.S. Knopman, D.L. Schacter. 1987. Neurochemical dissociation of memory systems. Neurol. 37:789-794.

Penatar, D.M. and J.H. McDonough, Jr. 1983. Effects of cholinergic drugs on delayed match-to-sample performance of rhesus monkeys. Pharmacol. Biochem. & Behav. 19:963-967.

Pirozzolo, F.J., K.J. Christensen, K.M. Ogle, E.C. Hansch and G.W. Thompson. 1981. Simple and choice reaction time in dementia: Clinical implications. Neurobiol. Aging 2:113-117.

Pirozzolo, F.J., R.K. Mahurin, D.W. Loring, S.H. Appel and G.J. Maletta. 1985. Choice reaction time modifiability in dementia and depression. Intern. J. Neuroscience 26:1-7.

Pontecorvo, M.J. and H.L. Evans. 1985. Effects of antiracetam on delayed matching-to-sample performance of monkeys and pigeons. Pharmacol. Biochem. & Behav. 22:745-752.

Presty, S.K., J. Bachevalier, L.C. Walker, R.G. Struble, D.L. Price, M. Mishkin and L.C. Cork. 1987. Age differences in recognition memory of the rhesus monkey (macaca mulatta). Neurobiol. of Aging 8:435-440.

Rossor, M.N., N.J. Garrett, A.L. Johnson, C.Q. Mountjoy, M. Roth and Iversen, L.L. 1982. A post-mortem study of the cholinergic and GABA systems in senile dementia. Brain 105:313-330.

Safer, D.J. and R.P. Allen. 1971. The central effects of scopolamine in man. Biol. Psychiat. 3:347-355.

Schoenberg, B.S., E. Kokmen, H. Okazaki. 1987. Alzheimer's disease and other dementing illnesses in a defined United States population:Incidence rates and clinical features. Ann. Neurol. 22:724-729.

Selkoe, D.J., D.S. Bell, M.B. Podlisny, D.L. Price and L.C. Cork. 1987. Conservation of brain amyloid proteins in aged mammals and humans with Alzheimer's disease. Science 235:873-877.

Struble, R.G., D.L. Price, L.C. Cork, and D.L. Price. 1985. Senile plaques in cortex of aged normal monkeys. Brain Res 361:267-275.

Ulrich, J. 1985. Alzheimer changes in nondemented patients younger than sixty-five:Possible early stages of Alzheimer's disease and senile dementia of Alzheimer type. Ann. Neurol. 17(3):273-277.

Whitehouse, P.J., D.L. Price, R.G. Struble, A.W. Clark, J.T. Coyle and M.R. DeLong. 1982. Alzhimer's disease and senile dementia: Loss of neurons in the basal forebrain. Science 215:1237-1239.

Wisniewski, H.M., A. Hirano. 1970. Neurofibrillary pathology. J. Neuropathol. Exp. Neurol. 29(2):163-176.

Wisniewski, H.M., B. Ghetti and R.D. Terry. 1973. Neuritic (senile) plaques and filamentous changes in aged rhesus monkeys. J. Neuropathol. Exp. Neurol. 32:566-583.

CHOLINESTERASE INHIBITORS INCREASE THE BRAIN'S NEED FOR FREE CHOLINE

Richard J. Wurtman[1], Jan Krzysztof Blusztajn[2],
John H. Growdon[3], and Ismail H. Ulus[4]

[1] Dept. of Brain and Cognitive Sciences, M.I.T., E25-604, Cambridge, MA;
[2] Dept. of Pathology, Boston Univ. School of Medicine, M1009, Boston, MA;
[3] Dept. of Neurology, Mass. General Hospital, ACC730, Boston, MA; [4] Dept. of Pharmacology, Univ. of Uludag Medical Faculty, Duacinari, Bursa, Turkey

Choline molecules undergo two very important metabolic fates within cholinergic nerve terminals (Fig. 1): They can be acetylated, forming the neurotransmitter acetylcholine (ACh), through the action of the enzyme choline acetyltransferase (CAT); or they can be phosphorylated, forming phosphocholine, an intermediate in phosphatidylcholine (PC) biosynthesis, through the action of the enzyme choline kinase (CK) (c.f., Blusztajn & Wurtman, 1983).

At normal tissue choline levels, neither CAT nor CK is saturated with its substrate, hence treatments that modify these levels can have marked effects on the syntheses of ACh or phosphocholine (Cohen and Wurtman, 1976; Haubrich et al., 1975; Millington and Wurtman, 1982; Wecker, 1985). Moreover, neuronal depolarization per se may selectively enhance CAT activity and suppress that of CK, thereby shunting choline into ACh synthesis and away from the formation of phosphocholine and, presumably, PC (Ando et al., 1987).

Three processes are known to provide free choline to cholinergic nerve terminals (c.f. Blusztajn & Wurtman, 1983): 1) enzymatic hydrolysis of intrasynaptic ACh, mediated by acetylcholinesterase (AChE), followed by high-affinity uptake of the free choline thus liberated; 2) transport of circulating choline across the blood-brain barrier, followed by its low-affinity uptake from the brain's extracellular fluid; and, 3) hydrolysis of choline-containing membrane phospholipids, like PC and sphingomyelin. Direct evidence that the latter process occurs in cholinergic cells has recently been attained using a neuroblastoma-derived cell line -- LA-N-2 -- in which the PC was isotopically labeled using 3H-methionine (which, as 3H-S-adenosylmethionine, methylated membrane phosphatidylethanolamine [PE]) (Blusztajn, et al., 1987). 3H-ACh, formed from the 3H-choline moiety in the 3H-PC, was shown to be released from the cells. It is not presently known whether only the PC formed from PE methylation constitutes a choline reservoir for ACh synthesis or whether the fatty acid composition of a PC molecule influences the likelihood that it will provide choline for ACh synthesis.

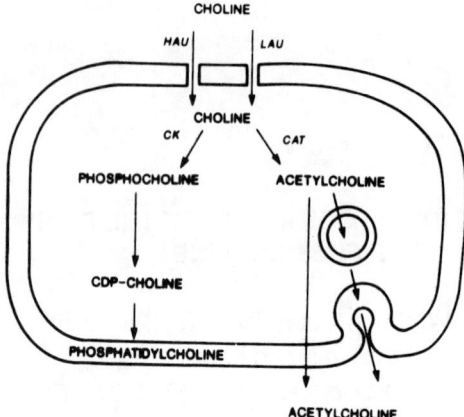

Figure 1. Metabolic fates of choline within cholinergic cerve terminals.
Choline is taken up by a high affinity- (HAU) or low affinity- (LAU) transport system. It may then be acetylated to acetylcholine by choline acetyltransferase (CAT) or phosphorylated to phosphocholine by choline kinase (CK). Phosphocholine is converted to membrane phosphatidylcholine via an intermediate, cytidinediphosphocholine (CDP-choline).

The observations that choline availability can limit the syntheses of both ACh and PC, and that the choline within membrane PC can serve as a precursor for ACh, have led us to examine what happens to membrane PC levels when the quantities of free choline available to cholinergic neurons are smaller than those needed to sustain ACh release. This sort of imbalance might arise as a consequence of prolonged neuronal depolarization or when drugs are given which, like presynaptic muscarinic receptor blockers, increase ACh release per firing. Alternatively, it might result from processes which diminish the amounts of exogenous choline that are available to the neuron -- such as, for example, old age itself (Mooradian, 1988), which apparently reduces the velocity of choline transport across the blood-brain barrier -- or from the chronic administration of drugs which, by inhibiting AChE, block the formation of free choline within the synapse.

To examine these relationships more closely, we have used an experimental system based on electrically-depolarized slices of rat corpus striatum; these are superfused with an artificial medium that lacks exogenous choline but contains an AChE inhibitor, physostigmine (Maire, et al., 1985). In initial studies in which the slices were stimulated for 30 minutes, we observed that, even though no choline had been added to the medium, ACh release continued unabated for the duration of stimulation (Fig. 2). This release could, of course, be enhanced by adding physiologic concentrations of choline to the medium; however, in the absence of such choline, not only were ACh production and release maintained, but tissue levels of choline and ACh failed to decline. However, when hemicholinium-3, a drug known to block choline's high-affinity uptake, was added to the medium, ACh synthesis and release rapidly diminished. We interpreted these findings as indicating that, even though no exogenous choline had been added to the medium, free choline was present just the same, and the most likely source of this free choline was the release from the large phospholipid pool present in cellular membranes. In support of this explanation, the total level of phosphatides (principally PC, PE and phosphatidylserine [PS]) in the slices, expressed per unit DNA, were found to have declined by about 20%, and this decline -- like the stimulation-induced release of ACh -- was blocked in the presence of tetrodotoxin, a drug that interferes with sodium channels (Fig. 3).

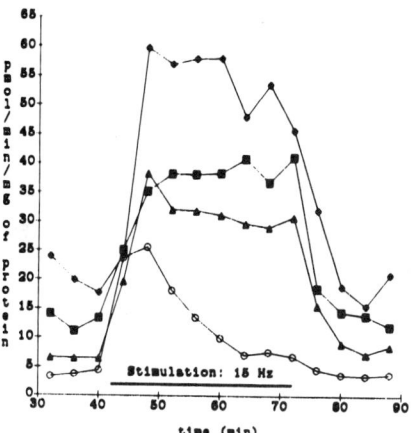

Figure 2. Acetylcholine release from striatal slices.
Slices of rat striatum were superfused at 0.5 ml/min with a physiological solution containing no choline (triangles); 5 µM choline (squares); or 20 µM choline (diamonds). 5 µM of hemicholinium-3 was added to choline-free medium (circles). After a 30 min equilibration, the superfusate was collected at 4 min intervals. Twelve minutes after the start of the collection period, the slices were electrically stimulated at 15 Hz for 30 min (horizontal bar) and then allowed to rest for a further 18 min period. ACh was determined in all fractions.

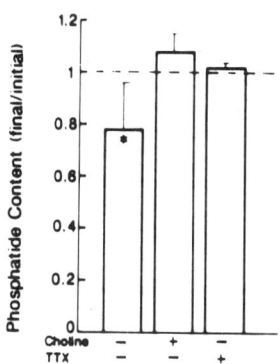

Figure 3. Phospholipids in brain slices.
Slices of rat striatum were superfused with a physiological solution containing: no choline or 20 µM choline or no choline plus 0.5 µM tetrodotoxin. After a 42 min equilibration period, the slices were stimulated electrically (15 Hz) for 30 min and then allowed to rest for an additional 18 min. Each bar represents the ratio of phospholipid phosphorus (in slices) at the end of the collection period ("final") to that at the beginning of the collection period ("initial"). This ratio was found to be 1.06 ± 0.06 in slices that were not subjected to electrical field stimulation under all conditions tested. The values are means $\pm$ S.E. of 2-6 perfusion studies. Statistical analyses were performed by the Wilcoxon matched-pairs signed ranks test comparing the electrically stimulated slices to those at rest. ($p<0.05$). TTX, tetrodotoxin.

We subsequently modified this experimental system to allow us to stimulate the striatal slices repeatedly, i.e., for as many as eight twenty-minute periods separated by rest periods of equal duration. Once again, repeated stimulation - now for as long as 180 minutes - in the absence of exogenous choline was associated with little or no reduction in ACh release per unit time. Free choline was released into the medium at a considerably greater rate than ACh (3 pmol/min/μg DNA of choline vs. 0.8 pmol/min/μg DNA of ACh during the periods of electrical stimulation); this rate declined with time but was unaffected by depolarization. Concurrently, the striatal slices underwent a dose-dependent (i.e., per number of stimulation periods) reduction in phosphatide levels, corrected for DNA (Fig. 4), amounting to 75% of initial levels, after eight stimulation periods. Slices of rat cerebellum, a tissue known to contain few cholinergic terminals, released too little ACh to measure, but relatively large quantities (1 pmol/min/μg DNA) of choline. In spite of this loss of choline, they failed to exhibit significant reductions in phosphatide levels, even with repeated stimulations.

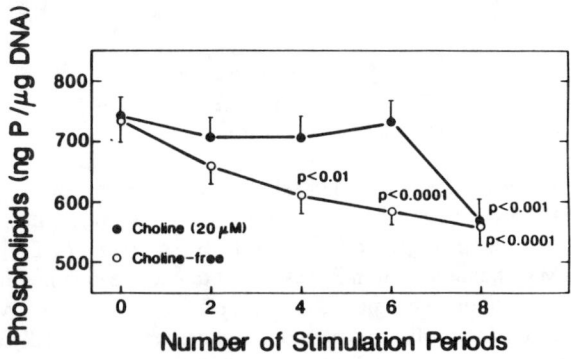

Figure 4. Effect of electrical stimulation with or without choline on total phospholipid contents of rat striatal slices. Slices of rat striatum were superfused (0.6 ml/min) with a physiological solution with or without added choline (20 μM). After a 20 min. equilibration period, the superfusate was collected for 20 min.; the slices were then stimulated electrically (15 Hz) for 20 min., after which they were allowed to rest for 20 min. This rest-stimulation cycle was repeated a total of 8 times. Samples of the slices were removed, for determination of phospholipid phosphate, 20 min. before the first stimulation and 10 min. after the 2nd, 4th, 6th and 8th stimulation periods. Phospholipids were extracted and quantitated by a phosphate assay, and corrected for the DNA contents of the slices. Each point represents the mean of 8-14 determinations. Vertical bars represent standard errors of the mean. Values indicate the significance of differences from the corresponding initial (0 stimulation) values. Linear regression analysis of the data indicate an inverse relationship between the number of stimulation periods and the phospholipid contents of the slices when superfused with the choline-free medium ($y = -20.5x + 717$; $r = -0.98$; $p < 0.0001$).

The stimulation-induced reductions in tissue phosphatides were found to be stoichiometrically distributed among all three of the major structural phosphatides, PC, PE, and PS (Fig. 5). Hence, the depletion of membrane PC associated with choline's utilization for ACh synthesis was not associated with a change in membrane composition (i.e., in the ratio of PC to other major constituents), but with an apparent change in the amount of membrane. This interpretation was supported by the finding that protein levels, corrected for DNA, also fell significantly, though not by so great a percentage as those of the phospholipids (probably reflecting the fact that protein, unlike PC, is distributed within both cell membranes and cytoplasm).

Addition of free choline to the medium, in a concentration range (10-40 μM) on the order of that present in blood and brain, largely or completely protected the slices against the expected stimulation-induced decreases in phosphatide (Fig. 5) (and protein) contents, while concurrently amplifying the release of ACh. Similar protection of membrane phospholipids was attained by omitting the cholinesterase inhibitor (i.e., physostigmine) from the medium (Ulus & Wurtman, 1988). Both the addition of choline and the omission of physostigmine would be expected to increase the availability of exogenous free choline to cholinergic nerve terminals.

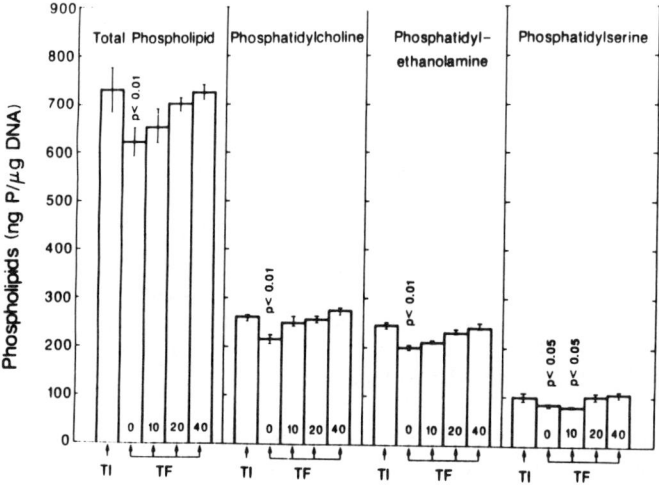

Figure 5: Effect of electrical stimulation with or without choline on phospholipid contents of rat striatal slices. Slices of rat striatum were superfused in a physiological solution containing no choline or choline (10-40 μM). After a 20 min. equilibration period, the superfusate was collected for 20 min.; the slices were then stimulated electrically (15 Hz) for 20 min., after which they were allowed to rest for 20 min., and the rest-stimulation cycle was repeated for a total of 4 times. Samples of the slices were removed, for determination of phospholipid phosphate, 20 min. before the first stimulation (TI) and 10 min after the 4th stimulation period (TF). Phospholipids were extracted, purified and quantified, corrected for μg of DNA contents of the slices. Each bar is the mean of 7 separate experiments. Vertical bars represent standard errors of the mean. Choline concentrations of 0, 10, 20 or 40 μM in the superfusion medium are given within the bars; p values indicate the significance of differences from the corresponding initial (TI)

These observations imply that within cholinergic neurons, there is a dynamic interrelationship between the choline in ACh and that in membrane PC, such that the excessive use of the quaternary amine to form ACh, its neurotransmitter product, occurs at the expense of tissue membrane levels. Whether significant changes in membrane levels also occur in vivo, if and when there are imbalances between choline's levels and its utilization, remains to be determined (as does the biochemical mechanism of the change in the phosphatides, i.e., decreased synthesis or accelerated degradation). If such imbalances actually occur, they seem most likely to do so in association with treatments that decrease the supply of exogenous choline. Clearly, one such treatment might be the administration of AChE inhibitors, since these drugs block the formation of choline within synapses, and thus reduce the amounts of choline that can enter presynaptic terminals via high-affinity choline uptake. Just as the addition of free choline to the medium could be shown, in the present studies, to protect the brain slices from the stimulation-induced phospholipid depletion that occurs in the presence of an AChE inhibitor, so also might the administration of a suitable choline source be protective in vivo.

REFERENCES

Ando, M., M. Iwata, K. Takahama, and Y. Nagata. 1987. Effects of extracellular choline concentration and K^+ depolarization on choline kinase and choline acetyltransferase activities in superior cervical sympathetic ganglia excised from rats. J. Neurochem. 48: 1448-1453.

Blusztajn, J.K. and Wurtman, R.J. 1983. Choline and cholinergic neurons. Science 221: 614-620.

Blusztajn, J.K., M. Liscovitch, and U.I. Richardson. 1987. Synthesis of acetylcholine from choline derived from phosphatidylcholine in a human neuronal cell line. Proc. Natl. Acad. Sci. USA 84: 5474-5477.

Cohen, E., and R.J. Wurtman. 1976. Brain acetylcholine synthesis: control by dietary choline. Science 191: 561-562.

Haubrich, D.R., P.F.L. Wang, D.E. Clody, and P.W. Wedeking. 1975. Increase in rat brain acetylcholine induced by choline or deanol. Life Sci. 17: 975-980.

Maire, J-C., and R.J. Wurtman. 1985. Effects of electrical stimulation and choline availability on release and contents of acetylcholine and choline in superfused slices from rat striatum. J. Physiol. (Paris) 80: 189-195.

Mooradian, A.D. 1988. Blood-brain barrier transport of choline is reduced in the aged rat. Brain Res. 440: 328-332.

Millington, W.R. and R.J. Wurtman. 1982. Choline administration elevates brain phosphorylcholine concentrations. J. Neurochem. 38: 1748-1752.

Ulus, I., and R.J. Wurtman. 1987. Depletion of membrane phosphatidylcholine by a cholinesterase inhibitor, and protection against this effect by supplemental choline. New Eng. J. Med. 318: 191.

Wecker, L. 1985. Neurochemical effects of choline supplementation, Can, J. Physiol. Pharmacol. 64: 329-333.

Part II

Physostigmine Treatment in Alzheimer's Disease

PHYSOSTIGMINE TREATMENT IN SDAT:
Type of Administration, Dose and Duration

Leon J. Thal[1], Bruce R. Lasker[1], David M. Masur[2],
Alan D. Blau[2], and Suzanne Knapp[3]

[1]Dept. of Neurology, VA Medical Center, San Diego, CA and Dept. of Neurosciences, U.C.S.D., La Jolla, CA; [2]Dept. of Neurology, Albert Einstein College of Medicine, Bronx, NY; [3]Dept. of Psychiatry, U.C.S.D., La Jolla, CA

INTRODUCTION

Until the twentieth century, diseases of the aged were not a major public health concern since average life expectancy was low. By 1900, however, with the introduction of significant public health measures, average survivorship had increased to approximately 65 years of age. With the introduction of antibiotics and other modern medical technologies survivorship increased even further, so that at the present time average survival for individuals born in the early portion of this century is approximately 65 years of age. For individuals born in the 1980s, 50% survivorship is estimated to be approximately 85 years of age (Katzman and Terry 1983).

During the aging process, approximately 15% of individuals will develop progressive cognitive impairment. This impairment is termed dementia and it is defined as "a deterioration in intellectual functioning of sufficient severity to interfere with occupational or social performances or both." The cognitive deficits always involve memory and usually involve impairment of abstract thinking, learning new skills, and problem solving. Additionally, a normal level of awareness persists until the late stages of the disease (DSM III 1980). The 1980 population census counted 232 million Americans of whom 11% or 25.5 million were over the age of 65. Fifteen percent of the individuals over the age of 65 have dementia resulting in approximately 3.8 million such individuals. Since one-half of all dementia is secondary to Alzheimer's disease (AD), in 1980 approximately 1.9 million individuals had this disorder. It is currently estimated that 2.7 to 2.8 million individuals are afflicted with this one disorder.

Until the 1960s and early 1970s the etiology of dementia was poorly understood. Most clinicians believed that dementia was due to "cerebrovascular disease" and "brain ischemia." It was therefore logical to postulate that treatment with vasodilator compounds might ameliorate symptoms of "senility." However, by the mid-1970s, several important concepts emerged. First, individuals dying with dementia most often had the neuropathological changes of AD with abundant plaques and tangles. In reality, the brains of individuals with AD show no more atherosclerosis than age-matched controls (Perry 1986). Second, the distinction between dementia beginning before and after the age of 65 seemed less compelling, since individuals in both age groups had similar neuropathological findings (Terry 1978). Third, modern neurochemical techniques were applied to postmortem pathological tissue and specific neurotransmitter changes were demonstrated. Major losses of choline acetyltransferase (CAT), the synthetic enzyme for the neurotransmitter acetylcholine, was detected in the cortex of individuals with AD (Davies and Malony 1976; Perry et al. 1977; White et al. 1977). The loss of cortical cholinergic innervation was subsequently shown to be secondary to degeneration of cholinergic neurons in the nucleus basalis of Meynert (NBM) (Whitehouse 1981). Subsequently, other investigators demonstrated smaller changes in the noradrenergic (Perry et al. 1981; Bondareff et al. 1982) and somatostatinergic (Davies et al. 1980; Rossor et al. 1980) systems. Smaller changes have also been demonstrated in the serotonin system (Arai et al. 1984; Frances et al. 1985) while changes in dopamine concentration remain controversial (Adolfsson et al. 1978; Yates et al. 1979). Thus, multiple neurotransmitter systems are abnormal in AD. However, only the changes in the cholinergic system correlate well with the memory loss of this disorder (Perry et al. 1978; Katzman et al. 1986). Additionally, the cholinergic deficit in AD appears to be largely presynaptic with most authors reporting either normal levels of muscarinic receptors (White et al. 1977; Davies and Verth 1978) or small decreases. Some have suggested that the small loss in receptor number may be due to loss of presynaptic muscarinic receptors (Mash et al. 1985) further strengthening the concept of a presynaptic cholinergic deficit.

CHOLINERGIC MANIPULATION IN AD

Because of the prominent cholinergic deficit that correlates with memory loss, enhancement of the cholinergic system was viewed as a logical step for the treatment of individuals with AD. Three strategies have been pursued. These include the use of precursor agents to increase the synthesis of acetylcholine, the use of cholinesterase inhibitors to inhibit the destruction of existing acetylcholine, and the use of direct-acting agonists to directly stimulate the presumably intact postsynaptic muscarinic receptor.

Precursor Therapy

Numerous clinical trials have been carried out utilizing both choline and lecithin as precursor agents. Studies utilizing choline have been uniformly disappointing with virtually all investigators reporting either minimal improvement or negative results on objective measures of learning (Boyd et al. 1977; Etienne et al. 1978; Fovall et al. 1980; Signoret et al. 1978; Smith et al. 1978; Renvoize and Jerram 1979; Thal et al. 1981). Studies using lecithin have similarly failed to demonstrate consistent improvement in memory or learning (Etienne et al. 1979; Brinkman et al. 1982; Dysken et al. 1982; Weintraub et al. 1983; Little et al. 1985). Failure to improve cognition with choline and lecithin does not appear to be caused by lack of absorption or failure to cross the blood brain barrier because oral choline clearly increases plasma and cerebrospinal fluid (CSF) levels of choline (Christie et al. 1979). It seems most likely that choline availability is not the rate-limiting step for the production of acetylcholine in the human brain.

Cholinergic Agonists

Few cholinergic agonists are available for testing in AD. Arecoline was reported to produce improvement at low doses in a small number of subjects (Christie et al. 1981). Oxotremorine has recently been tested and found to induce many side effects, including significant depression in 5 out of 7 treated patients (Davis et al. 1987). RS-86, a direct-acting cholinergic agonist, failed to improve memory in two studies (Wettstein and Spiegel 1984; Bruno et al. 1986) but improved 7 out of 12 AD patients on an Alzheimer Disease Assessment Scale (Hollander et al. 1987). Clinical improvement, however, was noted in only 2 subjects. A single study of pilocarpine in 2 patients with AD was likewise negative (Caine 1980). One investigator reported subjective improvement in a single blind study of 4 patients receiving intraventricular bethanechol (Harbaugh et al. 1984) but these findings have not been replicated by others. Many of the problems relating to the use of direct-acting cholinergic agonists relate to the induction of toxicity. Because of the many associated problems with the use of cholinergic agonists, most attention has focused on the use of cholinesterase inhibitors.

Cholinesterase Inhibition

Two cholinesterase inhibitors have been tested in AD: physostigmine and tetrahydroaminoacridine (THA). The latter compound will be dealt with in other sections of this volume.

Intravenous physostigmine. Numerous investigators have now completed studies using intravenous physostigmine. In 5 patients with early AD who received intravenous physostigmine, the three subjects given physostigmine while taking concomitant lecithin, consistently improved on a selective reminding task, a measure of verbal learning (Peters and Levin 1979). However, others have subsequently reported improvements in verbal learning on the same task in subjects treated with physostigmine without lecithin (Schwartz and Kohlstaedt 1986). Improvement using intravenous physostigmine has also been demonstrated on a recognition memory task (Christie et al. 1981; Davis and Mohs 1982) and on tests of construction such as figure copying (Muramoto et al. 1984). Two negative studies have been reported in patients with moderately severe dementia using only a single dose of intravenous physostigmine (Ashford et al. 1981; Franceschi et al. 1982). These negative results are not surprising since multiple investigators have reported improvement only when a moderate dose of physostigmine was used with failure to induce improvement at both low and high doses (Peters and Levin 1977; Berger et al. 1979). This phenomenon is probably due to the presence of an inverted U-shaped dose-response curve of improvement with physostigmine. This type of dose-response curve is quite common and has been observed in the psychopharmacological testing of many agents that affect memory. The preponderance of evidence suggests that the administration of intravenous physostigmine can transiently improve memory and other psychological functions to a small extent.

Oral physostigmine. Eight double-blind placebo-controlled trials have been reported using oral physostigmine. Four of these trials demonstrated significant improvement in patients with AD, including improvements in verbal memory assessed by a selective reminding task (Thal et al. 1986; Beller et al. 1985; Harrell et al. 1986) or an Alzheimer Disease Assessment Scale (Mohs et al. 1985). Forty-nine patients were treated in these four studies; improvement was noted in 32. A fifth study recently demonstrated improvement on digit symbol substitution and shape cancellation tasks, but not on the selective reminding task in a group of 22 AD patients (Stern et al. 1987). Three double-blind placebo-controlled negative studies have also been reported. In one study failure to improve was noted in patients with advanced dementia who were too demented to complete a selective reminding task (Wettstein 1983). Two additional studies also reported negative results in mild to moderate AD patients on tasks of learning and memory (Mitchell et al. 1986; Schmechel et al. 1984). There have also been a number of open and single-blind studies, some of which have reported improvement (Peters and Levin 1982) and some of which have failed to demonstrate positive results (Caltagirone et al. 1982, 1983; Jotkowitz 1983).

Response to oral physostigmine is clearly not a uniform phenomenon. Patients with mild to moderate dementia appear to respond somewhat better than those with severe dementia. Dose is also a major determinant of response, since in many studies patients tolerating the highest daily doses seemed to respond the best. This latter issue raises the question of absorption of drug after oral administration. Only two studies (Thal et al. 1986; Mohs et al.

1985) attempted to monitor the entrance of drug into the central nervous system (CNS) during treatment. A good correlation (r = 0.61) was demonstrated between retrieval from long-term storage on the selective reminding task and the degree of cholinesterase inhibition in CSF. A similarly robust correlation (r = 0.91) was demonstrated between the decrease in intrusions (a measure of incorrect responses) and cholinesterase inhibition in CSF in the same patient population (Thal et al. 1986). A correlation has also been demonstrated between changes on an Alzheimer Disease Assessment Scale score and mean percent change in serum cortisol levels (r = 0.88) in a separate cohort of patients (Mohs et al. 1985). These findings strongly suggest that failure to induce improvement in many of the patients reported in the literature may be due to failure to attain adequate CNS cholinesterase inhibition.

Controlled-release physostigmine. The major factor limiting the use of higher oral doses of physostigmine is the development of peripheral side effects. Additionally, the plasma and brain half-life of physostigmine is quite short, averaging only about 30 minutes (Sharpless and Thal 1985; Gibson et al. 1985; Weinstock et al. 1985; Hallak and Giacobini 1986). To overcome these difficulties, administration of small frequent doses of physostigmine has been used. However, dosing more frequently than every two hours is impractical, especially for demented subjects. A reasonable alternative would be the development of a form of physostigmine that would release a constant amount of drug over time. This would allow for the administration of a larger total amount of drug with the achievement of more constant blood levels with fewer side effects. A practical and logical approach to this problem is the development of a controlled-release tablet. To carry out such a study, the availability of a sensitive, specific and accurate assay for the measurement of physostigmine in blood was needed. An existing HPLC method (Whelpton 1983) was modified to increase sensitivity to 50 pg/ml. The presence of interfering peaks was investigated. It was determined that neostigmine, a preservative in many systems, and small amounts of hemolyzed blood, a possible contaminant during sample collection, did not produce interfering peaks. Similarly, theophylline, xanthine, nicotine and aspirin did not produce interfering peaks.

A healthy volunteer who ingested 2 mg of immediate-release physostigmine developed peak plasma concentrations of approximately 1 ng/ml 30 minutes after ingestion. Plasma levels were undetectable by 2 hr (Sharpless and Thal 1985). Repeat ingestions on five separate occasions by a single subject showed that peak plasma levels ranged from 0.5 to 1.0 ng/ml, and that the variance in the area under the curve was only 12%. CSF levels of physostigmine were also measured in two AD patients who demonstrated a good clinical response to 4 mg doses of oral physostigmine and were found to be 0.52 and 0.76 ng/ml.

Several controlled-release formulations of physostigmine tablets have recently been tested for pharmacokinetic availability. Tablets containing 9, 12 and 15 mg produced peak plasma levels of approximately 0.5, 1.6 and 2.4 ng/ml, respectively. Each preparation induced sustained elevations in plasma levels for 4 to 6 hours. Side effects, consisting of cholinergic hyperactivation, occurred only with the highest dose and when plasma concentrations approached 2 ng/ml (Thal et al. 1988). The availability of controlled-release preparations should prove to be invaluable for the clinical testing of oral physostigmine in AD.

CONCLUSIONS

At the present time, there is ample evidence that multiple neurotransmitter dysfunction occurs in AD. Nevertheless, only dysfunction of the cholinergic system has been strongly linked to the memory disorder of AD. The cholinergic system has been particularly difficult to manipulate with currently available pharmacological agents. Other neurotransmitter systems have been easier to manipulate. For example, administration of L-dopa to patients with Parkinson's disease resulted in an enormous increase in the quantity of free dopamine in brain and a more than 9-fold increase in homovanillic acid, the major metabolite of dopamine (Davidson et al. 1977). Similarly, the administration of 5-hydroxytryptophan to patients with myoclonus resulted in a 4- to 18-fold increment in CSF 5-hydroxyindoleacetic acid, the major metabolite of serotonin (Thal et al. 1980). By comparison, administration of physostigmine to normal rats resulted in a 2- to 3-fold increase in brain acetylcholine (Hallak and Giacobini 1986). Our failure to ameliorate the memory disorder in AD may be secondary to our inability to effectively increase central cholinergic neurotransmission with currently available agents.

Supported by NIH grant AG05386, a grant from the Medical Research Service of the Veterans Administration, and a grant from Forest Laboratories, Inc. The secretarial support of Barbara Reader and Marcia Strand is greatly appreciated.

REFERENCES

Adolfsson, R., C.G. Gottfries, L. Oreland, B.E. Roos, and B. Winblad. 1978. Reduced levels of catecholamines in the brain and increased activity of monoamine oxidase in platelets in Alzheimer's disease: Therapeutic implications. In Aging, Vol. 7, eds. R. Katzman, R.D. Terry, and K.L. Bick, 441-451. New York: Raven Press.

Arai, H., K. Kosaka, and R. Izuka. 1984. Changes of biogenic amines and their metabolites in postmortem brains from patients with Alzheimer-type dementia. J. Neurochem. 43: 388-393.

Ashford, J.W., S. Soldinger, J. Schaeffer, L. Cochran, and L. Jarvik. 1981. Physostigmine and its effect on six patients with dementia. Am. J. Psychiatry 138: 829-830.

Beller, S.A., J.E. Overall, and A.C. Swann. 1985. Efficacy of oral physostigmine in primary degenerative dementia. Psychopharmacology 87: 147-151.

Berger, P.A., K.L. Davis, and L.E. Hollister. 1979. Cholinomimetics in mania, schizophrenia and memory disorders. In Nutrition and the Brain, Vol. 5, eds. A. Barbeau, J.H. Growdon, and R.J. Wurtman, 425-441. New York: Raven Press.

Bondareff, W., C.Q. Mountjoy, and M. Roth. 1982. Loss of neurons of origin of the adrenergic projection to cerebral cortex (Nucleus locus ceruleus) in senile dementia. Neurology 32: 164-168.

Boyd, W.D., J. Graham-White, G. Blackwood, I. Glen, and J. McQueen. 1977. Clinical effects of choline in Alzheimer senile dementia. Lancet II: 711.

Brinkman, S.D., R.C. Smith, J.S. Meyer, G. Vroulis, T. Shaw, J.R. Gordon, and R.H. Allen. 1982. Lecithin and memory training in suspected Alzheimer's disease. J. Gerontol. 37: 4-9.

Bruno, G., E. Mohr, M. Gillespie, P. Fedio, and T.N. Chase. 1986. Muscarinic agonist therapy of Alzheimer's disease. Arch. Neurol. 43: 659-661.

Caine, E.D. 1980. Cholinomimetic treatment fails to improve memory disorders. N. Engl. J. Med. 303: 585-586.

Caltagirone, C., G. Gainotti, and C. Masullo. 1982. Oral administration of chronic physostigmine does not improve cognitive or mnesic performances in Alzheimer's presenile dementia. Int. J. Neurosci. 16: 247-249.

Caltagirone, C., A. Albanese, G. Gainotti, and C. Masullo. 1983. Acute administration of individual optimal dose of physostigmine fails to improve mnesic performances in Alzheimer's presenile dementia. Int. J. Neurosci. 18: 143-148.

Christie, J.E, I.M. Blackburn, A.I.M. Glen, S. Zeisel, A. Sherry, and C.M. Yates. 1979. Effects of choline and lecithin on CSF choline levels and on cognitive function in patients with presenile dementia of the Alzheimer type. In Nutrition and the Brain, Vol. 5, eds. A. Barbeau, J.H. Growden, and R.J. Wurtman, 377-387. New York: Raven Press.

Christie, J.E., A. Shering, J. Ferguson, and A.I.M. Glen. 1981. Physostigmine and arecoline: effects of intravenous infusions in Alzheimer presenile dementia. Br. J. Psychiatry 138: 46-50.

Davidson, D.L.W., C.M. Yates, C. Mawdsley, I.A. Pullar, and H. Wilson. 1977. CSF studies on the relationship between dopamine and 5-hydroxytryptamine in Parkinsonism and other movement disorders. J. Neurol. Neurosurg. Psychiatry 40: 1136-1141.

Davies, P., and A.J.F. Malony. 1976. Selective loss of central cholinergic neurons in Alzheimer's disease. Lancet II: 1403.

Davies, P., and A.H. Verth. 1978. Regional distribution of muscarinic acetylcholine receptor in normal and Alzheimer's-type dementia brains. Brain Res. 138: 385-392.

Davies, P., R. Katzman, and R.D. Terry. 1980. Reduced somatostatin-like immunoreactivity in cerebral cortex from cases of Alzheimer disease and Alzheimer senile dementia. Nature 288: 279-280.

Davis, K.L., and R. Mohs. 1982. Enhancement of memory processes in Alzheimer's disease with multiple-dose intravenous physostigmine. Am. J. Psychiatry 139: 1421-1424.

Davis, K.L., E. Hollander, M. Davidson, B.M. Davis, R.C. Mohs, and T.B. Horvath. 1987. Induction of depression with oxotremorine in patients with Alzheimer's disease. Am. J. Psychiatry 144: 468-471.

Diagnostic and Statistical Manual of Mental Disorders, American Psychiatric Association Task Force on Nomenclature and Statistics. 1980. 3rd ed., 494. Washington, D.C.

Dysken, M.W., P. Foval, C.M. Harris, J.M. Davis, and A. Noronha. 1982. Lecithin administration in Alzheimer dementia. Neurology 32: 1203-1204.

Etienne, P., S. Gauthier, G. Johnsen, B. Collier, T. Mendis, D. Dastoor, M. Cole, and H.F. Muller. 1978. Clinical effects of choline in Alzheimer's disease. Lancet 1: 508-509.

Etienne, P., S. Gauthier, D. Dastoor, B. Collier, and J. Ratner. 1979. Alzheimer's disease: clinical effects of lecithin treatment. In Nutrition and the Brain, Vol. 5, eds. A. Barbeau, J.H. Growdon, and R.J. Wurtman, 389-396. New York: Raven Press.

Fovall, P., M.W. Dysken, L.W. Lazarus, J.M. Davis, R.L. Kahn, R. Jope, S. Rinkel, and P. Rattan. 1980. Choline bitartrate treatment of Alzheimer-type dementias. Comm. Psychopharm. 4: 141-145.

Frances, T.P., A.M. Palmer, N.R. Sims, D.M. Bowen, A.N. Davison, M.M. Esiri, D. Neary, J.S. Snowden, and G.K. Wilcock. 1985. Neurochemical studies of early-onset Alzheimer's disease. N. Engl. J. Med. 313: 7-11.

Franceschi, M., O. Tancredi, G. Savio, and S. Smirne. 1982. Vasopressin and physostigmine in the treatment of amnesia. Eur. Neurol. 21: 388-391.

Gibson, M., T. Moore, C.M. Smith, and R. Whelpton. 1985. Physostigmine concentrations after oral doses. Lancet 1: 695-696.

Hallak, M., and E. Giacobini. 1986. Relation of brain regional physostigmine concentration to cholinesterase activity and acetylcholine and choline levels in rat. Neurochem Res. 11: 1037-1048.

Harbaugh, R.E., D.W. Roberts, D.W. Coombs, R.L. Saunders, and T.M. Reeder. 1984. Preliminary report: intracranial cholinergic drug infusion inpatients with Alzheimer's disease. Neurosurgery 15: 514-518.

Harrell, L., J. Falgout, D. Leli, R. Jope, C. McLain, M. Spiers, R. Callaway, and J. Halsey. 1986. Behavioral effects of oral physostigmine in Alzheimer's disease patients. Neurology 36: 269.

Hollander, E., M. Davidson, R.C. Mohs, T.B. Horvath, B.M. Davis, Z. Zemeshlany, and K.L. Davis. 1987. RS 86 in the treatment of Alzheimer's disease: cognitive and biological effects. Biol. Psychiatry 22: 1067-1078.

Jotkowitz, S. 1983. Lack of clinical efficacy of chronic oral physostigmine in Alzheimer's disease. Ann. Neurol. 14: 690-691.

Katzman, R., and R. Terry. 1983. In The Neurology of Aging, 1-14. Philadelphia: FA Davis Co.

Katzman, R., T. Brown, P. Fuld, L. Thal, P. Davies, and R. Terry. 1986. Significance of neurotransmitter abnormalities in Alzheimer's disease. In Neuropeptides in Neurologic and Psychiatric Disease, eds. J.B. Martin, and J. Barchas, 279-286. New York: Raven Press.

Little, A., R. Levy, P. Chuaqui-Kidd, and D. Hand. 1985. A double-blind, placebo controlled trial of high-dose lecithin in Alzheimer's disease. J. Neurol. Neurosurg. Psychiatry 48: 736-742.

Mash, D.C., D.D. Flynn, and L.T. Potter. 1985. Loss of M2 muscarine receptors in the cerebral cortex in Alzheimer's disease and experimental cholinergic denervation. Science 228: 1115-1117.

Mitchell, A., D.Drachman, B. O'Donnell, and G. Glosser. 1986. Oral physostigmine in Alzheimer's disease. Neurology 36: 295.

Mohs, R.C., B.M. Davis, C.A. Johns, A.A. Mathe, B.S. Greenwald, T.B. Horvath, and K.L. Davis. 1985. Oral physostigmine treatment of patients with Alzheimer's disease. Am. J. Psychiatry 142: 28-33.

Muramoto, O., M. Sugishita, and K. Ando. 1984. Cholinergic system and constructional praxis: a further study of physostigmine in Alzheimer's disease. J. Neurol. Neurosurg. Psychiatry 47: 485-491.

Perry, E.K., R.H. Perry, G. Blessed, and B.E. Tomlinson. 1977. Necropsy evidence of central cholinergic deficits in senile dementia. Lancet 1: 189.

Perry, E.K., B.E. Tomlinson, G. Blessed, K. Bergmann, P.H. Gibson, and R.H. Perry. 1978. Correlation of cholinergic abnormalities with senile plaques and mental test scores in senile dementia. Br. Med. J. 2: 1457-1459.

Perry, E.K., B.E. Tomlinson, G. Blessed, R.H. Perry, A.J. Cross, and T.J. Crow. 1981. Neuropathological and biochemical observations on the noradrenergic system in Alzheimer's disease. J. Neurol. Sci. 51: 279-287.

Perry, R.H. 1986. Recent advances in neuropathology. Br. Med. Bull. 42: 34-41.

Peters, B.H., and H.S. Levin. 1977. Memory enhancement after physostigmine treatment in the amnesic syndrome. Arch. Neurol. 34: 215-219.

Peters, B.H., and H.S. Levin. 1979. Effects of physostigmine and lecithin on memory in Alzheimer's disease. Ann. Neurol. 6: 219-221.

Peters, B.H., and H.S. Levin. 1982. Chronic oral physostigmine and lecithin administration in memory disorders of aging. In Aging, Vol. 19, eds. S. Corkin, K.L. Davis, J. Growdon, E. Usdin, and R.J. Wurtman, 421-426. New York: Raven Press.

Renvoize, E.G., and T. Jerram. 1979. Choline in Alzheimer's disease. N. Engl. J. Med. 301: 330.

Rossor, M.N., P.C. Emson, C.Q. Mountjoy, M. Roth, and L.L. Iversen. 1980. Reduced amounts of immunoreactive somatostatin in the temporal cortex in senile dementia of Alzheimer type. Neurosci. Lett. 20: 373-377.

Schmechel, D.E., F. Schmitt, J. Horner, W.E. Wilkinson, B.J. Hurwitz, and A. Heyman. 1984. Lack of effect of oral physostigmine and lecithin in patients with probable Alzheimer's disease. Neurology 34: 280.

Schwartz, A.S., and E.V. Kohlstaedt. 1986. Physostigmine effects in Alzheimer's disease: relationship to dementia severity. Life Sci. 38: 1021-1028.

Sharpless, N.S., and L.J. Thal. 1985. Plasma physostigmine concentrations after oral administration. Lancet 1: 1397-1398.

Signoret, J.L., A. Whiteley, and F. Lhermitte. 1978. Influence of choline on amnesia in early Alzheimer's disease. Lancet II: 837.

Smith, C.M., M. Swase, A.N. Exton-Smith, M.J. Phillips, P.W. Overstall, M.E. Piper, and M.R. Bailey. 1978. Choline therapy in Alzheimer's disease. Lancet II: 318.

Stern, Y., M. Sano, and R. Mauyex. 1987. Effects of oral physostigmine in Alzheimer's disease. Neurol. 22: 306-310.

Thal, L., N.S. Sharpless, L. Wolfson, and R. Katzman. 1980. Treatment of myoclonus with L-5-HTP and carbidopa: clinical, electrophysiological and biochemical observations. Ann Neurol 7: 570-576.

Thal, L.J., W. Rosen, N.S. Sharpless, and H. Crystal. 1981. Choline chloride fails to improve cognition in Alzheimer's disease. Neurobiol. Aging 2: 205-208.

Thal, L.J., D.M. Masur, N.S. Sharpless, P.A. Fuld, and P. Davies. 1986. Acute and chronic effects of oral physostigmine and lecithin in Alzheimer's disease. Prog. Neuropsychopharmacol. Biol. Psychiatry 10: 627-636.

Thal, L.J., B. Lasker, N.S. Sharpless, G. Bobotas, J.M. Schor, and A. Nigalye. Plasma physostigmine concentrations after controlled-release oral administration. Arch. Neurol. (In press).

Terry, R.D. 1978. Aging, senile dementia and Alzheimer's disease. In Alzheimer's disease: Senile Dementia and Related Disorders, Aging, Vol. 7, eds. R. Katzman, R.D. Terry, and K.L. Bick, 11-14. New York: Raven Press.

Weinstock, M., M. Razin, M. Chorev, and Z. Tashma. 1985. Pharmacological activity of novel anticholinesterase agents of potential use in the treatment of Alzheimer's disease. Paper presented at the 30th Oholo Biological Conference, Eliat, Israel.

Weintraub, S., M.-M. Mesulam, R. Auty, R. Baratz, B.N. Cholakos, L. Kapust, B. Ransil, J.G. Tellers, M.S. Albert, S. LoCastro, and M. Moss. 1983. Lecithin in the treatment of Alzheimer's disease. Arch. Neurol. 40: 527-528.

Wettstein, A. 1983. No effect from double-blind trial of physostigmine and lecithin in Alzheimer disease. Ann. Neurol. 13: 210-212.

Wettstein, A., and R. Spiegel. 1984. Clinical trials with the cholinergic drug RS 86 in Alzheimer's disease (AD) and senile dementia of the Alzheimer type (SDAT). Psychopharmacology 84: 572-573.

White, P., M.J. Goodhardt, J.K. Keet, C.R. Hiley, L.H. Carrasio, and I.E.I. Williams. 1977. Neocortical cholinergic neurons in elderly people. Lancet 1: 668-671.

Whitehouse, P.J., D. Price, A. Clark, J.K. Coyle, and M. DeLong. 1981. Alzheimer's disease evidence for a selective loss of cholinergic neurons in the nucleus basalis. Ann. Neurol. 10: 122-126.

Whelpton, R. 1983. Analysis of plasma physostigmine concentrations by liquid chromatography. J. Chromatogr. 272: 216-220.

Yates, C.M., Y. Allisin, J. Sampon, A.J.F. Maloney, and A. Gordon. 1979. Dopamine in Alzheimer's disease and senile dementia. Lancet 2: 851-852.

INTRACEREBROVENTRICULAR ADMINISTRATION OF CHOLINERGIC DRUGS:
Preclinical Trials & Clinical Experience in Alzheimer Patients

Ezio Giacobini, Robert Becker, Michael McIlhany, and Vinod Kumar
*Depts. of Pharmacology, Psychiatry and Surgery,
S.I.U. School of Medicine, Springfield, IL*

AN HISTORICAL OVERVIEW OF INTRAVENTRICULAR DRUG ADMINISTRATION

At the April, 1955 meeting of the Royal Society of Medicine in London, W. Feldsberg communicated that he and S.L. Sherwood during the last two years had studied the action of various drugs in the CNS by injecting them directly into the cerebral lateral ventricle of the conscious cat (Sherwood, 1952; Feldberg and Sherwood, 1954a,b, 1955). Feldberg specified that for this purpose they had "screwed a metal cannula into the skull and kept it permanently in position, sometimes for more than a year, without apparently producing any ill effects." The tip of the cannula rested in the lateral ventricle, and injections were made through a rubber diaphragm located outside the skull. A variety of substances had been administered intraventricularly, such as acetylcholine (ACh), DFP (diisopropylfluorophosphonate), atropine, eserine (physostigmine, Phy), noradrenaline and adrenaline. The cholinesterase (ChE) inhibitors produced motor effects such as changes in stance, gait and posture, hyperactivity and itching. The doses of eserine injected were high, from 20 ug to 100 ug. At the highest dose, agitation, anger and rage, as well as many autonomic effects, were visible (Feldberg and Sherwood, 1954a,b). This stage was followed in some cases by stupor and catatonia. According to Feldberg and Sherwood (1954a), "the effects observed on intracerebroventricular (i.c.v.) injection can, without any doubt, be attributed to central effects of the drugs; and their leakage into the systemic circulation, if it occurred at all, can be excluded as being responsible to a significant extent for the signs observed." It is interesting to note that the i.c.v. injection of ACh produced very similar effects to those of eserine.

Although Feldberg and Sherwood were the first to study systematically the effect of cholinergic drugs injected i.c.v., they were not first to use this route. Cushing (1931) had injected ACh into the ventricles of patients to study pituitary and hypothalamic responses, and Henderson and Wilson (1936) had injected ACh in humans to study autonomic effects.

In 1973, Beleslin et al. reported a detailed analysis of behavioral changes in conscious cats following i.c.v. injections of Phy, using even higher doses of Phy (0.1-2mg). When Phy was injected at the dose of 2.0 mg, pronounced behavioral, autonomic and motor phenomena were followed by clonic-tonic seizures. Convulsions appeared when ChE inhibition reached 33% in the thalamus and 61% in the caudate. Other experiments have stressed the antinociceptive effect of Phy and of other cholinomimetics (oxotremorine, carbachol) on rats and cats injected intrathecally (Yaksh et al., 1955).

USE OF INTRAVENTRICULAR ADMINISTRATION IN THERAPY

Intraventricular medication has been used when the administration of systemic agents has failed to influence the disease process. Opiate analgesics have been administered directly into the cerebrospinal fluid (CSF) for pain control (Harbaugh et al., 1982). Presently, delivery systems are used to provide continuous or intermittent infusion of potential therapeutic compounds including cholimomimetic drugs into cerebral ventricles or spinal subarachnoid space in humans (Coombs et al., 1983; Harbaugh et al., 1982, 1984; Penn et al., 1984, 1988; Gauthier et al., 1986). The rationale for intraventricular drug administration is: 1) better penetration of the drug in the brain than with systemic administration; 2) higher doses can be delivered directly to the brain; and 3) systemic toxic effects are reduced because the total dose is lower.

INTRAVENTRICULAR ADMINISTRATION OF CHOLINERGIC DRUGS IN HUMANS

Cholinergic drugs have been administered intraventricularly in humans since the experiments of Cushing (1931), for various indications and studies (Table 1).

A first preliminary report on intracranial cholinomimetic drug infusion in patients with SDAT has been published by Harbaugh et al. (1984). The clinical trial involved the use of an implantable infusion system connected to a Silastic intracranial catheter. Four patients were infused during three intervals with the cholinergic agonist bethanecol chloride (.05-0.7 mg/day). The results were encouraging with reported decreased confusion, increased initiative and improved memory. A second study of cholinomimetic intraventricular therapy on 12 patients was published by Fox et al. (1985), also using bethanechol chloride. During the insertion of the intraventricular catheter for long-term infusion, these authors performed a left frontal cortical biopsy, which confirmed the clinical diagnosis of senile dementia of Alzheimer type (SDAT) in all cases. A third study with bethanechol has been published by Penn et al. (1988). The first phase of this study was a 24-week double-blind crossover study of 3.5 mg/day bethanecol compared with placebo infusion. The authors found that the majority of assessment scores were "in the predicted direction" and that there was a non-significant trend ($p < 0.08$) toward decreased frequency of abnormal behaviors. In the second phase of the study, after an 8-week double-blind phase of low-dose i.c.v. bethanecol versus placebo, the subjects entered an unblinded escalating phase up to 1.75 mg/day. The authors' major conclusion is that bethanecol in moderate doses improves performance in AD patients and reduces the frequency of behavioral abnormalities whereas in high doses impairs cognitive abilities. The multicenter study of Dartmouth University (Harbaugh et al., 1984; Whitehouse, 1988) involved 11 clinics with a total of 49 patients. Preliminary data showed a statistically significant, small improvement in mental status score, but an increase in level of depression. All of these studies indicate that subject selection, experimental design, assessment, blinding and statistical analysis affect evaluation of drug studies in AD (Whitehouse, 1988). At this point of our search for a palliative therapy for Alzheimer, it is important to compare the clinical effects of two different strategies: direct ACh receptor stimulation (bethanecol studies) and ACh enhancement through acetylcholinesterase (AChE) inhibition. Both types of studies suggest that i.c.v. drug infusion is feasible in AD patients, and more extensive research on local delivery of drugs is granted (Cronin-Golomb, 1987).

PHYSOSTIGMINE I.C.V. ADMINISTRATION - PRECLINICAL TRIALS

Only a few animal studies have been performed to demonstrate advantages of i.c.v. administration as compared to other forms of administration of cholinergic drugs. This lack of basic knowledge has made therapeutic experimentation in humans more difficult to assess, and the application to therapy slow.

Table 1. Intracerebroventricular Administration of Cholinergic Drugs in Humans

CONDITION	DRUG	DOSE (ug)	AUTHOR/YEAR
ANTICHOLINERGICS			
Catatonic stupor	Banthine	500-1000	Sherwood (1952)
			Feldberg & Sherwood (1955)
Paranoid Schizophrenia	Acetylcholinesterase*	1000-5000	Sherwood (1952, 1955)
			Feldberg & Sherwood (1955)
Familial Myoclonic Epilepsy	Atropine	250-500	Sherwood (1952)
			Feldberg & Sherwood (1955)
Head Injury	Pentamethonium	150-300	Sherwood (1952)
			Feldberg & Sherwood (1955)
CHOLINOMIMETICS			
Pituitary gland dysfunction	Acetylcholine	--	Cushing (1931)
Alzheimer Disease	Bethanechol	50-700/day	Harbaugh et al. (1984)
Alzheimer Disease	Bethanechol	350/day	Fox et al. (1985)
Alzheimer Disease	Bethanechol	350-1750/day	Penn et al. (1988)
Alzheimer Disease	Bethanechol	350-2000/day	Harbaugh et al. (1984)
Alzheimer Disease	Physostigmine	.5-8/day	Giacobini et al. (1987a,b)

*from human erythrocytes or electric eel

Table 2. Cholinesterase Activity and Physostigmine Concentration in Areas of Brain After I.C.V. Administration of Physostigmine (4 ug) in the Rat

Brain Area	15 min		30 min	
	ChE* % act.	Phy* nmol/g	ChE % act.	Phy nmol/g
Striatum	69 $\pm$ 3.4	0.94 $\pm$ 0.25	70 $\pm$ 4.3	0.22 $\pm$ 0.05
Hippocampus	27 $\pm$ 6.9	1.56 $\pm$ 0.40	62 $\pm$ 4.9	0.36 $\pm$ 0.07
Medulla oblongata	42 $\pm$ 12	0.12 $\pm$ 0.007	70 $\pm$ 3.1	0.13 $\pm$ 0.007
Cerebellum	54 $\pm$ 4.0	0.15 $\pm$ 0.01	62 $\pm$ 4.3	0.18 $\pm$ 0.007
Cortex	55 $\pm$ 3.4	0.58 $\pm$ 0.11	77 $\pm$ 6.0	0.13 $\pm$ 0.03

* Expressed as mean $\pm$ S.E. of values of right and left hemispheres. n = 6 at 15 min, n = 12 at 30 min. Physostigmine was administered intraventricularly (i.c.v.) in the left ventricle.**
** From Hallak and Giacobini, 1987

Table 3. Acetylcholine Levels* in Three Different Areas of the Brain After I.C.V. Administration of Physostigmine (4 ug) 30 min in the Rat

Brain Area	Control	Phy	Percentage Change
Striatum	59.72 ± 2.18	73.11 ± 3.75*	18
Hippocampus	25.39 ± 1.74	31.60 ± 1.41*	22
Medulla oblongata	22.33 ± 1.74	22.32 ± 2.03	0

*Expressed as nmol/g mean ± S.E. of values of right and left hemispheres. n = 10. Mean values differ significantly from control (**p < 0.001) as determined by the t-test. Physostigmine was administered intraventricularly (i.c.v.) in the left ventricle.***
***From Hallak and Giacobini, 1987

To study effects of ChE inhibitors i.c.v. administration, we have conducted parallel studies of markers of ACh metabolism in AD patients in CSF and plasma (Giacobini et al., 1986a,b; 1987a) as well as studies of the effect of Phy on ChE activity and ACh in various species (rat, dog) (Hallak and Giacobini, 1987, 1988; Mattio et al., 1986).

In our preclinical experiments in rats and dogs we have determined Phy levels and ChE inhibition in plasma, brain and CSF after i.c.v. or i.v. administration of Phy (Mattio et al., 1986; Hallak and Giacobini, 1986, 1987). We have found that i.m. (650 mg/kg) and i.v. (100 mg/kg) administration of Phy produced a maximum 60% and 50% of AChE inhibition, respectively, in brain in these two animal models. In rats, very high doses (500-650 mg/kg) of Phy i.m. were required to produce a peak of 50% AChE inhibition in brain activity, and AChE activity returned to baseline after 42 min. The i.m. dose of 650 mg/kg is close to the LD_{50} value in rats and produces severe side effects. With i.v. administration (100 ug/kg), a peak of 50% AChE inhibition occurs in brain, but ChE activity returns to baseline after 7 min. These results may explain the short-lasting effect of Phy i.m. and i.v. obtained in therapy.

We have demonstrated that i.c.v. administration of very small doses (10-20 ug/kg) of Phy in dogs produces significant (50%) and long-lasting inhibition (up to 3 hrs) of AChE activity in selective brain regions including striatum, hippocampus and cortex. These areas contain cholinergic cells and terminals which are affected by Alzheimer disease (Whitehouse et al., 1981; Candy et al., 1986). The medulla oblongata and pons are much less affected.

Additional evidence for the effectiveness of the i.c.v. route is provided by our recent study in the rat (Hallak and Giacobini, 1987). The administration of 4 ug Phy i.c.v. produced at 15 min a 45% and 73% inhibition of AChE activity in cortex and hippocampus, respectively (Fig. 1, Table 2). At 30 min, ACh was increased by 18% and 22% in striatum and hippocampus, respectively (Table 3). No increase was seen at this time point in the medulla oblongata. At this dosage, however, the rat showed a higher frequency of side effects than the dog. Rats are known to be more sensitive to the effects of ChE inhibitors than most mammalian species.

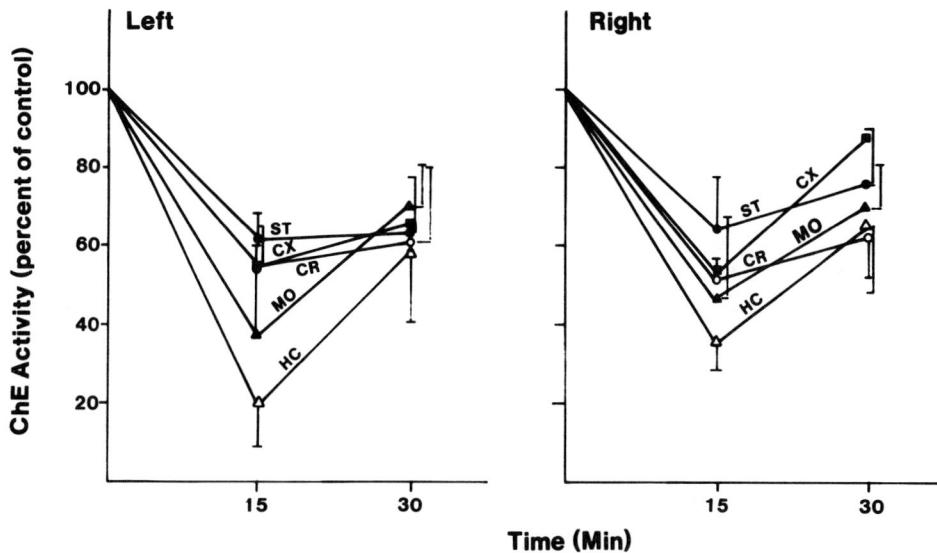

Figure 1. Cholinesterase activity (percentage of control) at 15 and 30 min in regions of rat brain of the right and left hemisphere after administration of physostigmine (4 ug i.c.v. in the left ventricle). ST = striatum; CX = cerebral cortex; CR = cerebellum; MO = medulla oblongata; HC = hippocampus. Each point is the mean $\pm$ SD of five animals. (Modified by Hallak and Giacobini, 1987)

CLINICAL TRIALS

Our clinical trials are based on results obtained on the rat and dog model. Our preclinical study in the dog (Mattio et al., 1986) showed that over 90% inhibition of CSF AChE activity was reached within 5 min of injection at all dosages (10-120 ug) and that even at the lowest dosage (10 ug), 50% inhibition was still present at 90 min. On the contrary, peripheral ChE inhibition in plasma was minimal (< 10%). Our data on Phy distribution in dog brain showed that with i.c.v. administration high concentrations of the drug were achieved, first in the striatum and the hippocampus and then in cortex (Mattio et al., 1986). A CSF inhibition of 90% at 15 min corresponded to a 64% inhibition in cortex. These data indicated that the drug rapidly inhibited (up to 90%) the AChE present in CSF and then diffused slowly into the brain. Based on these results in the dog, we suggested that the same changes could be true in humans, and peripheral side effects would be minimal with this route of administration. Our preliminary study on AD patients seems to confirm this hypothesis.

Our preliminary results are from an ongoing pilot study on six severely demented (CDR 2.5-3) patients who were treated acutely on several occasions with i.c.v. Phy infusion and followed for up to 6 hrs. Figure 2 summarizes the results obtained in the treatment of a 75-year-old patient with a 5-year history of the disease and a CDR score of 2.5. Several doses of Phy ranging from .5 ug to 8 ug were administered i.c.v. through the clinical reservoirs in single administrations at multiple sessions. Figure 2 shows the results of 9 consecutive experimental trials on the percent inhibition of CSF (upper curves) and plasma (lower curves) ChE activity. Five min from the starting of the injection, the lowest dose range (0.5-1 ug) inhibited CSF AChE by 60%; at 180 min AChE activity was

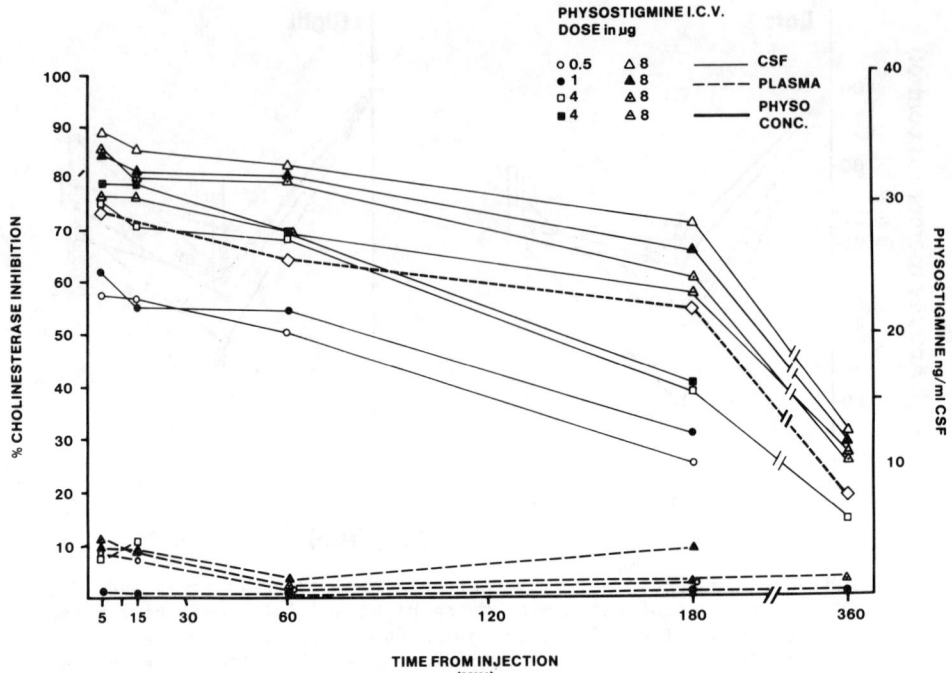

Figure 2. Effect of various doses of Phy administered i.c.v. to an AD patient on AChE activity in CSF (upper curves) and plasma (lower curves). The pattern of Phy concentrations in CSF is also shown following an 8 ug single dose i.c.v.

still 30% inhibited. The middle dose (4 ug) caused 75% and 45% AChE inhibition at 5 and 180 min, respectively. However, at 360 min AChE activity was inhibited by only 20%. The highest dose (8 ug), administered in 4 trials, resulted in 85% inhibition at 5 min and 70% inhibition at 180 min. At 360 min, AChE activity was still 30% inhibited. Contrary to the high degree of CSF AChE inhibition, plasma butyrylcholinesterase (BuChE) were maximally inhibited only 10% at the highest dose and 3% after 60 min. For comparison, in the same figure, the CSF concentration of Phy is reported following 8 ug dosage. Physostigmine levels were analyzed using the high pressure liquid chromatography (HPLC) method developed in our laboratory (Unni et al., 1988).

The concentration curve of Phy follows closely ChE inhibition, decreasing from the 5 min value of 20 ng/ml to 7 ng/ml at 360 min. We examined samples of 1.5 to 2.0 ml plasma for each time point for all i.c.v. studies. Physostigmine could not be detected in plasma at any time point following 8 ug dosage, indicating that the drug level is lower than the level of sensitivity of our method (0.25 mg/ml plasma). Larger volumes of plasma (> 2 ml) are necessary in order to determine such low drug levels at the present limit of sensitivity.

Figure 3 shows the effect on several cholinergic parameters and on Phy CSF concentration of an 8 ug i.c.v. dose administration of Phy on the same patient. The most significant effect is a five-fold increase of ACh in CSF from 0.9 nmol/ml to a peak value of 5 nmol/ml at 180 min. It is interesting to note that at 360 min, ACh levels are still elevated (4 nmol/ml) in spite of the low level (30%) of ChE inhibition in CSF. Inspite of the rapid and sustained increase in ACh, Ch does not seem to be utilized to synthesize the ACh recovered in CSF

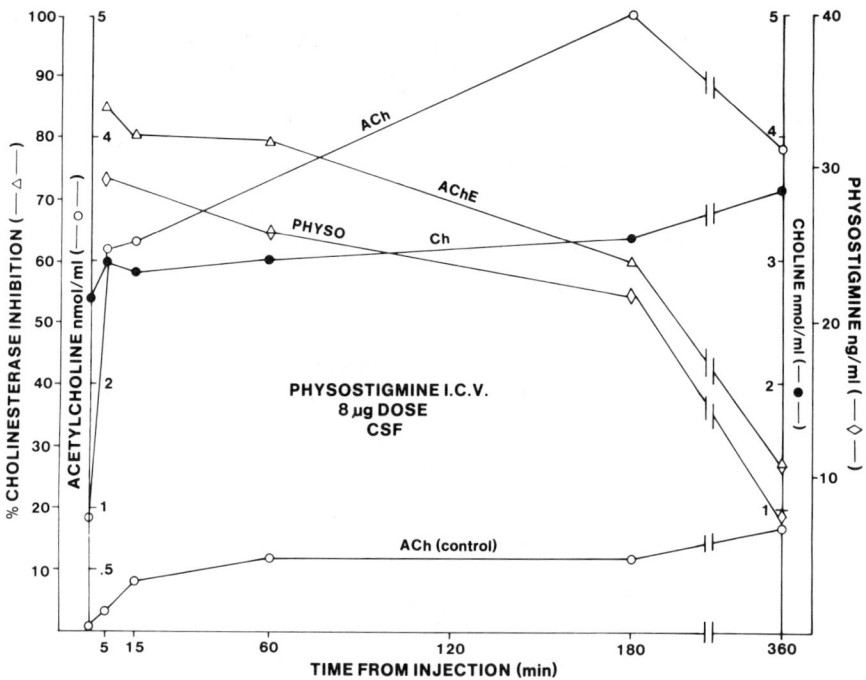

Figure 3. The effect of a single dose of 8 ug Phy i.c.v. on AChE activity - ACh, Ch and Phy concentrations in CSF of an AD patient. In the ACh control, AChE ws not inactivated immediately following collection of CSF.

since its levels remain stable. Inhibition of AChE activity and levels of Phy follow a parallel trend throughout the experiment. These results suggest that following a single dose i.c.v. injection of Phy, causing 85% inhibition of AChE in CSF, ACh is released from the CNS into the CSF and is maintained at a high steady state level during a period of at least 6 hours. If CSF ACh and Ch levels reflect closely extracellular fluid concentrations in the brain, as suggested by animal experiments (Tucek, 1984), then high and sustained cerebral ACh in CSF would be a direct result of Phy i.c.v. administration. This represents one of the major goals established for our therapy (Giacobini, 1987a; Becker and Giacobini, 1988). This result has not been achieved by any other route of administration, either in experimental animals or in humans (Hallak and Giacobini, 1986; Becker and Giacobini, 1988). As predicted by our dog model (Mattio et al., 1986), i.c.v. Phy should not produce significant peripheral side effects. Our patient, who was followed closely for cardiovascular, respiratory and gastrointestinal symptoms, did not show side effects. No central side effects could be recorded throughout the duration of the trial, not even at the highest Phy dose (8 ug). In particular, no peripheral side effect was observed, and the patient appeared to be bothered little by the Phy injection. This demonstrates that Phy can be safely injected i.c.v. into an elderly AD patient producing high levels of CNS AChE inhibition with little peripheral effect. Most significant is the long-lasting 5-fold increase in ACh. The patient was tested with our memory battery the day prior to the trial, the day of the trial before the injection, and at 1, 3 and 6 hours following Phy injection. The scores on a word and picture recognition test improved significantly in an open trial. However, at times, the patient seemed to be more irritable or sleepy.

Due to the slow mixing and low turnover rate of CSF ($t_{1/2}$ = 3.5 hours), the amount of Phy taken up by the brain from the CSF can be more easily estimated at late (> 180 min) time points than at early ones (Fig. 2). At 180 min, assuming uniform distribution, approximately 50% of the injected Phy is still present in CSF. Therefore, the CSF acts as a slow release system for Phy, causing a prolonged AChE inhibition in brain and preservation of high extracellular ACh levels, as reflected in CSF concentrations (Fig. 4). Another interesting observation is the 44% increase in baseline CSF AChE activity from 5.38 $\pm$ .84 to 9.48 + 67 nmol/ml/min at the higher doses (8 ug) (Fig. 2). The same effect has been observed in the dog with i.v. injections. This would be related to a stimulation of AChE release into the CSF as seen in dog and rat (Mattio et al., 1986; DeSarno et al., 1987).

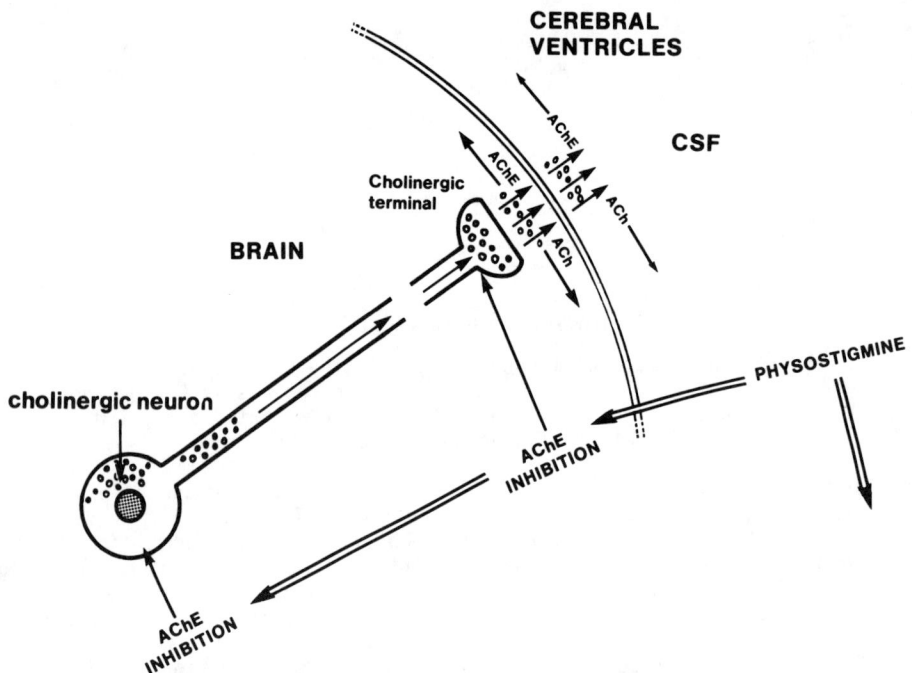

Figure 4. Physostigmine administered intraventricularly is distributed into the ventricular system and diffuses slowly into the brain. In the CNS it reaches first cholinergic synapses of regions which are located more closely to the ventricular surface (striatum and hippocampus) and then slowly diffuses to neurons deeper under and in the cerebral cortex. As a consequence of AChE inhibition, steady state acetylcholine levels are maintained in brain, and ACh diffuses into the CSF where it can be monitored. The release of AChE from cholinergic neurons is also represented in the diagram.

CONCLUSIONS

The attempt to modulate the central cholinergic system of animals and humans started with the pioneer findings of Cushing (1931), Feldberg and Sherwood (1954a,b) and Feldberg (1955), by using intraventricular drug administration. This approach has been followed by the use of ACh receptor blockers and cholinomimetic substances in several pathological conditions (Table 1). The use of the i.c.v. route for Phy for treatment of Alzheimer patients represents the most recent application of this methodology. It offers the unique possibility of affecting cholinergic nuclei deep in the brain and cortical synapses by increasing ACh levels and magnifying ACh effect on receptors with mimimal peripheral effects (Fig. 4). It also makes it possible to prolong the action of the drug by using a continuous infusion system using a procedure which could be well tolerated by the patient. Our preliminary results with Phy suggest the possibility of maintaining elevated ACh levels in brain for up to 6 hrs following single i.c.v. administration without producing intolerable side effects.

ACKNOWLEDGEMENTS

The authors wish to thank Diana Smith for typing and editing the manuscript. This work was supported in part by National Institute of Aging AG05416.

REFERENCES

Becker, R.E. and E. Giacobini. 1988. Mechanisms of cholinesterase inhibition in senile dementia of the Alzheimer type: clinical, pharmacological and therapeutic aspects. Drug Develop. Res. 12:163-195.
Beleslin, D.B., D. Andjelkovic and B.V. Vasic. 1973. An analysis of mechanism of the appearance of convulsions after intraventricular injection of eserine. Neuropharmacology 12:275-281.
Candy, J.M., E.K. Perry, R.H. Perry, J.A. Court, A.E. Oakley and J.A. Edwardson. 1986. The current status of the cortical cholinergic system in Alzheimer's disease and Parkinson's disease. In Progress in Brain Research, Vol. 70, ed. D.F. Swaab, E. Fliers, M. Mirmiran, W.A. Van Gool and F. Van Haaren, 105-132. Amsterdam: Elsevier.
Coombs, D.W., R.L. Saunders, M.S. Gaylor, A.R. Block, T. Colton, R.E. Harbaugh, M.G. Pageau and W. Mroz. 1983. Relief of continuous chronic pain by intraspinal narcotics infusion via an implanted reservoir. J. Amer. Med. Assoc. 250:2336-2339.
Cronin-Golomb, A. 1987. International study group on the pharmacology of memory disorders associated with aging. Neurobiol. Aging 8:277-282.
Cushing, H. 1931. Methods of action of pituitrin introduced into the ventricle. Proc. Natl. Acad. Sci., 17:239-249.
DeSarno, P., E. Giacobini and M. Downen. 1987. Release of acetylcholinesterase from the caudate nucleus of the rat. J. Neurosci. Res. 18:578-590.
Feldberg, W. 1955. Recent experiments with injections of drugs into the ventricular system of the brain, Proced. Royal Soc. Med. 48:853-864.
Feldberg, W. and S.L. Sherwood. 1954a. Injections of drugs into the lateral ventricle of the cat. J. Physiol. 123:148-167.
Feldberg, W. and S.L. Sherwood. 1954b. Behaviour of cats after intraventricular injections of eserine and DFP. J. Physiol. 125:488-500.
Feldberg, W. and S.L. Sherwood. 1955. Injections of bulbocaprine into cerebral ventricles of cats. Brit. J. Pharmacol. 10:371-374.
Fox, J.H., R. Penn, R. Clasen, E. Martin, R. Wilson and S. Savoy. 1985. Pathological diagnosis in clinically typical Alzheimer's disease. New Eng. J. Med. 313(22):1419- 1420.
Gauthier, S., R. Leblanc, R. Quirion, G. Carlsson, M. Beaulieu, R. Bouchard, D. Dastoor, F. ERvin, L. Gauthier, M. Gauvin, J. Henry, R. Palmour and Y. Robitaille. 1986. Transmitter-replacement therapy in Alzheimer's disease using intracerebroventricular infusions of receptor agonists. Can. J. Neurol. Sci. 13(4):394-402.
Giacobini, E., I. Mussini and T. Mattio. 1986a. Aging of cholinergic synapses: fiction or reality. In Dynamics of Cholinergic Function, Vol. 30, ed. I. Hanin, 177-190. New York: Plenum Press.

Giacobini, E., R. Becker, R. Elble, T. Mattio and M. McIlhany. 1986b. Acetylcholine metabolism in brain. Is it reflected by CSF changes. In Alzheimer's and Parkinson's Disease, ed. A. Fisher, 300-316. New York: Plenum Press.

Giacobini, E., R. Becker, R. Elble, T. Mattio, M. McIlhany and G. Scarsella. 1987a. Brain acetylcholine - a view from the cerebrospinal fluid. In Neurobiology of Acetylcholine, ed. N. Dun, 85-101. New York: Plenum Press.

Giacobini, E. 1987b. Models and strategies of cholinomimetic therapy of Alzheimer disease. In Cellular and Molecular Basis of Cholinergic Function, ed. M.J. Dowdall and J.N. Hawthorne, 882-901. England: Ellis Horwood.

Hallak, M. and E. Giacobini. 1986. Relation of brain regional physostigmine concentration to cholinesterase activity, and acetylcholine and choline levels in rat. Neurochem. Res. 11:1037-1048.

Hallak, M. and E. Giacobini. 1987. A comparison of the effects of two inhibitors on brain cholinesterase. Neuropharmacology 26(6):521-530.

Hallak, M.E. and E. Giacobini. 1988. The effects of repeated administration of acetylcholinesterase inhibitors in brain. Neuropharmacology (In Press).

Harbaugh, R.E., D.W. Coombs, R.L. Saunders, M. Gaylor and M. Pageau. 1982. Implanted continuous epidural morphine infusion system. J. Neurosurg. 56:803-806.

Harbaugh, R.E., D.W. Roberts, D.W. Coombs, R.L. Saunders and T.M. Reeder. 1984. Preliminary Report: Intracranial cholinergic drug infusion in patients with Alzheimer's disease. Neurosurgery 15(4):514-517.

Henderson, W.R. and W.C. Wilson. 1936. Intraventricular injection of acetylcholine and eserine in man. Quart. J. Exp. Physiol. 26:83-95.

Mattio, T., M. McIlhany, E. Giacobini and M. Hallak. 1986. The effects of Physostigmine on acetylcholinesterase activity of CSF, plasma and brain. A comparison of intravenous and intraventricular administration in beagle dogs. Neuropharmacology 25:1167-1177.

Penn, R.D., J.A. Paice, W. Gottschalk and A.D. Ivankovich. 1984. Cancer pain relief using chronic morphine infusion, early experience with a programable implanted drug pump. J. Neurosurg. 61:302-306.

Penn, R.D., E.M. Martin, R.S. Wilson, J.H. Fox and S.M. Savoy. 1988. Intraventricular bethanechol infusion for Alzheimer's disease: results of double-blind and escalating dose trials. Neurology 38:219-222.

Sherwood, S.L. 1952. Intraventricular medication in catatonic stupor. Brain 75:68-75.

Tucek, S. 1984. Problems in the organization and control of acetylcholine synthesis in brain neurons. Prog. Biophys. Molec. Biol. 44:1-46.

Unni, L., M. Hannant, R. Becker and E. Giacobini. 1988. Determination of physostigmine in plasma and CSF by high performance liquid chromatography with electrochemical detection. In Advances in Alzheimer Therapy, ed. E. Giacobini and R. Becker, New York: Taylor and Francis (In Press).

Whitehouse, P.J., D.L. Price, A.W. Clark, J.T. Coyle and M.R. DeLong. 1981. Alzheimer disease: evidence for selective loss of cholinergic neurons in the nucleus basalis. Ann. Neurology 10(2):122-126.

Whitehouse, P.J. 1988. Intraventricular bethanechol in Alzheimer's disease: A continuing controversy. Neurology 38:307-308.

Yaksh, T.L., R. Dirksen and G.J. Harty. 1985. Antinociceptive effects of intrathecally injected cholinomimetic drugs in the rat and cat. Eur. J. Pharmacol. 117:81-88.

TREATMENT OF ALZHEIMER DEMENTIA WITH STEADY-STATE INFUSION OF PHYSOSTIGMINE

Rodger J. Elble[1], Ezio Giacobini[2], Robert Becker[3], Ronald Zec[3], Sandra Vicari[3], Cindy Womack[3], Elizabeth Williams[2], and Constance Higgins[1]
[1]Dept. of Medicine, Div. of Neurology, [2]Dept. of Pharmacology, [3]Dept. of Psychiatry, S.I.U. School of Medicine, Springfield, IL

INTRODUCTION

The loss of cholinergic activity in the cerebral cortex of Alzheimer patients has been demonstrated in many laboratories and is correlated with the degree of dementia (White et al. 1977; Perry et al. 1977 and 1978; Wilcock et al. 1982; Francis et al. 1985; Near et al. 1986). This cortical deficit parallels the subcortical degeneration of cholinergic neurons in the nucleus basalis of Meynert (nbM) (Whitehouse et al. 1982; Koshimura et al. 1987) and tends to be greater in those patients who are afflicted at a younger age (Tagliavini and Pilleri 1983; Bird et al. 1983; Rossor et al. 1984; DeKosky et al. 1985; Hansen et al. 1988). However, there is only a limited correlation between the cortical cholinergic deficit and the neuronal loss in nbM (Perry et al. 1983; Etienne et al. 1986), and relative preservation of nbM may be seen in AD brains with significant cortical cholinergic loss (Perry et al. 1982; Pearson et al. 1983; Etienne et al. 1986).

Early psychopharmacologic studies in normal humans revealed memory enhancement with physostigmine (Davis et al. 1978; but see Drachman and Leavitt 1974) and memory impairment with scopolamine (Drachman and Leavitt 1974), supporting the importance of cholinergic transmission in normal memory function. Basal forebrain trauma in humans (Salazar et al. 1986) and in laboratory primates (Irle and Markowitsch 1987) produces anterograde amnesia which is partially reversed by physostigmine (Murray and Fibiger 1985). However, such lesions do not produce the global intellectual deficits of Alzheimer dementia (Salazar et al. 1986), suggesting that the cholinergic deficit is not sufficient to explain the many facets of Alzheimer dementia. Furthermore, cognitively impaired Parkinson patients have cortical cholinergic deficits comparable to those of Alzheimer patients, but Parkinson patients usually have less severe dementia (Perry et al. 1985).

Neuritic plaques and neurofibrillary tangles are found not only in the cerebral cortex but also in the basal forebrain (Rasool et al. 1986) and midbrain tegmentum (Yamamoto and Hirano 1985), suggesting that the same disease process simultaneously affects subcortical and cortical structures. In the basal forebrain, both cholinergic and noncholinergic neurons are affected (Rasool et al. 1986). Alzheimer disease therefore results in several neurochemical deficits other than acetylcholine (Ellison et al. 1986; Yamamoto and Hirano 1985; Rasool et al. 1986; Bondareff et al. 1987). Even though plaques and tangles clearly involve cholinergic neurons (Struble et al. 1982; Arendt et al. 1985; Mesulum et al. 1987; Tago et al. 1987), AD is not simply a disease of cholinergic transmission.

Therefore, the functional importance of the cholinergic deficit is still a matter for debate (Collerton 1986) and must be explored in the context of numerous pathological and behavioral complexities. We continue to examine physostigmine not only as a potential treatment but also as an experimental probe into the clinical importance of the cholinergic deficit. In this report, we present data from an ongoing intravenous physostigmine study that is patterned after the study of Mohs and Davis (1982). In contrast to their study, we have also compared the effects of physostigmine with those of low-dose scopolamine in individual patients, hoping to gain insight into the variable response of AD patients to these drugs. In particular, we have examined the usefulness of scopolamine as a tool for identifying cholinergic responders.

METHODS

Sixteen patients, 6 male and 10 female, with dementia of Alzheimer type were selected using the diagnostic criteria of the National Institute of Neurological and Communicative Disorders and Stroke and the Alzheimer's Disease and Related Disorders Association (McKhann et al. 1984), as outlined in a previous report (Elble et al. 1987). All patients and their next-of-kin signed informed consent. The age range of our patients was 55 to 81 yr (mean 71.8 ± 7.5 S.D.), and their Mini-Mental State (MMS) scores (Folstein et al. 1975) ranged from 5 to 28 (mean 17.9 ± 6 S.D.). Only one patient had a MMS score less than 10, and the patient with a score of 28 was a retired college professor who subsequently required nursing home placement.

Memory function was assessed using the experimental protocol of Mohs and Davis (1982). A sequence of 12 words or 12 black-and-white pictures were presented to the patient, and the patient was required to read or name aloud each word or picture, respectively. These words or pictures were then presented pseudorandomly with 12 additional distractor words or pictures of equal difficulty, and the patient was asked to identify the original words or pictures by responding "yes" and the distractors by responding "no". This process was repeated twice, each time with a new sequence of distractors.

In an initial screening session, both the word recognition task and the picture task were given to each patient. Then, each patient was assigned to the word task or picture task based upon the requirement of exceeding chance recognition of the words or pictures. The more demented patients (4 of 16) could not exceed chance performance on the word task and were therefore assigned to the picture recognition task.

Repeat memory testing with new sequences of words or pictures was then performed once each week for six weeks. Throughout the six weeks of testing, our patients took 12.4 gm of lecithin (270 mg choline) per day. During the first three weeks of testing, 300, 600 and 900 micrograms/m^2 physostigmine were given intravenously (IV) in pseudorandom order, blind to the examiner and patient, to determine the optimum dose, defined as that dosage resulting in the greatest total number of correct word or picture identifications (sum of three trials). During this dose-finding phase of the study, no attempt was made to identify and exclude cholinergic <u>nonresponders</u>.

Physostigmine was administered as a 2 hr and 15 min infusion (Figure 1). One-third of the total dosage was given during the first 15 min, with the remainder given over the subsequent 2 hr. This method of physostigmine administration was determined, in a preliminary trial-and-error study of 6 normal volunteers, to produce steady-state plasma butyrylcholinesterase (BuChE) inhibition during the final hour of infusion. Memory testing was performed during the last 30 min of infusion (Figure 1).

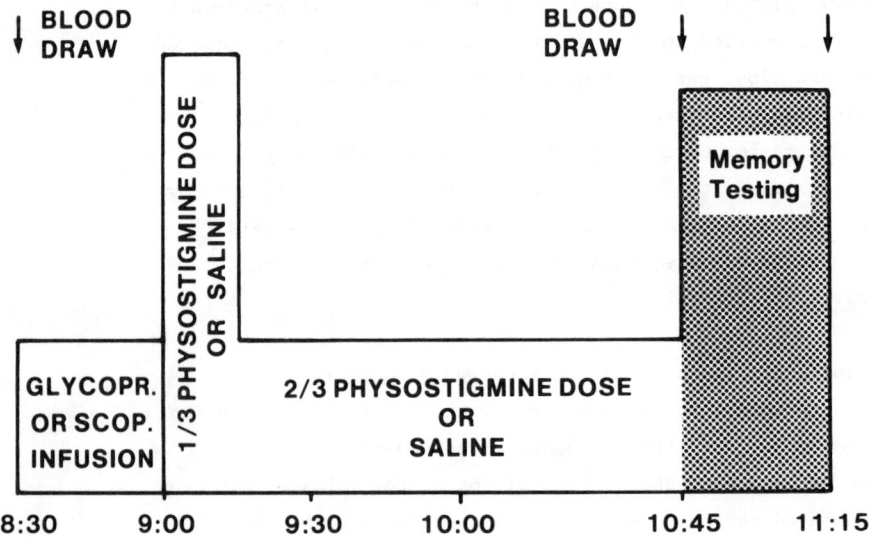

Figure 1: Flow diagram of the weekly experimental sessions. See text for discussion. Glycopr = glycopyrrolate and scop = scopolamine.

During the final three weeks, the best-dose of physostigmine was tested against saline placebo and 0.161 mg/m^2 IV scopolamine in a randomized, double-blind study. The entire dose of scopolamine was given as a 30-min infusion, prior to the infusion of normal saline (Figure 1). This dosage of scopolamine is equivalent to 0.3 mg for an adult of average weight and height and is known to produce detectable memory impairment in AD patients without producing incapacitating side effects (Sunderland et al. 1987). In this regard, we have seen dosages as low as 0.4 mg produce incapacitating catatonia in moderately-demented AD patients. Even with the dosage used in our study, one of our patients was not testable after scopolamine infusion, due to acute confusion and restlessness. Consequently, memory data for scopolamine infusion are reported for fifteen patients.

Prior to receiving physostigmine, all patients received 0.1 mg glycopyrrolate IV (Figure 1). Additional 0.1 mg dosages were given as needed to control side effects, but this drug had limited efficacy in this regard. All patients were monitored continuously but unobtrusively by a nurse, with the aid of a portable electrocardiograph. Side effects were charted throughout the procedure.

An initial baseline blood sample was obtained at 8:30 a.m., prior to the administration of any drugs, and two additional blood samples were obtained at 10:45 and 11:15 a.m., shortly before and after memory testing (Figure 1). Plasma BuChE activity was determined using the radiometric method of Johnson and Russell (1975). The substrate was acetylcholine (ACh) chloride at concentrations of 2.5 mM in the presence of ACh iodide [acetyl-^{3}H spec. act. 100 mCi/mM, New England Nuclear].

The memory test results were analyzed trial by trial and as the sum of three trials using repeated-measures analysis of variance (ANOVA) and the Tukey test for multiple comparison of means (Zar 1984). The test results were also examined using signal detectability theory (Parks 1966), employed for this purpose by Mohs and Davis (1982). In this analysis, two parameters are computed, d' and C. The parameter d' is a measure of the information stored in memory about the sequence of words or pictures, and C is a measure of the word or picture familiarity that is required before the patient replies "yes, I studied it". Thus, when C is low, patients tend to mistakenly say "yes" when distractors are presented. All means are reported ± one standard deviation (S.D.).

RESULTS

A steady-state BuChE inhibition was established during the final 30 min of physostigmine infusion for all three dosages (Figure 2), and there was a linear relationship between BuChE inhibition and physostigmine dosage (Figure 2).

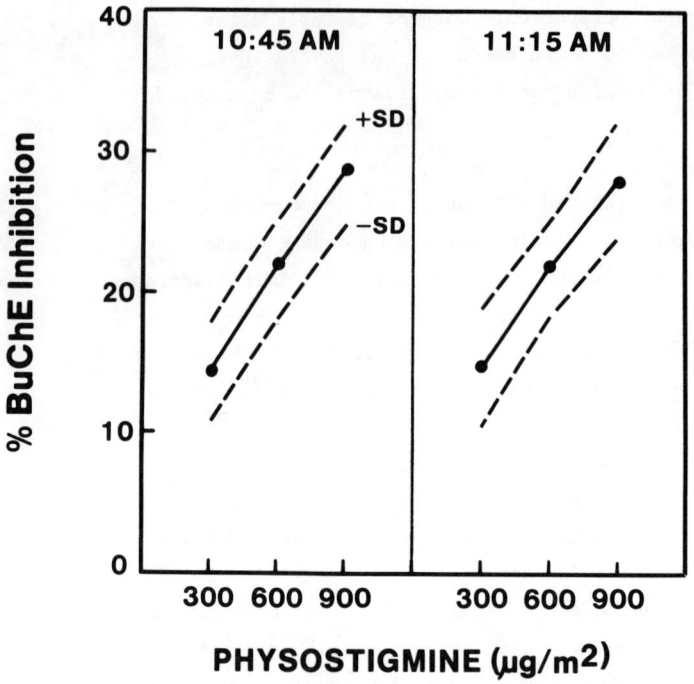

Figure 2: Plasma BuChE inhibition increased linearly with physostigmine dosage and was constant during the period of memory testing (10:45 to 11:15 am).

Four patients experienced nausea and vomiting with the 900 µg/m^2 dose of physostigmine, and seven patients experienced lightheadedness. Only one patient experienced nausea and vomiting during the 600 µg/m^2 infusion, but six patients experienced lightheadedness. No nausea or vomiting occurred during the 300 µg/m^2 infusion, but four patients experienced lightheadedness. The less bothersome side effects of dry mouth (11 patients; presumably due to the glycopyrrolate), belching (7 patients), abdominal cramping (2 patients), and blurred vision (2 patients) occurred during the 900 µg/m^2 infusion and less frequently during the 600 and 300 µg/m^2 infusions.

The selection of "best" physostigmine dose was made purely on the basis of memory test performance, and it was never necessary to resort to a lower dose because of untoward side effects. During the replication phase of our study, physostigmine side effects were never so severe as to preclude meaningful memory testing. Seven patients received 900 $\mu g/m^2$, 4 received 600 $\mu g/m^2$ and 5 received 300 $\mu g/m^2$ physostigmine infusion.

All patients experienced some drowsiness following scopolamine infusion, but this side-effect resolved by the time of memory testing. One patient became confused and restless and could not be tested. Minor side effects of dry mouth (3 patients), lightheadedness (7 patients), abdominal cramping (1 patient), tremor (1 patient) and headache (1 patient) also occurred, but except for the mild headache, these side effects had resolved by the time of memory testing.

Scopolamine impaired memory by an average 9.0 ± 11.8% ($p < 0.05$), as measured by comparing the total correct answers for all three trials with the total for saline placebo. Physostigmine produced a statistically-insignificant mean improvement of 4.3 ± 13%. In a trial-by-trial analysis, scopolamine-induced memory impairment was statistically significant for trials 1 ($p = 0.015$) and 2 ($p = 0.037$) but not for trial 3 ($p = 0.2334$).

These results were largely confirmed by signal detection analysis. The memory parameter d' was significantly reduced for scopolamine in trial 2 ($p = 0.046$) but not in trials 1 and 3 (Figure 3). Physostigmine had no significant effect on d'. However, for all three memory trials, there was a consistent trend for increased d' with physostigmine and for reduced d' with scopolamine (Figure 3).

Physostigmine and scopolamine also had opposite effects on parameter C although these effects failed to reach statistical significance (Figure 4). There was a consistent trend for increased C with physostigmine and for reduced C with scopolamine. These results suggest that physostigmine increases the familiarity threshold for responding "yes", thus accounting for a statistically insignificant trend toward fewer incorrect "yes" responses during physostigmine infusion (mean 10.6 ± 9.9) as compared to saline (13.5 ± 9.9). These trends are consistent with the previous findings of Mohs and Davis (1982). By contrast, the lower C with scopolamine indicates a reduced familiarity threshold for responding "yes". Consequently, there were more incorrect "yes" answers under scopolamine (17.9 ± 7.5) as compared to placebo and physostigmine (p < 0.025).

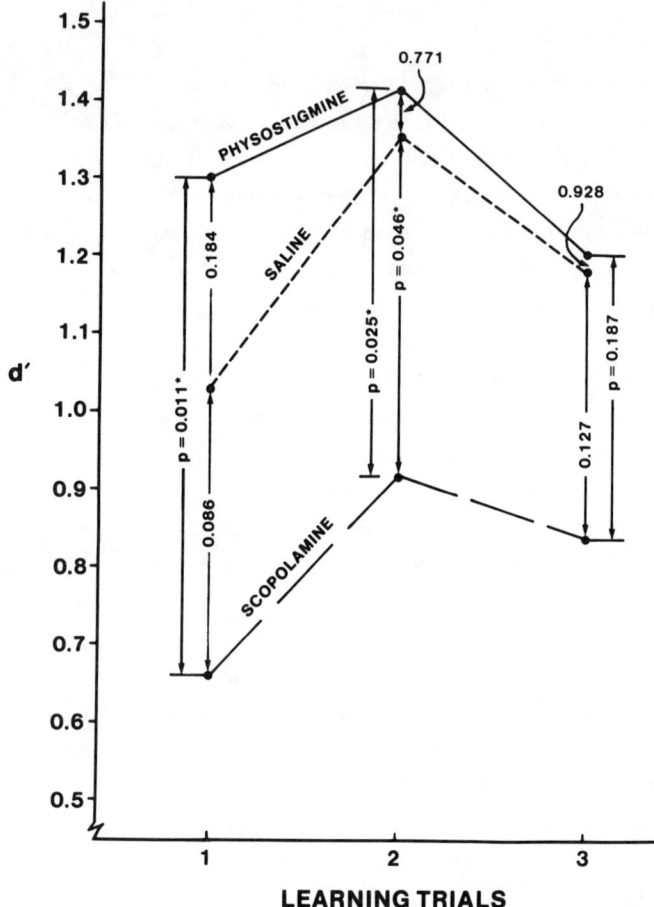

Figure 3: A trial-by-trial graph of the mean d' values for physostigmine, saline and scopolamine. The p values for comparison for means are also given.

The memory scores (total correct answers over three trials) for best-dose physostigmine during the dose-finding phase were significantly correlated with the physostigmine memory scores during the replication phase (r = 0.69; p < 0.005). Similarly, the correlation between the d' scores for the two physostigmine trails was 0.65 (p =

0.006). Significant correlation is to be expected since patients with a wide range of dementia were studied, and those patients with greatest dementia should achieve lower scores than those with less dementia, independent of any physostigmine effect. When interpreted in this context, these correlations are weaker than expected. Fifty to 60% of the total variance is unexplained, indicating considerable variability in patient performance.

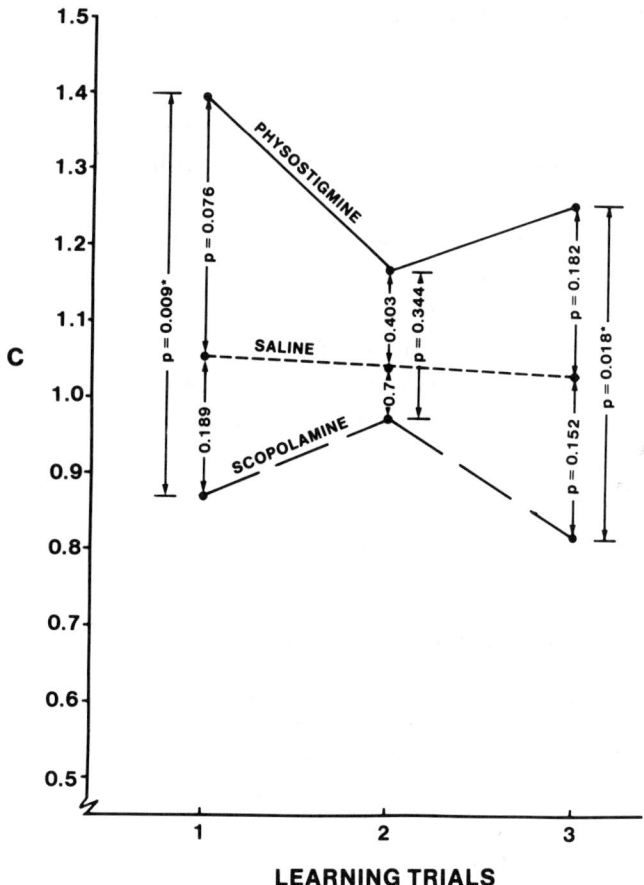

Figure 4: A trial-by-trial graph of the mean C values for physostigmine, saline and scopolamine. The p values for comparison of means are also given.

Of primary interest is the reproducibility of memory improvement
under physostigmine since memory scores are empirically used to
distinguish cholinergic responders from nonresponders. Accurate
assessment of this reproducibility would require replication of the
replication phase of this study in a large number of patients. This
has not been done. However, to obtain a rough estimate of this
reproducibility, baseline memory scores were subtracted from the
dose-finding best-dose physostigmine scores, and these _improvement_
scores were then regressed with the difference between the replication-phase physostigmine and saline scores. A correlation of only
0.41 ($p > 0.10$) was found, again indicating a highly variable patient
response to physostigmine.

The response to scopolamine was not a predictor of physostigmine
response. Eleven of fifteen patients exhibited memory deterioration
with scopolamine, ranging from -2 to -32 percent. There was no
correlation between the scopolamine-induced memory decrement (as
compared to saline placebo) and the improvement with best-dose
physostigmine ($r = 0.371$, $p > 0.10$). There was also no correlation
between physostigmine response and the magnitude of plasma BuChE
inhibition ($r = 0.331$, $p > 0.10$), Mini-Mental State score ($r = 0.100$,
$p > 0.50$) or age ($r = 0.105$, $p > 0.50$). Similarly, scopolamine-
induced memory impairment was not correlated with Mini-Mental state
($r = 0.043$, $p > 0.5$) or age ($r = 0.104$, $p > 0.5$).

DISCUSSION

We had hoped that steady-state cholinesterase inhibition might lead
to greater memory enhancement than has been found in previous studies
using rapid IV physostigmine infusion. Although, we could not
measure brain cholinesterase inhibition, laboratory studies in rats
(Hallak and Giacobini 1986; Somani and Khalique 1987) have
demonstrated that parenteral physostigmine administration produces
peak cholinesterase inhibition in less than 10 minutes. It is
therefore very likely that steady-state central inhibition was
achieved in our patients. If so, we may conclude that the attainment
of steady-state cholinesterase inhibition during acute IV
physostigmine adminstration produces little if any memory enhancement
in a general population of Alzheimer patients. However, it is

possible that chronic drug administration may lead to better results (Thal et al. 1986).

The use of a dose-finding phase to determine the best dose of physostigmine and to eliminate patients who might be insensitive to this and similar drugs has been popularized by Thal and colleagues (1986) and by Mohs and colleagues (1985). However, this method of identifying cholinergic responders is purely empirical, and no neurochemical or behavioral markers have been identified by which one can predict drug response to cholinergic enhancement therapy. Consequently, the concept of cholinergic response has been questioned, and there is justifiable concern that the exclusion of cholinergic nonresponders is a source of statistical bias (Stern et al. 1987). Nevertheless, it is noteworthy that physostigmine studies which incorporated a dose-finding phase were generally successful in demonstrating a small beneficial effect (Brinkman and Gershon 1983), and we know of no negative study which excluded nonresponders prior to the placebo-controlled replication phase.

However, the reliability of identifying cholinergic responders has been questioned by Stern et al. (1987) who found that memory performance in a dose-finding phase was a poor predictor of replication-phase performance. Our data are in agreement with those of Stern et al. (1987). Despite the use of a very simple memory task, the performance of our patients was so variable and the effect of physostigmine so small as to cast considerable doubt on the validity of the dose-finding phase in this and similar studies. Previous studies have involved patient populations as small or smaller than ours. Therefore, the reliability of this study design, using a large population of Alzheimer patients, has yet to be adequately tested.

Scopolamine has been used as a pharmacologic probe into the cholinergic mechanisms of memory (Sunderland et al. 1986), and we had hoped that scopolamine infusion might be a useful tool for identifying cholinergic responders. Unfortunately, we found no correlation between the responses to scopolamine and physostigmine. Similarly, physostigmine performance did not correlate with patient age or Mini-Mental State score. However, more patients should be studied, using more comprehensive dementia scales, before drawing firm conclusions.

The question remains as to why cholinergic enhancement therapy does not improve memory in all patients. The <u>cholinergic responder</u> phenomenon is also said to occur in healthy young adults (Davis et al. 1978) and is therefore not peculiar to Alzheimer patients. Consequently, it is possible that cholinergic responsiveness in Alzheimer patients is primarily due to a life-long, premorbid biological trait and only secondarily due to age- and disease-related factors. This hypothesis would explain the failure of all investigators to correlate physostigmine response with a variety of clinical factors. Furthermore, laboratory primates with basal forebrain lesions (Aigner et al. 1987) exhibit less memory improvement with cholinesterase inhibition than do aged controls (Bartus 1979). Therefore, a small and variable response to physostigmine in Alzheimer patients is perhaps what should be expected.

SUMMARY

We present a double-blind, placebo-controlled study of intravenous physostigmine (Phy) in the treatment of memory loss in 16 Alzheimer patients. Phy was administered as a 2-hour infusion with one-third of the total dose given during the first 15 minutes. This method of Phy infusion produced steady-state inhibition of plasma butyrylcholinesterase (BuChE) during the final hour of drug administration, and our memory tests were administered during this period of drug infusion. The study began with a "best dose" finding phase in which Phy dosages of 300, 600 and 900 micrograms/m^2 were administered at weekly intervals. These dosages produced an average plasma BuChE inhibition of 14.5, 21.8 and 28.5 percent, respectively. The best dose was then tested against 0.161 mg/m^2 intravenous scopolamine (Scop) and saline placebo. Memory was assessed in mild-moderately demented patients (12) with three trials of a 12-word recognition task. A 12-picture recognition task was used in advanced patients (4). Scop reduced memory scores by an average 9 ± 11.8% ($p < 0.05$) while best dose Phy produced a statistically insignificant 4.3 ± 13% improvement. Memory changes during the administration of these drugs did not correlate with degree of dementia nor was there any relationship between the responses to these drugs. Furthermore, patient performances exhibited considerable variability from trial to trial. These results illustrate the difficulties in identifying cholinergic drug responders and add to an already sizeable body of literature describing little or no clinical response to physostigmine. While these results may reflect negatively on the cholinergic hypothesis, they are equally indicative of the complexities of Alzheimer dementia.

ACKNOWLEDGMENTS

Supported by grants from the National Institute on Aging (AG05416), the Southern Illinois University Alzheimer Research Project Fund, and the E.F. Pearson Foundation. We gratefully acknowledge the valuable assistance with statistical analysis provided by Drs. Jerry Colliver and Paul Kolm.

REFERENCES

Aigner, T.G., S.J. Mitchell, J.P. Aggleton et al. 1987. Effects of scopolamine and physostigmine on recognition memory in monkeys with ibotenicacid lesions of the nucleus basalis of Meynert. Psychopharmacol. 92: 292-300.

Arendt, T., V. Bigl, A. Tennstedt, et al. 1985. Neuronal loss in different parts of the nucleus basalis is related to neuritic plaque formation in cortical target areas in Alzheimer's disease. Neurosci. 14: 1-14.

Bartus, R.T. 1979. Physostigmine and recent memory: effects in young and aged nonhuman primates. Science 206: 1087-1089.

Bird, T.D., S. Stranahan, S.M. Sumi, and M. Raskind. 1983. Alzheimer's disease: choline acetyltransferase activity in brain tissue from clinical and pathological subgroups. Ann. Neurol. 14: 284-293.

Bondareff, W., C.Q. Mountjoy, M. Roth, et al. 1987. Age and histopathologic heterogeneity in Alzheimer's disease. Arch. Gen. Psychiat. 44: 412-417.

Brinkman, S.D. and S. Gershon. 1983. Measurement of cholinergic drug effects on memory in Alzheimer's disease. Neurobiol. Aging 4: 139-145.

Collerton, D. 1986. Cholinergic function and intellectual decline in Alzheimer's disease. Neurosci. 19: 1-28.

Davis, K.L., R.C. Mohs, J.R. Tinklenberg, et al. 1978. Physostigmine: improvement of long-term memory processes in normal humans Science 201: 272-274.

Dekosky, S.T., S.W. Scheff, and W.R. Markesbery. 1985. Laminar organization of cholinergic circuits in human frontal cortex in Alzheimer's disease and aging. Neurology 35: 1425-1431.

Elble, R., E. Giacobini, and G.F. Scarsella. 1987. Cholinesterases in cerebrospinal fluid: a longitudinal study in Alzheimer disease. Arch. Neurol. 44: 403-407.

Ellison, D.W., M.F. Beal, E.D. Bird, et al. 1986. A postmortem study of amino acid neurotransmitters in Alzheimer's disease. Ann. Neurol. 20: 616-621.

Etienne, P., Y. Robitaille, P. Wood, et al. 1986. Nucleus basalis neuronal loss, neuritic plaques and choline acetyltransferase activity in advanced Alzheimer's disease. Neurosci. 19: 1279-1291.

Folstein, M.F., S.E. Folstein, and P.R. McHugh. 1975. "Mini-Mental State": a practical method for grading the cognitive state of patients for the clinician. J. Psychiatr. Res. 12: 189-198.

Francis, P.T., A.M. Palmer, N.R. Sims, et al. 1985. Neurochemical studies of early-onset Alzheimer's disease: possible influence on treatment. NEJM 313: 7-11.

Hallak, M. and E. Giacobini E. 1986. Relation of brain regional physostigmine concentration to cholinesterase activity and acetylcholine and choline levels in rat. Neurochem. Res. 11: 1037-1048.

Hansen, L.A., R. DeTeresa, P. Davies, and R.D. Terry. 1988. Neocortical morphometry, lesion counts, and choline acetyltransferase levels in the age spectrum of Alzheimer's disease. Neurology 38: 48-54.

Irle, E. and H.J. Markowitsch. 1987. Basal forebrain-lesioned monkeys are severely impaired in tasks in association and recognition memory. Ann. Neurol. 22: 735-743.

Johnson, C.D. and R.L. Russell. 1975. A rapid, simple radiometric assay for cholinesterase, suitable for multiple determinations. Anal. Biochem. 64: 229-238.

Koshimura, K., T. Kato, I. Yohyama, et al. 1987. Correlation of choline acetyltransferase activity between the nucleus basalis of Meynert and the cerebral cortex. Neurosci. Res. 4: 330-336.

McKhann, G., D. Drachman, M. Folstein, et al. 1984. Clinical diagnosis of Alzheimer's disease: report of the NINCDS-ADRDA work group under the auspices of the Department of Health and Human Services Task Force on Alzheimer's Disease. Neurology 34: 939-944.

Mesulam, M.M., C. Geula, and A. Moran. 1987. Anatomy of cholinesterase inhibition in Alzheimer's disease: effect of physostigmine and tetrahydroaminoacridine on plaques and tangles. Ann. Neurol. 22: 683-691.

Mohs, R.C. and K.L. Davis. 1982. A signal detectability analysis of the effect of physostigmine on memory in patients with Alzheimer's disease. Neurobiol. Aging 3: 105-110.

Mohs, R.C., B.M. Davis, A.A. Mathe, et al. 1985. Intravenous and oral physostigmine in Alzheimer's disease. Interdiscipl Topics Geront. 20: 140-152.

Murray, C.L. and H.C. Fibiger. 1985. Learning and memory deficits after lesions of the nucleus basalis magnocellularis: reversal by physostigmine. Neurosci. 14: 1025-1032.

Neary, D., J.S. Snowden, D.M.A. Mann, et al. 1986. Alzheimer's disease: a correlative study. J. Neurol. Neurosurg. Psychiat. 49: 229-237.

Parks, T.E. 1966. Signal-detectability theory of recognition memory performance. Psychol. Rev. 73: 46-58.

Pearson, R.C.A., M.V. Sofroniew, A.C. Cuello, et al. 1983. Persistence of cholinergic neurons in the basal nucleus in a brain with senile dementia of the Alzheimer's type demonstrated by immunohistochemical staining for choline acetyltransferase. Brain Res. 289: 375-379.

Perry, E.K., R.H. Perry, G. Blessed, and B.E. Tomlinson. 1977. Necropsy evidence of central cholinergic deficits in senile dementia. Lancet 1: 189.

Perry, E.K., B.E. Tomlinson, G. Blessed, et al. 1978. Correlation of cholinergic abnormalities with senile plaques and mental test scores in senile dementia. Br. Med. J. 2: 1457-1459.

Perry, R.H., J.M. Candy, E.K. Perry, et al. 1982. Extensive loss of choline acetyltransferase activity is not reflected by neuronal loss in the nucleus of Meynert in Alzheimer's disease. Neurosci. Lett. 33: 311-315.

Perry, E.K., M. Curtis, D.J. Dick, et al. 1985. Cholinergic correlates of cognitive impairment in Parkinson's disease: comparisons with Alzheimer's disease. J. Neurol. Neurosurg. Psychiat. 48: 413-421.

Rasool, C.G., C.M. Svendsen, and D.J. Selkoe. 1986. Neurofibrillary degeneration of cholinergic and noncholinergic neurons of the basal forebrain in Alzheimer's disease. Ann. Neurol. 20: 482-488.

Rossor, M.N., L.L. Iverson, G.P. Reynolds, et al. 1984. Neurochemical characteristics of early and late onset types of Alzheimer's disease. Br. Med. J. 288: 961-964.

Salazar, A.M., J. Grafman, S. Schlesselman, et al. 1986. Penetrating war injuries of the basal forebrain. Neurology 36: 459-465.

Somani, S.M. and A. Khalique. 1987. Pharmacokinetics and pharmacodynamics of physostigmine in the rat after intravenous administration. Drug Metab. Disposition 15: 627-633.

Stern, Y., M. Sano, and R. Mayeux. 1987. Effects of oral physostigmine in Alzheimer's disease. Ann. Neurol. 22: 306-311.

Struble, R.G., L.C. Cork, P.J. Whitehouse, et al. 1982. Cholinergic innervation in neuritic plaques. Science 216: 413-415.

Sunderland, T., P.N. Tariot, R.M. Cohen, et al. 1987. Anticholinergic sensitivity in patients with dementia of the Alzheimer type and age-matched controls. Arch. Gen. Psychiatr. 44: 418-426.

Tagliavini, F., and G. Pilleri. 1983. Basal nucleus of Meynert: a neuropathological study in Alzheimer's disease, simple senile dementia, Pick's disease and Huntington's chorea. J. Neurol. Sci. 62: 243-260.

Tago, H., P.L. McGeer, and E.G. McGeer. 1987. Acetycholinesterase fibers and the development of senile plaques. Brain Res. 406: 363-369.

Thal, L.J., D.M. Masur, N.S. Sharpless, P.A. Fuld, and P. Davies. 1986. Acute and chronic effects of oral physostigmine and lecithin in Alzheimer's disease. Prog. Neuropharmacol. Biol. Psychiat. 10: 627-636.

White, P., M.J. Goodhard, J.P. Keet, et al. 1977. Neocortical cholinergic neurons in elderly people. Lancet 1: 668-670.

Whitehouse, P.J., D.L. Price, R.G. Struble, and D.L. Price. 1982. Alzheimer's disease and senile dementia: loss of neurons in the basal forebrain. Science 215: 1237-1239.

Wilcock, G.K., M.M. Esiri, D.M. Bowen, et al. 1982. Alzheimer's disease: correlation of cortical choline acetyltransferase activity with the severity of dementia and histological abnormalities. J. Neurol. Sci. 57: 407-417.

Yamamoto, T., and A. Hirano. 1985. Nucleus raphe dorsalis in Alzheimer's disease: neurofibrillary tangles and loss of large neurons. Ann. Neurol. 17: 573-577.

Zar, J.H. 1984. Biostatistical Analysis (2nd edit.), Englewood Cliffs, New Jersey: Prentice Hall.

MEASURING CHANGES IN MEMORY & COGNITIVE FUNCTIONING IN ALZHEIMER'S DISEASE WITH ADMINISTRATION OF ORAL PHYSOSTIGMINE

David M. Masur[1], Alan D. Blau[1], Leon J. Thal[2], and Paul A. Fuld[1]

[1] Dept. of Neurology, Albert Einstein College of Medicine, 1300 Morris Park Ave., Bronx, NY; [2] Dept. of Neurology, San Diego VA Medical Center and Neurosciences, Univ. of Calif., San Diego, La Jolla, CA

INTRODUCTION

Current efforts to develop an effective drug treatment for Alzheimer Type Dementia (ATD) are in large part based upon the consistent finding that individuals with this disease suffer from marked reductions of acetylcholine in the brain, and that these reductions have been associated with changes in memory (Davies and Maloney, 1976; Katzman, Brown, Fuld, Thal, Davies, and Terry, 1986; Perry, Tomlinson, Blessed, Perry, Cross, and Crow, 1978). Physostigmine, an agent that increases the availability of acetylcholine through the inhibition of acetylcholinesterase has been considered a promising therapeutic approach for the treatment of ATD. While a number of recent studies with physostigmine have reported varying degrees of improvement in memory functioning (Peters and Levin, 1979; Davis and Mohs, 1982; Thal, Fuld, Masur, and Sharpless, 1983; Schwartz and Kohlstaedt, 1986), others have failed to demonstrate a positive effect (Wettstein, 1983; Mitchell, Drachman, O'Donnell, and Glosser, 1986; Stern, Sano, and Mayeux, 1987). The difficulty in obtaining consistently encouraging results with physostigmine has been generally attributed to pharmacological factors such as the variability of the rate of absorption of the drug into the CNS (Thal, et al., 1983), the permeability of the blood-brain barrier (Schwartz, and Kohlstaedt, 1986), or differences in drug metabolism (Sharpless and Thal, 1985).

A related issue of extreme significance revolves around the type of patient that is likely to demonstrate a positive response to physostigmine. While it has been thought that subjects with more intact cholinergic systems and thus milder dementia would benefit most from drug treatment, improvement of memory in both mildly impaired and severely impaired ATD subjects has been observed (Thal, et al., 1983; Schwartz and Kohlstaedt, 1986). In addition, a recent study suggested relatively greater improvement in moderately impaired Alzheimer subjects on some non-memory measures of cognitive functioning when compared to subjects with mild dementia (Stern, Sano, and Mayeux, 1987). One way in which to clarify the relationship between severity of dementia and response to the drug is to further examine the criteria by which ATD subjects are labelled mild, moderate, or severely impaired for purposes of participation in a physostigmine study. The heterogeneous nature of ATD in terms of the manifestation of different patterns of neuropsychological deficits has received increasing attention (Crystal, Horoupian, Katzman, and Jotkowitz, 1982; Mayeux, Stern, and Spanton, 1985; Becker, Huff, Nebes, Holland,

and Boller, 1988). These differences may be subtle and escape detection by screening instruments which form the basis for judgement of severity, thus resulting in wide variations of performance capability by ATD subjects. This in turn may mask the true nature of the effect of physostigmine on cognitive functioning. The focus of this paper will be to discuss the psychometric parameters involved in subject selection and its subsequent effect on the results of a recently completed double-blind outpatient study of oral physostigmine (Thal, Masur, Blau, and Fuld, manuscript submitted for publication).

MATERIALS AND METHODS
Subjects

Sixteen subjects with presumed early to moderate ATD were enrolled in a 12 week double-blind parallel study of the effects of oral physostigmine on memory and social functioning (Thal, et al., submitted). Subjects were diagnosed based on NINCDS criteria for probable ATD (McKhann, Drachman, Folstein, Katzman, Price, and Stadlin, 1984). Efforts were made to include only those subjects who were mildly to moderately impaired. This was done in part based upon the following clinically derived criteria: 1) a Blessed Mental Status Test error score of 15 or less, 2) the ability to sustain performance during the administration of a 12 item, 12 trial memory test using the procedure of Selective Reminding (SR, Buschke, 1973; Buschke and Fuld, 1974), and 3) the ability to place at least one item into long term storage on the SR test. The demographic data for the entire subject sample is illustrated in Table 1.

TABLE 1

DEMOGRAPHIC CHARACTERISTICS OF THE SUBJECT SAMPLE

SUBJECT	GROUP	SEX	AGE	EDUCATION (Years)
1	DRUG	M	76	19
2	DRUG	M	62	18
3	DRUG	F	66	8
4	DRUG	M	71	8
5	DRUG	M	80	20
6	DRUG	M	57	18
7	DRUG	F	66	12
8	DRUG	F	70	12
9	DRUG	M	68	13
10	DRUG	M	60	14
		MEAN	67.6	14.2
		S.D.	7.1	4.4
1	PLACEBO	F	67	8
2	PLACEBO	M	74	4
3	PLACEBO	M	62	16
4	PLACEBO	F	71	7
5	PLACEBO	M	80	16
6	PLACEBO	F	62	16
		MEAN	69.3	11.2
		S.D.	7.1	5.5
TOTALS	MALES 10	MEAN	68.3	13.1
	FEMALES 6	S.D.	6.9	4.9

Procedure

Subjects who satisfied the entrance criteria for early to moderate ATD were then randomized so that two-thirds of the subjects received physostigmine (10 subjects) and one-third received placebo (6 subjects). Subjects underwent extensive baseline neuropsychological testing which consisted of the Fuld Adaptation of the Blessed Mental Status Test, (Fuld, 1978), the SR test, the Mattis Dementia Rating Scale (Mattis, 1976), Information, Vocabulary, Similarities, Digit Span, Digit Symbol, Block Design, and Object Assembly subtests of the Wechsler Adult Intelligence Scale (Wechsler, 1955), and the Rosen Construction Task (Rosen, 1984). Verbal fluency was measured by the total number of items retrieved from five semantic categories, with 60 seconds of response time allowed for each category (Howard, 1979). In addition, subjects were administered a depression scale (Raskin and Crook, 1976) as well as a number of scales designed to assess capability in the activities of daily living (Kuriansky and Gurland, 1976; Lawton and Brody, 1969). Both the drug and placebo group subjects were titrated on either physostigmine or placebo for a period of three weeks, (2 to 4 mg. q.i.d.) and response to the increasing dose was measured by the administration of two different forms of the SR test on a weekly basis. The first test was given approximately 30 minutes after ingestion of the medication (drug or placebo), and the second test was given about 1 hour later. This procedure was adhered to as closely as possible for each clinic visit. The dose yielding the best response for each subject was then administered for six weeks, and subjects were evaluated every two weeks with 2 SR tests. At the end of six weeks, all subjects were placed on placebo for two additional weeks, and underwent a final test session at the end of this period. All tests were administered by experienced examiners who had no knowledge of the treatment condition of the subjects.

Scoring

Although a considerable amount of data regarding both short term and long term memory functions can be obtained from the SR test (Masur, Fuld, Blau, Thal, Levin, and Aronson, in press), improvement was defined as an increase in consistent retrieval (CR), a reliable measure of long term storage (Masur, et al., in press), and a reduction of intrusions, or inappropriate verbal responses (Thal, et al., 1983). Calculation of CR and intrusions was accomplished according to the usual method (Fuld, 1980). For each clinic visit, the scores of the two SR tests for each of these components were averaged. Each subject's total baseline score was computed by averaging the tests obtained during the initial baseline (prior to titration) with those obtained at final baseline (all subjects on placebo). Similarly, the best dose scores were calculated by averaging the individual scores from all SR tests administered during the six week period of best dose (3 test sessions, 6 SR tests).

RESULTS

Initially, the results from the SR test were analyzed using a 2 factor ANOVA comparing baseline vs. best dose score and drug group vs. placebo group. A separate ANOVA was performed for CR and for the number of intrusions. The results demonstrate a significant group difference in CR (p .05), but no significant difference in the degree of reduction of intrusions. In addition, there was no interaction between group and dose level, suggesting that the significant group difference in CR may have been present at baseline. While these results did not appear to be encouraging, inspection of Table 2 revealed that the standard deviations for both CR and for intrusions were extremely large regardless of group or test condition. Despite these observations, the scores for CR and for intrusions showed some degree of improvement in the drug group (<u>reduction</u> of intrusions is defined as improvement), and decline in the placebo group. Thus, it appeared that a trend toward a positive drug effect was present, but was obscured by the unexpected variability in the subject's test performance.

TABLE 2

<u>ANALYSIS OF VARIANCE FOR EACH COMPONENT OF MEMORY</u>

DRUG GROUP (N = 10)

MEMORY VARIABLE	MEAN BASELINE	S.D.	MEAN BEST DOSE	S.D.
CR *	14.7	(11.0)	17.1	(12.1)
INTRUSIONS	7.4	(5.8)	6.8	(7.1)

PLACEBO GROUP (N = 6)

MEMORY VARIABLE	MEAN BASELINE	S.D.	MEAN BEST DOSE	S.D.
CR	6.9	(7.3)	7.8	(8.2)
INTRUSIONS	4.3	(2.3)	5.5	(2.8)

* $p < .05$ (DRUG > PLACEBO)

In an effort to lessen the effect of the variability in test performance, a case by case analysis was performed for each subject. Percent improvement or decline for CR and intrusions was computed by subtracting the average baseline score from the verage best dose score and dividing the result by the baseline score. A positive response to physostigmine was defined as improvement in CR <u>and</u> a reduction in intrusions (see Table 3).

TABLE 3

EFFECTS OF PHYSOSTIGMINE ON MEMORY: NON-PARAMETRIC ANALYSIS

SUBJECT	GROUP	% CHANGE CR	% CHANGE INTR.	*PSYCHOMETRIC RESPONSE
1	DRUG	+ 20	+ 24	
2	DRUG	+ 44	- 24	YES
3	DRUG	- 38	- 53	
4	DRUG	+ 22	- 06	YES
5	DRUG	+ 63	- 58	YES
6	DRUG	- 36	- 20	
7	DRUG	+ 29	- 10	YES
8	DRUG	+ 173	- 05	YES
9	DRUG	+ 37	- 29	YES
10	DRUG	+ 18	- 25	YES
1	PLACEBO	- 07	+ 04	
2	PLACEBO	+ 10	0	
3	PLACEBO	0	- 20	
4	PLACEBO	+ 15	+ 73	
5	PLACEBO	- 10	+ 67	
6	PLACEBO	+ 167	+ 83	

* Psychometric response - Improvement in CR and a reduction in the number of Intr.

** 7/10 Drug treated patients demonstrated psychometric response and 0/6 in the placebo treated group ($p < 0.01$ by Fisher Exact Probability test for the proportion of responding patients).

The results demonstrate a positive response in 70 percent of the drug group (7 of 10 subjects), while the placebo group did not produce a positive responder (Fisher Exact Probability; p .001). Graphic display of these results highlights both the trend toward improvement in memory in the drug group and the variability of performance by each subject regardless of group. Figures 1A and 1B depict the effects of physostigmine on CR and intrusions, respectively for each subject in the treatment group. Similarly, Figures 2A and 2B illustrate the performance of subjects in the placebo group.

146   D. M. Masur et al.

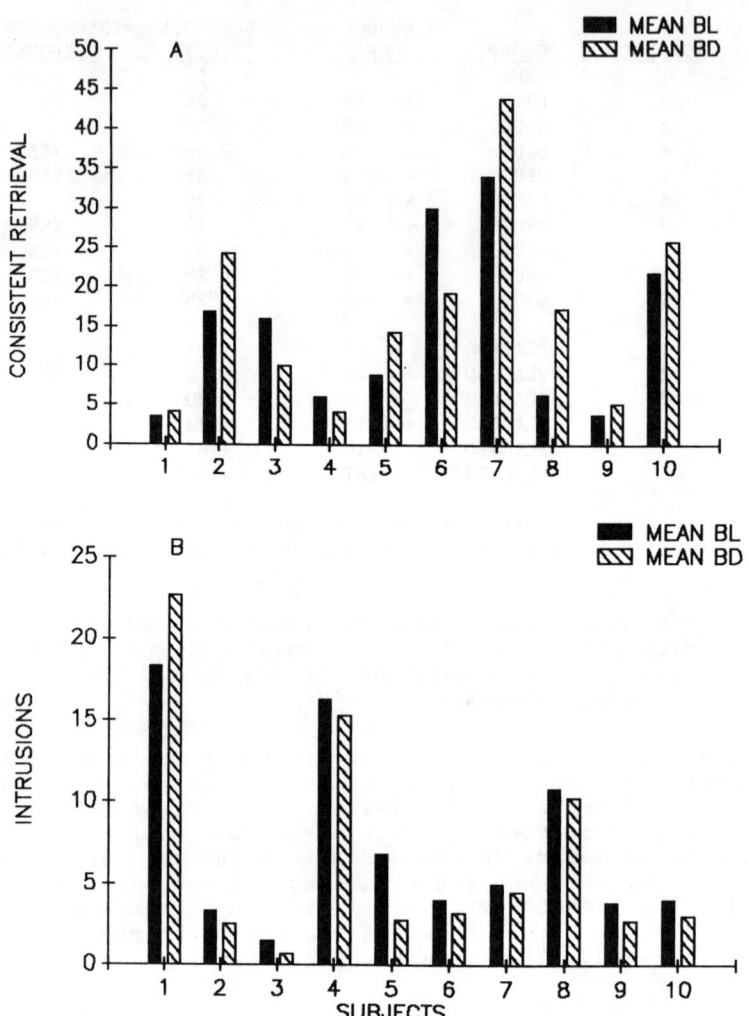

FIGURES 1A AND 1B
EFFECTS OF PHYSOSTIGMINE ON MEMORY — DRUG TREATMENT GROUP

NOTE: BL indicates Baseline
       BD indicates Best Rate

FIGURES 2A AND 2B

EFFECTS OF PHYSOSTIGMINE ON MEMORY – PLACEBO TREATMENT GROUP

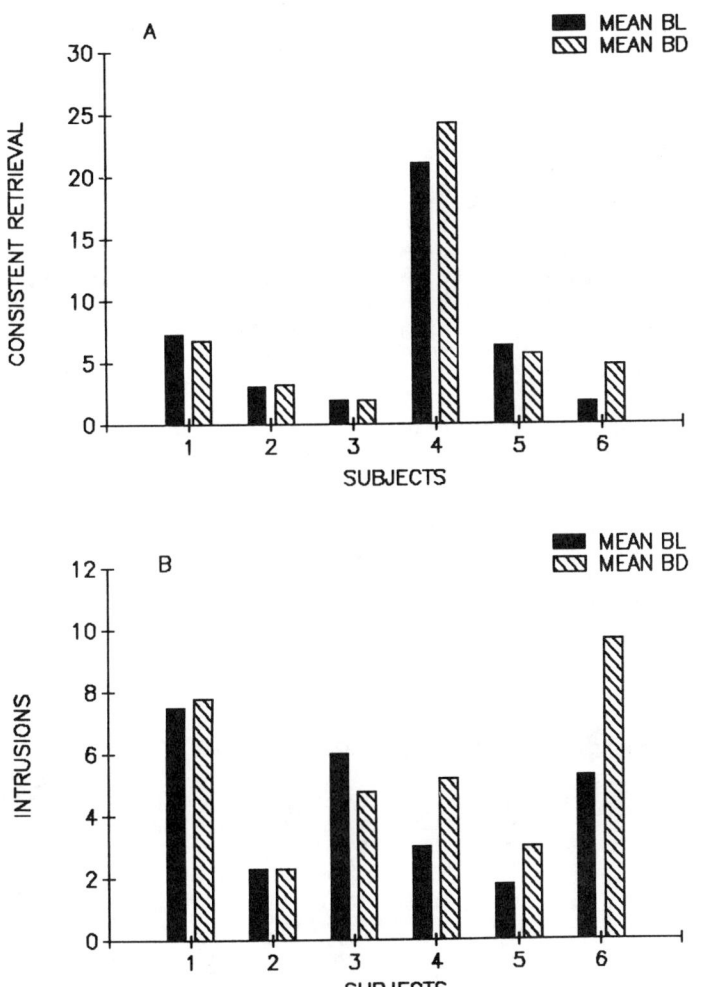

NOTE: BL indicates Baseline
       BD indicates Best Rate

There was no significant improvement in either constructional bility as measured by the Rosen Construction Task, or in activities of daily living. However, subjective reports from the families of six of the subjects who demonstrated a positive response to physostigmine indicated some degree of improvement of functioning in the home.

DISCUSSION

The present study demonstrates improvement on two components of memory functioning in 70% of the subjects who received oral physostigmine, while every subject who received placebo failed to meet the criteria for a positive response. While these results confirmed our previous inpatient study in which 10 of 16 subjects demonstrated improvement with oral physostigmine and lecithin (Thal, et al., 1983), the current study also highlights the difficulties encountered in consistently determining the effects of oral physostigmine on memory. The relatively small sample size of our study may inhibit the statistical demonstration of a more powerful effect; however, research in this area is generally performed with small samples (between 10 and 20 subjects). This is likely due to the time required to screen large numbers of potential subjects in order to obtain a homogeneous sample that meets rigit diagnostic and psychometric criteria for study entry. Despite our efforts to recruit mildly impaired subjects who would be homogeneous in terms of cognitive functioning, our study sample revealed huge individual differences in memory functioning. Thus, both intergroup and intragroup variability in performance was manifested on SR testing, which prevented the detection of a potential overall group effect with administration of oral physostigmine. (see Table 2).

In order to further understand why we obtained such a heterogeneous group of ATD subjects despite the use of what we though were rigid screening criteria, we studied each subject's cognitive functioning in greater detail. We examined the individual Blessed scores obtained at baseline along with the results of other neuropsychological tests that were not part of the entry criteria. These data are presented in Table 4 and provide some insight into the unexpected range of cognitive capability in these subjects with ostensibly mild impairment.

TABLE 4

BASELINE NEUROPSYCHOLOGICAL MEASURES FOR EACH SUBJECT

SUBJECT	GROUP	BLESSED ERRORS	MDRS	VIQ	PIQ	CR	INT	FLUENCY	PSYCHOMETRIC RESPONSE
1	DRUG	11	123	122	110	3.5	16.0	35.5	
2	DRUG	13	109	103	N/A	19.5	3.5	43.5	YES
3	DRUG	4	103	75	71	19.0	0.5	33.5	
4	DRUG	15	114	93	77	5.5	11.5	40.0	YES
5	DRUG	8	127	123	124	9.0	9.0	46.5	YES
6	DRUG	8	103	80	101	32.0	3.5	19.0	
7	DRUG	6	139	97	98	38.5	6.5	53.0	YES
8	DRUG	13	122	106	92	5.5	11.0	48.0	YES
9	DRUG	8	115	106	88	2.5	3.0	31.5	YES
10	DRUG	12	136	100	75	29.5	5.5	32.0	YES
	MEAN	9.8	119.0	101.0	92.9	16.4	7.0	38.3	
	S.D.	3.5	12.6	14.4	17.4	13.2	4.8	9.9	
1	PLACEBO	15	106	80	79	10.0	6.0	32.0	
2	PLACEBO	11	103	68	71	3.0	2.5	20.0	
3	PLACEBO	11	101	100	N/A	3.5	6.5	25.0	
4	PLACEBO	13	107	88	98	21.5	2.5	38.5	
5	PLACEBO	6	116	103	87	10.5	1.0	21.0	
6	PLACEBO	13	119	83	65	1.0	3.5	50.5	
	MEAN	11.5	108.7	87.0	79.8	8.3	3.7	29.5	
	S.D.	3.1	7.2	13.1	12.7	7.6	4.3	11.3	
TOTALS	MEAN	10.4	115.2	95.8	88.2	13.3	5.8	35.6	
	S.D.	3.4	11.8	16.4	16.7	11.8	4.3	10.9	

As can be seen from inspection of Table 4, subjects varied dramatically in their performance on the Dementia Rating Scale (MDRS) and in their Verbal and Performance I.Q. scores, despite Blessed scores which indicated mild to moderate impairment. A striking example of this variability is the performance of subject 3 in the drug group. Her very low Blessed score strongly suggested an individual with very mild impairment who was likely to be early in the course of her dementia. However, she had thepoorest performance of all the subjects in the drug group on virtually all other measures of cognitive functioning. Examination of these scores displayed much greater impairment in this subject than was thought to be possible. Tne surprising degree of heterogeneity of cognitive ability was also reflected in the baseline scores for both CR and intrusions, which were the major dependent measures. Further inspection of Table 4 demonstrates suprisingly little relationship between presumed level of impairment and memory functioning at entry into the study. For example, subject 1 in the drug group, despite having Verbal and Performance I.Q. scores within the bright average to superior range, scored extremely low on CR and made a much larger number of intrusions than would normally be expected. By contrast, subject 10 had a Blessed virtually identical to subject 1 and much lower I.Q. scores, yet had a far less memory impairment as suggested by his score on CR. Subject 1 did not reach the criteria for a positive response to physostigmine, while subject 7 demonstrated a clear improvement (see Table 1).

Our data illustrate the wide range of cognitive ability in a well chosen sample of ATD subjects with ostensibly mild impairment. Recent studies have suggested the presence of subgroups of ATD that can be distinguished on the basis of differing profiles of cognitive dysfunction (Mayeux, et al., 1985; Becker, et al., 1988). Of particular interest is the recent evidence for the formation of ATD subgroups characterized by _either_ a disproportionate impairment in lexical/semantic and syntactic processes, or a predominant visuoconstructional disturbance (Becker et al., 1988). Identification of such subgroups may have major implications for the way in which the behavioral response to physostigmine is measured. It is reasonable to speculate that disruption of the lexical/semantic system may compromise the ability to perform consistently on a challenging and sensitive verbal memory test like SR, despite the fact that the performance may be adequate on this test for entry intro the study. This situation could result in the failure to detect improvement when it is expected. Such may have been the case in the current study. Referring again to Table 4, two of the three subjects in the drug group (subjects 3 and 6) who did not demonstrate a positive response had the lowest Verbal I.Q. scores, and among the lowest scores in verbal fluency. This may have interfered with the detection of improvement in memory as hypothesized above. Alternatively, it is conceivable that the amount of cholineresterase inhibition attained with physostigmine and thus its efficacy may be affected more by the presence of subgroups of ATD that manifest a specific cluster of cognitive impairments, than by the overall severity of ATD. Thus, the selection of ATD subjects on the basis of the subtype as well as the stage of this complex disorder may provide new information regarding the benefits of treatment with physostigmine.

REFERENCES

Becker, J.T., J. Huff, R.D. Nebes, A. Holland, A. and F. Boller. 1988. Neuropsychological function in Alzheimer's disease. Pattern of impairment and rates of progression. Arch Neurol 45: 263-8.

Buschke, H. 1973. Selective reminding for analysis of memory and learning. Jour Verb Learn Verb Beh 12: 543-50.

Buschke, H., and P.A. Fuld. 1974. Evaluation of storage, retention and retrieval in disordered memory and learning. Neurol 24: 1019-25.

Crystal, H.A., D.S. Horoupian, R. Katzman, and S. Jotkowitz. 1982. Biopsy-proved Alzheimer's disease presenting as a right parietal lobe syndrome. Ann Neurol 12: 186-8.

Davies, P., and A.J.F. Maloney. 1976. Selective loss of central cholinergic neurons in Alzheimer's disease. Lancet II: 1403.

Davies, K.L., and R.C. Mohs. 1982. Enhancement of memory processes in Alzheimer's disease with multiple-dose intravenous physostigmine. Am Jour Psych 139: 1421-4.

Fuld, P.A. 1978. Psychological testing in the differential diagnosis of the dementias. In: Alzheimer's disease: senile dementia and related disorders Aging, Vol. 7, eds. R. Katzman, R.D. Terry and K.L. Bick, 185-193. New York: Raven Press.

Fuld, P. 1980. Guaranteed stimulus-processing in the evaluation of memory and learning. Cortex 16: 255-72.

Howard, D.V. 1979. Category Norms for Adults Between the Ages of 29 and 80. Technical Report NIA-79-1, Studies of Aging And Semantic Structure. Cognition Laboratory, Georgetown University, Washington, D.C.

Katzman, R., T. Brown, P. Fuld, L. Thal, P. Davies, and R. Terry. 1986. What is the significance of neurotransmitter abnormalities in Alzheimer's disease? Assn Res Nerv Ment Dis 63 :

Kuriansky, J., and B. Gurland. 1976. The performance test of activities of daily living. Int Jour Aging Human Dev 7 : 343-51.

Lawton, M.P., and E.M. Brody. 1969. Assessment of older people: self-maintaining and instrumental activities of daily living. Gerontologist 9 : 179-86.

Masur, D.M., P.A. Fuld, A. Blau, L.J. Thal, H.S. Levin, and M.K. Aronson. In press. Distinguishing normal and demented elderly with the selective reminding test. Jour Clin Exp Neuropsy.

Mattis, S. 1976. Dementia Rating Scale. In: Dementia and aging, eds. L. Bellack and B. Karsau. New York: Grune & Stratton.

Mayeux, R., Y. Stern, and S. Spanton. 1985. Heterogeneity in dementia of the Alzheimer type: evidence of subgroups. Neurol 35: 453-61.

McKhann, G., D. Drachman, M. Folstein, R. Katzman, D. Price, and E.M. Stadlan. 1984. Clinical diagnosis of Alzheimer's disease: report of the NINCDS-ADRDA Work Group under the auspices of the Department of Health and Human Services Task Force on Alzheimer's disease. Neurol 34: 939-44.

Mitchell, A., D. Drachman, B. O'Donnell, and G. Glosser. 1986. Oral physostigmine in Alzheimer's disease. Neurol 36 (Suppl. 1): 280.

Perry, E., Tomlinson, B., Blessed, G., Perry, R., Cross, A., and Crow, T. 1981. Neuropathological and biochemical observations on the noradrenergic system in Alzheimer's disease. J Neuro Sci 51: 279-287.

Peters, B., and H.S. Levin. 1979. Effects of physostigmine and lecithin on memory in Alzheimer's disease. Ann Neurol 6: 219-21.

Raskin, A., and T. Crook. 1976. Sensitivity of rating scales completed by psychiatrists, nurses, and patients to antidepressant drug effects. Jour Psychia Res 13: 31-41.

Rosen, W. 1984. Rosen construction task. Paper presented at the International Neuropsychological Society, Houston, Texas.

Schwartz, A.S., and E.V. Kohlstaedt. 1986. Physostigmine effects in Alzheimer's disease: relationship to dementia severity. Life Sci 38: 1021-8.

Sharpless, N.S., and L.J. Thal. 1985. Plasma physostigmine concentrations after oral administration. Lancet I: 1397-8.

Stern, Y., M. Sano, and R. Mayeux. 1987. Effects of oral physostigmine in Alzheimer's disease. Ann Neurol 22: 306-10.

Thal, L.J., P.A. Fuld, D.M. Masur, and N.S. Sharpless. 1983. Oral physostigmine and lecithin improve memory in Alzheimer's disease. Ann Neurol 13: 491-6.

Thal, L.J., D.M. Masur, A.D. Blau, and P.A. Fuld. Submitted for publication. Chronic oral physostigmine without lecithin improves memory on Alzheimer's disease.

Wechsler, D. 1955. Manual for the Wechsler Adult Intelligence Scale. New York: The Psychological Corporation.

Wettstein, A. 1983. No effect from double-blind trial of physostigmine and lecithin in Alzheimer's disease. Ann Neurol 13: 210-2.

EFFECTS OF PHYSOSTIGMINE ON RECOGNITION MEMORY IN AD PATIENTS:
A Comparative Review

Ronald F. Zec
Dept. of Psychiatry, S.I.U., School of Medicine, Springfield, IL

INTRODUCTION

As discussed in another chapter (Elble et al. 1988), we at SIU did not find memory enhancement in our acute intravenous physostigmine study. Our study was modelled after a study by Mohs and Davis (1982) in which intravenous physostigmine was found to improve performance on a recognition memory test. In this chapter these two studies will be compared and relevant methodological issues will be discussed.

The rationale for developing cholinergic therapies is based on the cholinergic hypothesis of Alzheimer's disease (AD). Although there is a major loss in several neurotransmitter systems in AD, the loss of cholinergic neurons is the most severe and correlates most closely with memory impairment (Thal et al. 1988, Davies 1983, Greenwald and Davis 1983). Cholinergic therapy is thought to be theoretically possible because the cholinergic loss is largely pre-synaptic with little or no change in post-synaptic muscarinic receptors (Thal et al. 1988).

Precursor loading with choline, however, has yielded uniformly negative results (Becker and Giacobini 1988). Only 3 of 29 studies have reported positive findings. Apparently, the cholinergic neurons in the Alzheimer brain are not driven by the precursor. More promising results have been obtained with physostigmine, a cholinesterase inhibitor (Becker & Giacobini 1988). Twenty-three studies (62%) have reported positive effects of physostigmine on memory, whereas 14 (38%) have reported negative findings. The magnitude of the positive effects, however, were generally too small to be considered therapeutic. On the other hand, unlike choline, the majority of the studies using physostigmine do report positive findings. Thus, the question arises why were positive findings not obtained in our study at SIU.

METHODS

Comparison of Subject Characteristics. In the Mount Sinai (MtS) study (Mohs and Davis 1982) there were 10 subjects (6 male, 4 female) whose ages ranged from 50-68 years. In the Southern Illinois University (SIU) study there were 16 patients (6 male, 10 female) whose ages ranged from 55-82 years (mean age = 72). The mean age of the patients was not specified in the MtS study, but it appears that their patient sample averaged about a decade younger than our sample. Since age has not been demonstrated to be a determinant of response to physostigmine, it is unlikely that this age difference is a critical variable in these studies.

In the two studies the degree of dementia of the patient samples appears to be approximately in the same range. The mean score of the SIU patient group on the Mini Mental State Exam (MMSE) was 17.9, i.e., moderate dementia. The MMSE scores ranged from 5 (severe) to 28 (very mild). There was only one patient with a MMSE score below 10. In the MtS study the Memory Information Test (MIT) was used to assess degree of dementia. Their criteria for subject selection was a MIT score of less than 11 and a Dementia Rating Scale score greater than 4. The mean level of dementia in their study was not indicated, so it is difficult to compare the severity of impairment of the patient samples in the two studies. It appears, however, that their patients were mild to moderately demented.

The patient samples in the two studies had similar levels of memory impairment. On the recognition memory test the mean level of performance and the standard error of the mean (S.E.M.) were almost identical for the two patient samples (Table 1). Furthermore, the ratio of patients in each study assigned the word versus the picture recognition test were approximately equivalent.

Comparison of design and procedure. There were both similarities and differences in the designs of the two studies. All patients were pretested to determine whether the word or the picture recognition test was more appropriate. This test requires the patient to read aloud and remember 12 target words or identify and remember 12 pictures which are visually presented one at a time. These 12 target words/pictures are then mixed with 12 distractor items and the patient is asked to identify which are "old" versus "new" items. Each patient was assigned to the version of the test on which he/she scored approximately midway between chance (=12) and a perfect score (=24). A mid-range baseline score would avoid "ceiling" and "floor" effects and thus would allow detection of improvements and decrements in performance. The mean score for the groups in both studies was slightly below the midway point. The picture recognition task usually is more appropriate for the more severely demented patients. One-third of the patients in the MtS study were assigned the picture recognition test compared to one-fourth of the patients in the SIU study.

In the MtS study three doses of physostigmine and a placebo were administered during the dose-finding phase. Each dose was evaluated separately over four consecutive days and had been given in a random order. In the SIU study three doses of physostigmine were administered in random order on separate days, at weekly intervals, and evaluated in a double-blind fashion. It is doubtful that the daily versus weekly intervals between test sessions in the two studies can account for the different results during the replication phases.

In the MtS study physostigmine (either .125, .25 or .5 mg) was administered intravenously over a 30-minute period. In the SIU study physostigmine (either 300, 600 or 900 $\mu g/m^2$) was infused over a 2.25-hour period. The mean milligram doses used in the SIU study were .60 mg, 1.19 mg and 1.79 mg. (i.e., these doses were 4.8, 4.8 and 3.6 times greater, respectively, than the doses in milligrams used in the MtS study). Since the SIU infusion was over 2.25 hours rather than over 30 minutes, (i.e., over a 4.5 times longer period), drug dosages used in the two studies were actually similar per unit of time.

A double-blind replication phase followed the dose-finding phase in both studies. In the MtS study "best dose" physostigmine and placebo were administered. In the SIU study a relatively low dose of scopolamine (.3 mg) was given in addition to "best dose" physostigmine and placebo. This low dose of scopolamine is known to produce cognitive impairments without appreciable side-effects (Sunderland et al. 1987). Lecithin was administered to the patients in the SIU study to maximize our chances of getting a positive effect with physostigmine. The dose of lecithin used in our study may have been too low to supply sufficient levels of choline (Wurtman 1988). Glycopyrrolate was also administered to lessen the side-effects of physostigmine. However, the lecithin and glycopyrrolate did not appear to affect the results of the SIU study.

The dose of physostigmine used during the replication phase was the dose on which the patient displayed the greatest number of total correct responses during the dose-finding phase. In both studies the highest dose was the "best dose" for the greatest proportion of patients. In the MtS study the highest dose was the "best dose" for six patients, the intermediate dose for 2 patients, and the lowest dose for another two patients. In the SIU study the highest dose was the "best dose" for seven patients, the intermediate dose for 4 patients, and the lowest dose for 5 patients. Thus, the proportion of patients in each study receiving the low, middle and high doses of physostigmine were similar.

In the MtS study the memory test was given over the last 20 minutes of a 30-minute infusion. In the SIU study one-third of the total dose was administered in the first 15 minutes of a 2.25-hour infusion and the other two-thirds of the dose was administered over the subsequent 2-hour period. Our plasma butylcholinesterase inhibition (BuChE) data indicated that a steady-state inhibition approaching 30% for the highest dose was produced over the last hour when the recognition memory test was admininstered (Elble et al. 1988). We had hoped that the steady-state cholinesterase inhibition in our study during the last hour, when the memory test was administered, would increase our chances of finding positive effects with physostigmine.

RESULTS

The mean response measures and the S.E.M.s for the placebo and physostigmine conditions were very similar in the SIU and MtS studies (Table 1). But unlike the SIU study, the MtS study found a positive, albeit modest, effect of physostigmine on recognition memory (p <.03). The mean improvement in the MtS study was 1 to 1.5 items per trial (chance=12 items, maximum correct=24) (Table 1). The error rates decreased by 12%, 19% and 13% on trials 1, 2, and 3, respectively. However, there was no difference in the "hit" rate (i.e., number of correct "yes" responses) between physostigmine and saline (Tables 1 and 2). On the other hand, physostigmine did decrease the number of "false alarms", (i.e., incorrect "yes" responses) (Tables 1 and 2). In summary, physostigmine did not increase the number of correct "yes" responses, it only decreased the number of incorrect "yes" responses in the MtS study.

Unlike the MtS study, we did not find any statistically significant effects of physostigmine compared to saline on any of our response measures (Table 2). There was a nonsignificant 6% trend for improvement (p=.32) with physostigmine compared to saline on trial 1 (Figure 1). There was no trend for improvment, however, on trial 2 (p=.87) or trial 3 (p=1.0). By contrast, we found a statistically significant 15% worsening in memory performance with scopolamine compared with saline (Figure 1, Table 2). There was a nonsignificant 14% increase in error rate with scopolamine on trial 1 (p=.05), a statistically significant 17% increase on trial 2 (p=.02), and a nonsignificant 15% increase on trial 3 (p=.06) (Figure 1). Performance with scopolamine was significantly worse compared with physostigmine on trial 1 (p=.01) and trial 2 (p=.01), while there was a nonsignificant trend for poorer performance on trial 3 (p=.10) (Figure 1). In summary, we did not find a memory-enhancing effect of physostigmine on the recognition memory test, but we did find an amnestic effect of scopolamine. The only positive trend for physostigmine was a nonsignificant

6% improvement on the first trial compared to the saline condition. Analyses using total correct or the d' (signal detection memory index) showed almost identical results.

Table 1. Comparison of SIU and MtS Studies: Mean Response Measures for Replication Study (± S.E.M.)

	Placebo			Physostigmine		
	Trial			Trial		
	1	2	3	1	2	3

Total Correct Responses

SIU:	16.4±0.79	17.7±0.95	17.1±0.83	17.2±0.87	17.8±0.81	17.1±0.88
MtS:	15.6±1.07	16.4±0.82	17.3±0.83	17.1±0.92	18.7±0.92	18.9±0.99

Correct Yes Responses ("Hits")

SIU:	8.8±0.71	10.0±0.38	9.9±0.45	8.8±0.56	9.3±0.68	8.8±0.73
MtS:	7.9±0.96	9.3±0.63	10.2±0.51	8.2±1.00	9.8±0.51	9.2±0.57

Incorrect Yes Responses ("False Alarms")

SIU:	4.4±0.87	4.3±0.94	4.8±0.93	3.6±0.99	3.4±0.89	3.7±0.97
MtS:	4.3±1.15	4.9±1.01	4.9±1.01	3.1±1.06	3.1±1.12	2.3±0.91

Table 2. ANOVAs (p values) for Replication Phases of the SIU and MtS Studies.

(SIU Study)	Physostigmine vs. Saline	Scopolamine vs. Saline	Physostigmine vs. Scopolamine
Total Correct	0.538	0.004*	0.008*
Hits	0.283	1.000	0.424
False Alarms	0.211	0.005*	0.009*

(MtS Study)			
Total Correct	<0.03*		
Hits	>0.05		
False Alarms	<0.01*	(* = statistically significant)	

In the SIU study there were no statistically significant differences in the "hit" rate for the three drug conditions (physostigmine, saline and scopolamine) (Figure 2, Table 2). Thus, as in the MtS study, there was no effect of physostigmine on "hit rate" compared to saline. Furthermore, we extended the findings of the MtS study by demonstrating that scopolamine had no effect on the number of hits. In summary, we found that physostigmine did not increase hit rate and scopolamine did not decrease hit rate.

Physostigmine and scopolamine did, however, affect the "false alarm" rate (Table 1, Figure 3). Scopolamine significantly increased the number of false alarms compared to saline (Table 2), whereas physostigmine tended to decrease the incorrect "yes" responses (Table 1, Figure 3). The number of false alarms were significantly greater with scopolamine compared with saline on each of the three learning trials (p=.03, .02, .03 respectively).

False alarms were also significantly greater with scopolamine compared with physostigmine on each trial (p=.03, .01, .02 respectively). Although the effect of physostigmine on false alarms was not statistically significant when compared with saline (Table 2), the trend was in the same direction as in the MtS study. We extended that finding by showing that scopolamine increased the number of false alarms. In other words, increased cholinergic function decreased false alarms, while decreased cholinergic function increased false alarms. Thus, manipulation of the cholinergic system in opposite directions has opposite effects on false alarms.

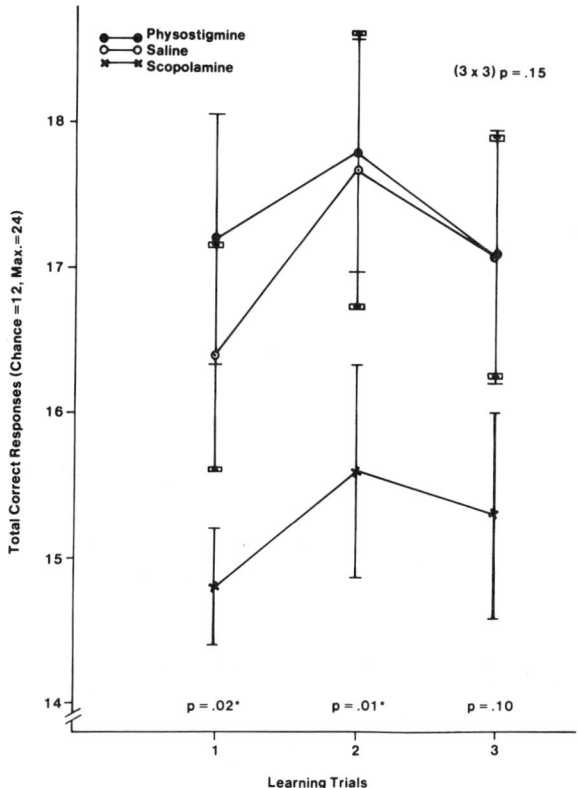

Figure 1: Mean total correct (+/- S.E.M.) for three learning trials and three drug conditions (n=16).

Although we did not find a group effect, perhaps there was a subgroup of patients who responded to physostigmine. To determine if there were "responder" individuals, three different criteria were applied to the data. A patient's score had to increase or decrease by 4, 5, or 6 total correct responses to meet one of these three criteria (Table 3). For example, using a criterion of ±4, a patient must have an increase or decrease of 4 or more correct on the drug trial compared to the saline trial. If the percentage of patients showing a change in one direction (e.g., +4) is greater than the other direction (e.g., -4) then this would be evidence that there was a subgroup of patients who responded to the drug.

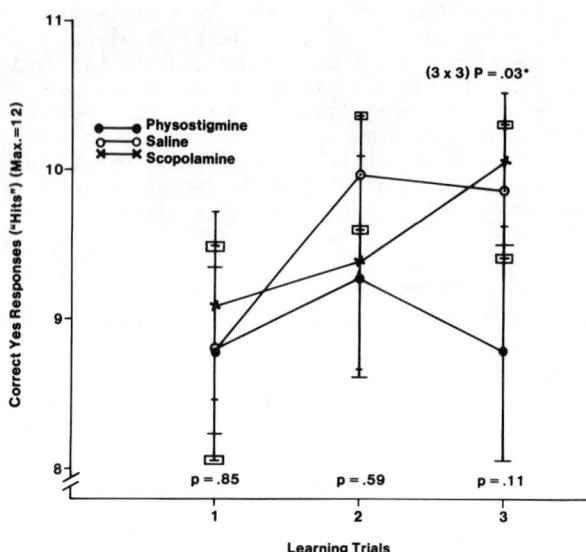

Figure 2: Mean "hits" (+/- S.E.M.) for three learning trials and three drug conditions (n=16).

A majority of the patients using each of these 3 criteria showed a response to scopolamine, i.e., an amnestic effect (Table 3). Only one patient under the ±4 criterion, one patient with the ±5 criterion, and 0 patients with the ±6 criterion showed an effect in the opposite direction. Applying the same criteria to physostigmine reveals very different results. Using the strictest criterion for change (i.e., 6 words or more across the three trials) there is no evidence for any responders to physostigmine (Table 2). Just as many patients responded in a negative direction as in a positive direction.

Using the more lenient criteria of ±4 or ±5 total correct responses, there is evidence for a weak effect of physostigmine in a subgroup of patients (Table 1). For example, using the ±4 criterion, 56% of the patients exhibited improved memory while only 25% showed a decrement in memory. Subtracting 25% from 56% suggests that only a quarter of our patients were showing a positive, albeit marginal, response to physostigmine.

Table 3: Percentage of Patients Showing a Drug Response Using Three Different Criteria.

CRITERION	SCOPOLAMINE	PHYSOSTIGMINE
+4	7%	56%
-4	69%	25%
+5	6%	44%
-5	56%	19%
+6	0%	19%
-6	56%	19%

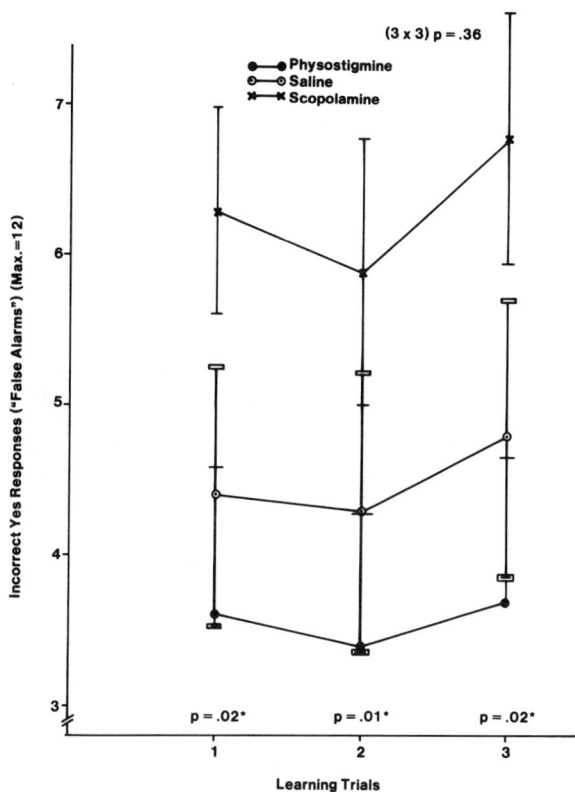

Figure 3: Mean "false alarms" (+/- S.E.M) for three learning trials and three drug conditions (n=16).

There was a statistically significant correlation between butylcholinesterase (BuChE) inhibition and physostigmine dosage (see Elble et al. 1988). The higher the dose of physostigmine the greater the inhibition of BuChE in the plasma and by inference the greater the cholinesterase inhibition in the CNS. There was no correlation between the effects of scopolamine and physostigmine on recognition memory. In other words, those patients who responded negatively to scopolamine were not the same people who showed a positive response to physostigmine (Table 4). There was no correlation between physostigmine response and the magnitude of the plasma butylcholinesterase inhibition. Furthermore, there was no correlation between response to physostigmine and scores on the MMSE. Nor was there any correlation between physostigmine response and the age of the patient. In summary, the few patients who appeared to be "responders" to physostigmine could not be identified in terms of their age, degree of dementia, or the degree of BuChE inhibition in the plasma. Also no correlation was found between the effects of scopolamine and scores on the MMSE.

Table 4: Difference scores for scopolamine-saline (Sc-Sa) and for physostigmine-saline (Ph-Sa).

Patient	1	2	3	4	5	6	7	8	9	10	11	12	13	14	15	16
AGE	55	72	76	78	60	71	58	76	71	64	74	77	71	71	71	76
Sc-Sa	-22	-11	-11	-8	-7	-7	-7	-6	-6	-4	-4	-1	0	+1	+3	+5
Ph-Sa	-8	+8	+4	+5	+5	-1	-1	+7	+5	+5	-3	-9	+4	-4	-6	+8

DISCUSSION

There is no clear methodological reason why phosostigmine improved recognition memory in the MtS study but not in the SIU study. One potential methodological issue concerns the notion of "individualized optimal dosing". It has been argued by Mohs and Davis (1982) and other researchers (Davis et al. 1978) that there is a narrow therapeutic range in which physostigmine can have positive effects on memory. Higher doses of physostigmine can impair memory rather than enhance it (Davis et al. 1976). Thus, it is possible that the doses used in our study were either too high or too low. Drug dosages used in the SIU and MtS studies were similar when duration of infusion was taken into consideration.

The second issue is that there may be individual differences in the degree of absorption of physostigmine into the CNS (Thal et al., 1988). This may determine who is a "responder" or "nonresponder" to the drug. Another methodological issue is the characteristics of the patients being studied. If the patient samples in the two studies differed significantly in terms of their level of dementia, then this could have differentially affected the findings. Degree of dementia was approximately equivalent in the SIU and MtS patient samples and more importantly level of performance on the recognition memory test was nearly identical.

Another issue is the test-retest variability in performance on memory tests. In most studies only one or two baseline or placebo measurements are taken. To determine test-retest variability we administered both a 10-word recall test and a word recognition test to two patients with moderate dementia and two with mild dementia once a week for six weeks. Performance on the recall test was quite consistent from week to week for each patient. By contrast, the recognition memory paradigm that we used in our study showed considerable test-retest variability. The problem of variability in memory performance needs to be taken into consideration in future studies. A single measurement may not provide a reliable baseline for assessing the effects of a drug for an individual patient. On the other hand, we did detect the amnestic effects of low dose scopolamine using the recognition memory test. This finding suggests that despite test-retest variability, the recognition memory paradigm should have been capable of detecting an improvement with physostigmine if one existed.

The fifth issue is the existance of "responder" and "nonresponder" patients to physostigmine. Although we did not find a statistically significant effect of the drug on the patient group, there was evidence that a subgroup of our patients displayed a modest improvement with physostigmine. Thus, individual responsiveness to the drug must be examined in addition to analyzing group data. A related issue is reproducibility of results for individual patients. If a given patient does not consistently have the same response to a particular drug dose, the drug will not be clinically useful. The reliability of the "responder-nonresponder" classification has not yet been empirically validated. In summary, there is no clear methodological difference for our failure to replicate the positive effects on recognition memory in AD patients reported by Mohs and Davis (1982). Our failure to replicate suggests that the memory-enhancing effect of physostigmine is small in magnitude, only occurs in a minority of AD patients, and consequently is not a robust phenomenon.

The use of physostigmine to enhance the cholinergic system and scopolomine to inhibit the cholinergic system produced opposite effects on false alarm rates in our study. Neither drug had any effect on "hit" rate. The implication of these findings is that the cholinergic system may play an important role in the inhibition of inappropriate responses. This finding is in agreement with Mohs & Davis (1982) who reported that physostigmine significantly decreased the number of false alarms but only slightly increased the number of hits. False alarms on a recognition memory test are comparable to the intrusion errors on recall tests. AD patients make more intrusion errors on recall tests than normal control subjects (Fuld et al. 1982). Thus, intrusion errors or false alarms are a characteristic sign of AD and have proven useful in making differential diagnoses (Fuld et al. 1982). Our findings are compatible with the current literature stating the importance of intrusion errors as a sign of AD. Furthermore, the number of intrusion errors that AD patients make on recall tests correlate with decreased cholinergic function measured at autopsy (Fuld et al. 1982). Future studies on the effects of cholinergic manipulations should pay particular attention to false alarm or intrusion rates.

Physostigmine, a cholinesterase inhibitor, has a relatively modest effect on memory in AD patients. L-dopa, the precursor of dopamine, has a dramatic effect on tremors in Parkinson patients (Adams and Victor 1981). If the cholinergic deficit plays a critical role in AD, why does cholinergic enhancement not produce a clinically significant effect? Is the fault with our technique or with our theory? Perhaps we have not yet perfected cholinergic therapy. On the other hand, it may be that the cholinergic hypothesis is too simplistic. Given the disappointingly small effects of cholinergic therapy, a re-evaluation of the cholinergic hypothesis is needed.

Currently, there is no evidence that physostigmine or any other cholinergic therapy will retard the progression of Alzheimer's disease. Perhaps this should not be expected since dopaminergic therapy does not slow down the progression of Parkinson's disease (Adams and Victor 1981). Thus, the most that can be expected is that cholinergic therapies may be a useful but limited palliative. A palliative is nonetheless a worthwhile aim. Researchers must not, however, lose sight of the ultimate aim which is to find either a prevention or cure for AD. Our penultimate aim is to stop or at least slow down the progression of the disease. It is also important to realistically appraise the potential effectiveness of cholinergic therapy for AD patients. Researchers and the media must be very careful about not misleading the public and creating false hope about cholinergic therapy.

An effective pharmacotherapy for AD has not yet been developed. A multimodal cholinergic strategy and/or a multimodal pharmacological strategy may be more effective than a single approach alone. Clinicians have observed considerable benefits of psychosocial and environmental interventions on the quality of life for AD patients (Cohen and Eisdorfer 1986, Mace and Rabins 1981). The combination of both pharmacological and psychosocial interventions is likely to be the best approach to management.

SUMMARY

In our study at SIU "best dose" physostigmine did not have a statistically significant effect on a recognition memory test in AD patients. A "low dose" of scopolamine, on the other hand, significantly impaired performance on the same test. Physostigmine had no group effect, but there was some evidence that a minority of patients may have had a modest improvement with physostigmine. Manipulating the cholinergic system in opposite directions had no effect on hit rate but produced opposite effects on the false alarm rate. Thus, the cholinergic system appears to play an important role in the inhibition of inappropriate responses.

ACKNOWLEDGEMENTS

I gratefully acknowledge the valuable editing assistance of Eleanor Feldman and Tami McDaniel, the statistical assistance of Steve Markwell and Dr. Jerry Colliver, and the word processing assistance of Gayle Jones and Jackie Wilmot.

REFERENCES

Adams, P.A., R. Katzman, P. Davis and R.D. Terry. 1982. Intrusions as a sign of Alzheimer dementia:Chemical and pathological verification. Ann. Neurol. 11: 155-159.

Becker, R.E. and E. Giacobini. 1988. Mechanisms of cholinesterase inhibition in senile dementia of the Alzheimer type: Clinical, pharmacological, and therapeutic aspects. Drug Dev. Res. 12: 163-195.

Cohen, D., and C. Eisdorfer. 1986. The Loss of Self: A Family Resource for the Care of Alzheimer's Disease and Related Disorders. New York: W.W. Norton & Company.

Davis, K.L., L.E. Hollister, J. Overall, A. Johnson and K. Train. 1976. Physostigmine: Effects on cognition and affect in normal subjects. Phychopharmacology 51: 23-27.

Davis, K.L., R.C. Mohs, J.R. Tinklenberg, L.E. Hollister, A. Pfefferbaum and B.S. Kopell. 1978. Physostigmine: Enhancement of long-term memory processes in normal subjects. Science 201: 272-274.

Davies, P. 1983. An Update on the Neurochemistry of Alzheimer Disease. In The Dementias, eds. R. Mayeux and W.R. Rosen, 75-86, New York:Raven Press.

Elble, R.J., E. Giacobini, R. Zec, S. Vicari, C. Womack, E. Williams, and C. Higgins. 1988. Treatment of Alzheimer dementia with steady-state infusion of physostigmine. In Current Research in Alzheimer Therapy: Cholinesterase Inhibitors, ed. E. Giacobini, R. Becker, New York:Taylor and Francis.

Fuld, P.A., R. Katzman, P. Davis and R.D. Terry. 1982. Intrusions as a sign of Alzheimer dementia: Chemical and pathological verification. Ann. Neurol. 11: 155-159.

Greenwald, B.S., K.L. Davis. 1983. Experimental pharmacology of Alzheimer disease. In The Dementias. eds. R. Mayeux and W.R. Rosen, 87-102, New York:Raven Press.

Mace, N., and P. Rabins. 1981. The 36-Hour Day: A Family Guide to Caring for Persons with Alzheimer's Disease, Related Dementing Illnesses, and Memory Loss in Later Life. Baltimore: John Hopkins University Press.

Mohs, R.C. and K.L. Davis. 1982. A sign detectability analysis of the effect of physostigmine on memory in patients with Alzheimer's disease. Neurobiol. Aging 3: 105-110.

Thal, L.J., B.R. Lasker, D.M. Masur, A.D. Blau, S. Knapp. 1988. Physostigmine treatment in SDAT: Type of administration, dose & duration. In Current Research in Alzheimer Therapy: Cholinesterase Inhibitors, ed. E. Giacobini, R. Becker, New York:Taylor and Francis.

Wurtman, R.J., J.K. Blusztajn, J.H. Growdon and I.H. Ulus. 1988. Cholinesterase inhibitors increase the brain's need for free choline. In Current Research in Alzheimer Therapy: Cholinesterase Inhibitors, ed. E. Giacobini, R. Becker, New York:Taylor and Francis.

COGNITIVE AND NEUROENDOCRINE CHANGES WITH PHYSOSTIGMINE IN ALZHEIMER'S DISEASE PATIENTS

V. Kumar[1], J. Murphy[2], K.A. Sherman[3], J.W. Ashford[1], and S.J. Markwell[4]
[1]Dept. of Psychiatry, [2]Laboratory Medicine, [3]Dept. of Pharmacology, [4]Dept. of Statistics, S.I.U. School of Medicine, P.O. Box 19230, Springfield, IL

INTRODUCTION

Dementia of the Alzheimer's type (DAT) is a progressive degenerative brain disorder characterized by impairment of memory, orientation, language, and judgement. There are a wide range of psychological, neurological and neurochemical deficits (Gottfries et al. 1982) associated with this disorder. Among the neurochemical deficits, the decrease of acetylcholine and ChAT (Bowen et al. 1976, Davis et al. 1976, Perry et al. 1977) have been widely reported. A number of investigators have attempted to correct this cholinergic deficit by administering various cholinergic agents to Alzheimer's patients to improve their behavior and other cognitive functions (Davis et al. 1982, Mohs et al. 1985 a,b,c, Drachman et al. 1980, 1982, Summers et al. 1986).

During the administration of cholinomimetic agents to Alzheimer's patients, there appears to be two problems: (1) to know whether the cholinomimetic agent is really increasing the central cholinergic activity or not; (2) to find biological markers to indicate that this increased central cholinergic activity is in fact producing the cognitive changes seen in these patients after the administration of the cholinomimetic agent.

The rationale for physostigmine therapy is simple, it is supposed to inhibit the enzyme acetylcholinesterase, which is involved in the metabolism of brain acetylcholine. Effects on Acetylcholine are believed to mediate the cognitive and behavioral effects of physostigmine. It is also well established (Janowsky et al. 1985) that an increase in ACh in the central nervous system has a direct effect on the hypothalamus-pituitary-adrenal axis. Increased acetylcholine in the CNS produces an increase in corticotropic releasing factor, A.C.T.H., and cortisol.

Recently Mohs et al. (1985) reported that oral physostigmine improved cognitive function in 11 of 12 patients in a dose finding phase and 7 of 10 in a replication phase of their study. The improvement during treatment with oral physostigmine was correlated with cortisol increase, and these authors suggested that the change in cortisol induced by the drug may reflect the central effects of physostigmine. Hollander et al. (1987), from the same group, has reported similar findings with RS86 a long acting and specific muscarinic agonist. RS86 produced a significant increase in peak nocturnal cortisol levels and this increase in cortisol correlated with improvement measured by the Alzheimer's Disease

Assessment scale. These two studies indicate that plasma cortisol elevation may be an indirect indication of increase in central cholinergic activity following administration of cholinomimetic drugs.

Since both of these studies involved the use of multiple doses of physostigmine or RS86 and several blood samples were drawn over a 24 hour period, we decided to explore the relationship between the change in central cholinomimetic activity, endocrine secretion and cognitive improvement in acute administration of physostigmine by oral, intravenous and intracerebral ventricular administration to Alzheimer's disease patients. In our studies we also studied other parameters (plasma cholinesterase inhibition and CSF acetylcholinesterase inhibition) wherever it was possible. Our objectives for these studies were to address:

(1) Is there a relationship between the endocrine secretion and the central cholinomimetic activity after acute administration of cholinomimetic drugs?
(2) Do the cognitive changes produced by cholinomimetic agents correlate with a change in endocrine secretions?

METHOD

We conducted our studies by using physostigmine through various routes of administration and comparing the endocrine changes with the patients' own baseline and with the values obtained after the administration of a placebo.

<u>Study A</u> - This study involved administering a fixed dose of physostigmine (Antilirium; 60 µg/kg) and neostigmine bromide 7.5 mg orally to ten patients meeting the NINCDS/ADRDA diagnostic criteria for probable Alzheimer's disease (McKhann et al. 1986) in a double-blind crossover design. Seven of these patients were transferred to experiment B of this study where patients were given I.V. scopolamine 0.1 mg, or I.V. saline 0.25 ml in a double-blind crossover design. Blood samples were collected for plasma cortisol and plasma cholinesterase at predrug baseline and at 30, 60, 90, and 120 minutes post drug administration; in some experimental sessions blood was also drawn at later time points. Some blood sampling points were missed because of technical problems. These patients were also administered a battery of psychological tests during predrug as well as postdrug periods. Cognitive functions were assessed using a battery of neuropsychological tests lasting approximately 30 minutes, administered prior to and during each hour of drug testing. Eight counter-balanced forms of the tests (4 versions used on 1 drug day and 4 on the other) were used. Each test included assessment of short-term memory (Rey word list learning, noun recognition), followed by long-term memory (Boston Naming category fluency), short-term visual memory (post card recognition, design recognition), attention (digit span, Smith symbol digit), motor coordination (tapping) and perceptual speed (7's cancellation test).

Out of ten patients who received physostigmine, only eight were able to complete the neuropsychological testing. Nine of these patients also completed the neostigmine session. Of the patients who participated in the experiment there were only seven who completed all of the parts of the studies, four experimental conditions, physostigmine neostigmine, scopolamine, and saline. The cholinesterase inhibition, plasma cortisol were measured as described in our previous reports (Sherman et al. 1987, Kumar et al. 1988). Repeated measures analyses of variance (ANOVA'S) were used to compare the different drug conditions and their changes over time. Paired t-tests were used as follow-up tests.

Study B - We entered 14 DAT patients into this study. Each patient met NINCDS/ADRDA criteria (McKhann et al. 1986) for Alzheimer's disease. Each patient was given lecithin 12.5 gm/day for a week before the infusion of physostigmine and saline as described by Elble et al. in this book. The plasma ACTH by radioimmunoassay method and cholinesterase activity were measured by method described by Sherman et al. (1987). All these patients were given a word recognition test about 105 minutes after the start of the infusion. Paired t-tests were used to make comparisons between the physostigmine and saline condition.

Study C - Four and eight µg physostigmine was administered by ICV route to one DAT patient as described by Giacobini et al. (1988) in several repeated sessions and plasma cortisol, plasma ACTH, CSF acetylcholine and acetylcholinesterase were measured at baseline, 5 minutes, 15 minutes, 60 minutes, 180 minutes, and 360 minutes. These patients were also given a battery of neuropsychological tests which included word recall commands, constructional praxis, ideational praxis, orientation, language, comprehension of spoken language, word finding difficulty and remembering test instruction at various times after injection of physostigmine.

RESULTS

Study A - As evident from Figure 1, the increase in plasma cortisol was significantly higher with physostigmine as compared to neostigmine at 90, 120, 180 minutes after administration of these drugs. The cortisol peaked between 90-120 minutes, after the administration of physostigmine 30 minutes after the peak of cholinesterase inhibition. There was no increase in plasma cortisol after the administration of neostigmine.

In the seven patients who completed experiment A, B and all of the neuropsychological testing, the cortisol secretion under physostigmine, neostigmine, scopolamine, and saline was in the same range at baseline 30 and 60 minutes. However, the cortisol level was significantly higher under physostigmine at 120 minutes than with the other three conditions (Table 1). There was no significant relationship of neuropsychological performance and any of the biological parameters in these patients.

Study B

As evident in Table II, ACTH secretion was quite variable in the 14 patients. Only 5 patients showed increase in ACTH compared to saline sessions (the increase ranged from 6.5% to 79.5%). The improvement on neuropsychological performance with physostigmine compared to saline session was observed in only 7 of the 14 patients (ranged from 5.6% to 26.7%). For the fourteen patients as a group there was no significant relationship between cognitive improvement and the change in ACTH levels from the saline to the physostigmine conditions. However, there was a nearly significant higher ACTH level in physostigmine sessions as compared to saline. There were only 4 patients (2, 3, 4, 13) who showed positive response in cognitive function and increased plasma ACTH. Two of these 4 patients were on 900 µg/m2, one on 600 µg/m2, and the remaining one on 300 µg/m2. It is difficult to establish a dose response relationship in these four subjects.

Study C (ICV)

The increase in plasma cortisol and ACTH was noted with 4 µg physostigmine, but was more with 8 µg physostigmine. After administration of 8 µg physostigmine we observed CSF acetylcholinesterase inhibitions up to 90% and acetylcholine in CSF was also increased up to 5 fold. The increase in acetylcholine peaked at 180 minutes and it is interesting to note that plasma cortisol also peaked at 180 minutes. To our surprise there was no noticeable change in cognitive functions before 360 minutes after the administration of physostigmine. At 360 minutes there was 20% improvement in word recognition compared to placebo session, but there was no change in word recall, commands, constructional praxis, ideational praxis, orientation, language, comprehension of spoken language, word finding difficulty and remembering test instructions.

TABLE I

Mean plasma cortisol (Ng/ml) following administration of Physostigmine (oral 60 µg/kg) Neostigmine (oral 7.5 mg) Scopolamine (I.V. 0.1 mg) and Saline (I.V. 0.25 ml)

	Physostigmine Oral	Neostigmine Oral	Scopolamine	Saline	P Value
Baseline	14.25 ± 3.86	13.26 ± 2.52	13.27 ± 3.54	15.64 ± 4.38	N.S.
30 minute	11.66 ± 3.10	10.99 ± 2.15	14.95 ± 5.00	13.15 ± 2.33	N.S.
60 minute	10.72 ± 2.52	10.46 ± 2.83	13.92 ± 4.64	12.62 ± 6.20	N.S.
120 minute	21.72 ± 8.25**	9.99 ± 2.96*	12.43 ± 3.18*	12.01 ± 6.55*	<0.05

DISCUSSION

Our results from the oral study, I.V. study and I.C.V. study suggest that changes in plasma cortisol and ACTH are indirect indicators of central cholinomimetic activity produced by physostigmine. The absence of a positive correlation between cortisol increase and cognitive changes in the oral study could be because: (1) there is no relationship between cortisol increase and cognitive change; (2) we might have missed this relationship because we could not delineate a significant improvement in cognitive function as a group or in any individual; (3) there may be a complex relationship between the cognitive effects and the endocrine responses because it is difficult to exclude the effect of stress on the secretion of cortisol and ACTH altogether.

In the I.V. study, again there appears to be a tendency for greater secretion of ACTH with physostigmine than with saline. In seven responders who showed improvement in their neuropsychological test performance from 5.6% to 26.7% change and in ACTH from -5.7% to 79.5%, there was no significant relationship between the cognitive changes and the ACTH secretion.

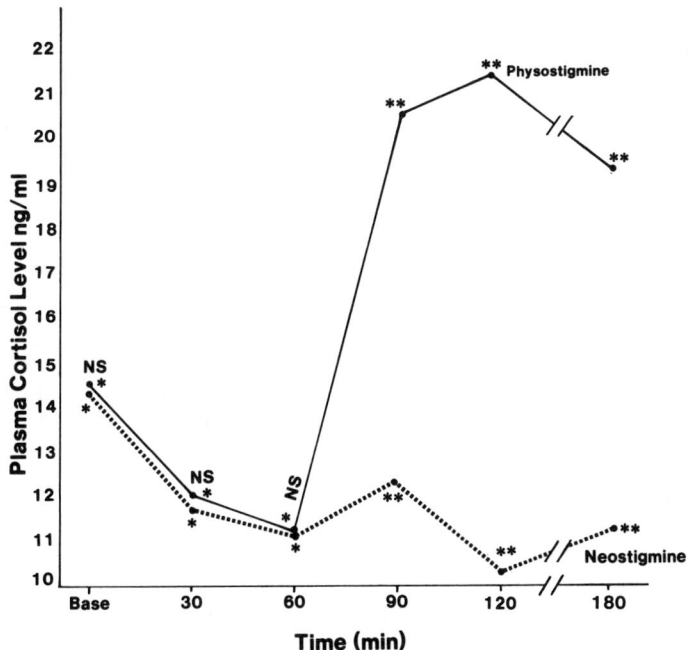

Figure 1. Mean plasma cortisol levels, plotted as a function of time for physostigmine and neostigmine. Significant difference (P <0.05) between the two drug conditions at 90, 120, 180 minutes.

TABLE II

Cognitive and Neuroendocrine change after I.V.
infusion of Physostigmine in DAT patients

Patient	Age	Sex	Best Dose	% Δ from Saline	from Saline
1	77	M	900	-10.5	-8.5
2	62	M	900	11.1	20.6
3	77	F	300	21.4	11.2
4	77	M	900	15.8	79.5
5	76	M	600	0	-27.2
6	76	F	600	0	-6.9
7	72	F	900	-15.4	-17.0
8	72	M	900	0	-15.0
90	55	F	300	-8.7	-7.0
10	58	F	300	14.3	-3.6
11	77	M	900	5.6	-5.7
12	74	F	300	14.3	-2.3
13	81	F	600	26.7	6.5
14	72	F	600	-33.3	28.9

It appears from these data that it is very difficult to predict the cognitive response by measuring ACTH or vice versa. Because of our relatively small sample size it is difficult to speculate further on the relationship between ACTH and cognitive changes. However, we should be aware of the fact that we do not have the knowledge of the time interval between IV physostigmine administration and the possible increase in acetylcholine in the human brain tissue and the time required to modulate the cognitive process in Alzheimer's disease patients. It was interesting to note that in the ICV study, the CSF acetylcholine peaked at 180

TABLE III
Mean Plasma ACTH (Pg/ml) and Cortisol (Ng/ml) Levels for
I.C.V. Physostigmine

Time	8 µg Physo		4 µg Physo	
	ACTH	Cortisol	ACTH	Cortisol
5	47.20	13.90	49.22	17.20
15	53.89	13.40	44.78	14.70
60	55.65	15.70	49.36	14.40
180	55.51	16.80	50.34	14.75
360	53.44	9.46	46.24	12.20

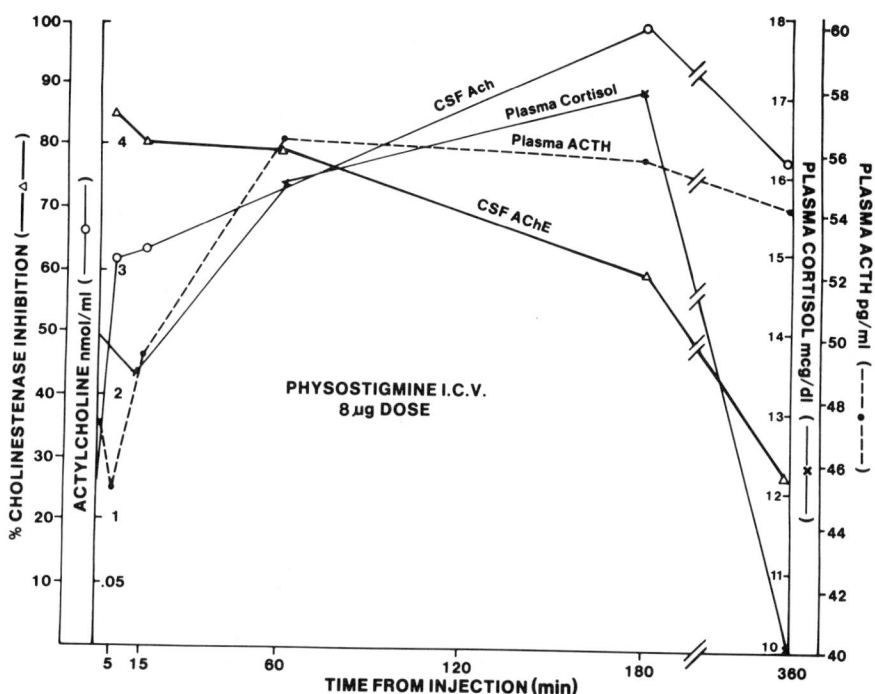

Figure 2. Intracerebro-venticular administration of 8 ug physostigmine in an A.D. patient and its effect on CSF Acetylcholine (ACh), CSF Acetylcholinesterase (AChE) plasma cortisol and plasma ACTH across time.
(Modified from Giacobini et al., this volume)

minutes, but it nearly took three hours more to show 20% improvement on word recognition test. This observation underscores the complexity of the cholinergic system and its direct role in producing cognitive changes. The most interesting observation was that the plasma cortisol peaked at the time when the highest concentration of acetylcholine was detectable in the CSF. One conclusion which could be drawn from this observation is that the time of increase in acetylcholine may be the same as the time for changes in endocrine secretions after the administration of physostigmine but there may be a latency period before we observe cognitive changes. The time course of cortisol changes may be similar to acetylcholine changes in the central nervous system and not to changes in cognitive functions. Therefore, we did not find a time course relationship between the ACTH, cortisol secretion and the cognitive performance in our acute studies. Mohs et al. (1985) and Hollander et al. (1987) have observed a positive correlation in increase in basal cortisol secretion and the cognitive improvement in Alzheimer's disease patients. Unfortunately, we cannot compare our results with the above studies because they used more than one dose of these drugs and they did not study the time course of the changes in cortisol secretion and the cognitive changes as we did in our experiments.

Our data answer the first of the two proposed questions in the beginning by showing a consistent rise in plasma cortisol, and ACTH on administration of physostigmine and this rise seems to be due to increased central cholinomimetic activity. The second question of possible relationship between cognitive change and increase in plasma cortisol and ACTH could not be answered adequately. The available data from our acute studies suggest that there is no relationship between these two parameters. We are trying to study this relationship in our ongoing studies with chronic administration of various cholinomimetic drugs.

ACKNOWLEDGEMENTS

We would like to thank Dr. R.E. Becker, M.D., Dr. E. Giacobini, M.D., Ph.D., Dr. R. Zec, Ph.D., Dr. R. Elble, M.D., Ph.D., Sandy Vicari, and Constance Higgins for all of their help during these studies.

REFERENCES

Bowen, D.M., C.B. Smith, P. White and A.M. Davison. 1976: Neurotransmitter related enzymes and indices of hypoxia in senile dementia and abiotrophies. Brain 99:459.
Davis, K.L. and R.C. Mohs. 1982: Enhancement of memory processes in Alzheimer's disease with multiple-dose intravenous physostigmine. Am. J. Psychiatry 139:1421-1424,.
Drachman, D.H., B.J. Sohakian. 1980: Memory and cognitive functions in the elderly: a preliminary study. Arch Neurology 37:674-675,.
Drachman, D.A., G. Glosser, P., Fleming and G. Longenecker. 1982: Memory decline in the aged: Treatment with lecithin and physostigmine. Neurology 32:944-950,.
Gottfries, C.G. 1982. The metabolism of some neurotransmitters in ageing and dementia disorders. Gerontology Supplement 2:11-19.
Hollander, E., M. Davidson, R.C. Mohs, T.B. Horvath, B.M., Davis, Z. Zemishlony, K.L. Davis. 1987. RS86 in the treatment of alzheimer's disease: cognitive and biological effects. Biol Psychiat 22:1067-1078.
Janowsky, D.S., S.C. Risch, L.L. Judd, L.Y. Huey, and D.C. Parker. 1985. Brain cholinergic system and the pathogenesis of affective disorders. In: Central Cholinergic Mechanisms and Adaptive Dysfunctions (ed. Mon Sinyl, D.M. Warburtan and H. Lab). Plenum Press, New York/London.
Kumar, V., R.C. Smith, K. Sherman, J.W. Ashford, J. Murphy, E. Giacobini, J. Colliver. 1988: Cortisol responses to cholinergic drugs in Alzheimer's Disease. Clinical Pharmacology Therapy and Toxicology, in press.
McKhann, G., D. Drachman, M. Folstein, R. Katzman, D. Price, E.M. Stadlan, Clinical diagnosis of Alzheimer's disease: report of the NINCDS-ADRDA work group under the auspices of Department of Health and Human Services task force on Alzheimer's disease. Neurology 34: 939-944.
Mohs, R.C., B.M. Davis, C.A. Johns, A.A. Mathe, B.S. Greenwald,

T.B. Horvath, and K.L. Davis. 1985a: Oral physostigmine treatment of patients with Alzheimer's disease. Am. J. Psychiatry 142:28-33.

Mohs, R.C., B.M. Davis, A.A. Mathe, W.G. Rosen, C.A. Johns, T.B. Horvath, and K.L. Davis. 1985b: Intravenous and oral physostigmine in Alzheimer's disease. Interdiscipl. Top. Geronol. 20:140-152,.

Mohs, R.C., B.M. Davis, B.S. Greenwald et al. 1985c. Clinical studies of the cholinergic deficit in Alzheimer's disease II Psycopharmacological Studies. Amer. J. Geriatr. Soc. 33:749-757.

Perry, E.K., R.H. Perry, G. Blessed et al. 1977. Neurotransmitter enzyme abnormalities in senile dementia. Neurol. Sci. 34:247.

Sherman, K.A., V. Kumar, J.W. Ashford, J.M. Murphy, R. Elble, R.C. Smith and E. Giacobini. 1987: Effect of oral physostigmine in senile dementia patients: Utility of blood cholinesterase inhibition and neuroendocrine responses to define pharmakinetics and pharmacodynamics - In: CNS Disorders of Aging: Strategies for Intervention. (Aging Vol.33) Ed. by R. Strong, WG Wood & WJ Burke. Raven Press, NY. pp. 71-90.

Summers, W.K., L.V. Majovski, G.M. Marsh, K. Tachiki, A. Kling. 1986: Oral tetrahydroaminoacridine in long-term treatment of senile dementia, Alzheimer type. New Engl. Med. 315:1241-1245.

HIGH PERFORMANCE LIQUID CHROMATOGRAPHY-ELECTROCHEMICAL DETECTION: Effect of Electrode Surface Area on Sensitivity of Physostigmine Determination

L.K. Unni[1], R.E. Becker[1], and E. Giacobini[2]
[1]Dept. of Psychiatry, [2]Dept. of Pharmacology,
S.I.U. School of Medicine, Springfield, IL

Introduction:
Physostigmine (Phy), an alkaloid from the calabar bean, is a reversible cholinesterase inhibitor and has long been used in the treatment of glaucoma (Argyll,1863), as an antidote for anticholinergic drugs and also to treat overdosage of tricyclic antidepressants (Daunderer (80), Duvoisin and Katz (68) and Aquilonius and Hedstrand (78)). Phy, a tertiary amine readily crosses the blood-brain barrier unlike other reversible quarternary amine cholinesterases such as neostigmine and pyridostigmine and has therefore gained prominence as a potential drug in the treatment of Alzheimer disease. Phy is rapidly metabolized by plasma and liver as indicated by the short half life after i.v. or oral administration (Aquilonius and Hartvig, 86, Unni and Somani, 86, Hartvig et al, 86). Clinical pharmacokinetics of Phy have been lacking even though the drug has been used since the turn of the century. The drug concentration in plasma falls rapidly in the range 10 ng/ml to 1 ng/ml after i.v., s.c. or oral administration of the drug in humans (Hartvig et al, 86). Only in the last few years sensitive analytical methods have become available for the determination of Phy levels in plasma. These methods have lacked either chromatographic conditions that permit extended usage of the column or extraction procedures with linear recovery for low Phy concentrations. Whelpton and Moore (85) and Brodie et al (87) reported a high performance liquid chromatographic (HPLC) method using electrochemical detector (EC) and fluorescence detector respectively with a detection limit of 0.1 ng/ml using silica-based columns and basic buffer. Silica based columns should be used under acidic conditions (pH 2 to 6.5) since basic buffers would disrupt the column packing and thereby reduce column life and reproducibility between injections. Isaksson and Kissinger (87) reported an HPLC-EC method by achieving chromatographic separation with a biophase octyl column and an acidic buffer. The same authors followed a solid phase extraction method with Sep-pak resulting in nonlinear recovery with a detection limit of 0.5 ng/ml. Liquid-liquid extraction methods (Whelpton and Moore, 85 and Brodie et al, 87) are more reliable in obtaining consistant recovery which is critical due to the low levels of Phy normally encountered in patients after drug administration. Chromatographic methods sensitive enough to detect low levels of

plasma Phy encountered in Alzheimer patients after drug administration have often resulted in extraneous peaks thereby making the extraction process very difficult.

We have developed an alternate method for the determination of Phy in plasma and CSF by HPLC with EC detector (Unni et al, 88) using a silica column and an acidic buffer. The drug level determination in CSF has not been reported earlier. The advantage of our method over some of the previously reported HPLC-EC methods is the chromatographic conditions which permit the separation of Phy from plasma or CSF contaminants without causing column deterioration during extended usage which makes the method economical for routine monitoring of patient plasma drug levels. Moreover, the liquid-liquid extraction method gives consistant recovery with day-to-day reproducibility for the wide range of Phy concentrations normally encountered in Alzheimer patients. A sensitivity limit of 0.5 ng/ml was achieved using 2 ml plasma or 0.5 ml CSF although lower concentrations can be detected with larger sample volumes. Since Phy is rapidly metabolized by liver, the plasma drug concentration can be expected to fall below the detection limit of the presently available methods after oral administration of the drug. The response from an electrochemical detector is directly proportional to the surface area of the working electrode. However an increase in surface area can result in a proportionate increase in background noise. Under constant chromatographic conditions, the possibility of increasing the sensitivity with increase in the surface area of the working electrode surface was tested by comparing liquid chromatographic systems from Bioanalytical System, West Lafayette, IN and that from Sekisui Chemical Co., Japan.

Methods:
The HPLC system was either from Bioanalytical System (BAS) or from Sekisui Chemical Co. Ltd. (SCC). The BAS system consisted of a PM 30 pump, LC-4B electrochemical detector, Ag/AgCl reference electrode, glassy carbon working electrode, RYT recorder and a Rheodyne injector. The SCC system consisted of a LCP 320 pump, ECD 120 electrochemical detector, Ag/AgCl reference electrode, a glassy carbon working electrode, BAS RYT recorder and Rheodyne injector. The chromatographic separation for both the systems was achieved with a spherisorb silica column and 0.01M sodium acetate in 90% methanol, pH 4.6. The sensitivity for both the systems was set at 5 nA full scale deflection. An oxidation potential of +0.95V was applied to the working electrode of both the systems. The surface area of SCC working electrode was about 5 times that of BAS System.

Results and Discussion:
The retention time of Phy was 4.33 min. with BAS system. A slight increase in retention time to 5.5 min. due to change in the tubing length was seen with SCC system. A linear relationship between peak height and Phy concentration was observed in the range of 0.5 to 10 ng Phy/injection. The chromatogram obtained by injecting 10 ng standard Phy in the BAS system is shown in figure 1A, for comparison with the chromatogram for 1 ng standard Phy injected onto the SCC system (Figure 1B). Both figures represent a 34% reduction from actual size. A peak height of 114 mm for 1 ng Phy was obtained with SCC system while that for 10 ng Phy in BAS system was 130 mm. The

ratio of peak heights obtained from SCC system to BAS system was 8.8 for 1 ng Phy.

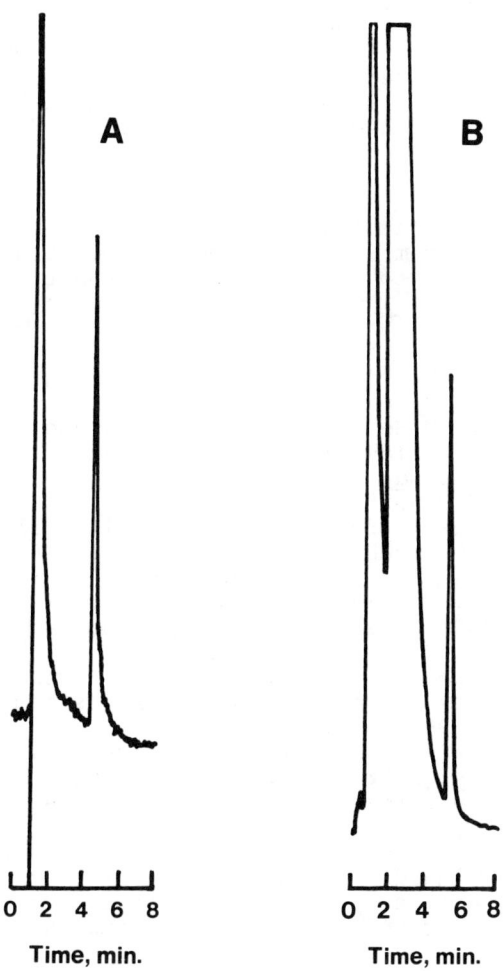

Figure 1. Chromatogram of
(A) 10 ng Phy injected into BAS system
(B) 1 ng Phy injected into SCC system

The signal to noise ratio for 1 ng Phy injection was 6.5 with BAS system whereas it was 114 for SCC system. The working electrode surface area of SCC system was approximately 5 times that of BAS system. The increase in surface area of SCC system did not increase the background noise. The SCC system had a voltage regulator in addition to a pulse dampener which probably reduced the background noise considerably whereas the BAS system had only a pulse dampener. Increasing the sensitivity of BAS detector from 5 nA to 2 nA full scale deflection resulted in a

2.5 fold increase in baseline noise whereas a 10 fold increase in sensitivity from 5 nA to 0.5 nA resulted in a 1.5 fold increase in background noise with SCC system. The HPLC-EC method for the determination of Phy in plasma and CSF that was developed in our laboratory used a dual electrode BAS system. At present we do not have a dual electrode set up for SCC system in our laboratory. The SCC system can increase the sensitivity for Phy by about 17 to 18 times as indicated by signal to noise ratio values. Theoretically, a dual electrode SCC system could increase the sensitivity of our method to about 5-10 pg Phy/ml using 2 ml plasma or 0.5 ml CSF.

Acknowledgements:
Sekisui Liquid Chromatographic System with Electrochemical Detector was a gift from Sekisui Chemical Co. Ltd., Japan. The authors would like to thank M.E. Hannant for technical assistance.

References:
Argyll, R.D. 1863. The calabar bean as a new agent in ophthalmic practice. Edin.Med.J. 8: 815-820
Aquilonius,S.M. and U.Hedstrand. 1978. The use of physostigmine as antidote in tricyclic antidepressant intoxication. Acta Anaesthesio. Scand. 22: 40-45.
Aquilonius,S.M. and P.Hartvig. 1986. Clinical pharmacokinetics of cholinesterase inhibitors.Clin. Pharmacokinet.11: 236-249.
Brodie, R.R., L.F. Chasseaud and A.D.Robbins. 1987. Determination of physostigmine in plasma by high performance liquid chromatography. J. Chromatogr. 415: 423-431.
Daunderer, M. 1980. Physostigmine salicylate as an antidote. J. Clin. Pharm. Ther. Toxicol. 18: 523-535.
Duvoisin, R.C. and R. Katz. 1968. Reversal of central anticholinergic syndrome in man by physostigmine. J. Amer. Med. Assoc. 206: 1963-1965.
Hartvig, P., L. Wiklund and B. Lindstgrom. Pharmacokinetics of physostigmine after intravenous, intramuscular and subcutaneous administration in surgical patients. Acta Anaesthesiol. Scand. 30: 177-182.
Isaksson, K. and P.T. Kissinger. Determination of physostigmine in plasma by liquid chromatography with dual electrode amperometric detection. J. Liq. Chromatogr. 10: 2213-2229.
Unni, L.K. and S.M. Somani. 1986. Hepatic and muscle clearance of physostigmine in the rat. Drug Metab. Disp. 14: 183-189.
Unni, L.K., M.E. Hannant, R.E. Becker and E. Giacobini. 1988. Determination of physostigmine in plasma and CSF by high performance liquid chromatography with electrochemical detection. J. Chromatogr. Submitted.
Whelpton, R. and T. Moore. Sensitive liquid chromatographic thod for physostigmine in biological fluids using dual electrode electrochemical detection. J. Chromatogr. 341: 341-361.

Part III

THA: Preclinical Studies and Mechanism of Action of Cholinesterase Inhibitors

EFFECTS OF CHOLINERGIC AND ADRENERGIC ENHANCING DRUGS ON MEMORY IN AGED MONKEYS

Raymond T. Bartus, and Reginald L. Dean III
Lederle Laboratories, Pearl River, NY

INTRODUCTION

Awareness and interest in the cognitive problems associated with aging and age-related neurodegenerative diseases has increased rapidly during the past decade. During this same period attempts to treat these problems with appropriate CNS-active drugs has also accelerated. Until sufficient understanding of the phenomena responsible for the physical damage to the CNS is achieved, most investigators recognize that much benefit could be gained from simply reducing the severity of the cognitive symptoms of the patients affected. As more has been learned about some of the neurotransmitter deficiencies that contribute to the cognitive disturbances, increased emphasis has been placed on so-called "replacement therapy", to try to compensate for specific chemical defects. Although a number of neurochemicals have been implicated in the cognitive defects, including certain neuropeptides and amine analogs, the greatest effort to date has been directed toward drugs which enhance cholinergic activity, or the activity of neurons which utilize acetylcholine as one of their neurotransmitters (Bartus, et al 1982; 1984).

Despite this effort, research has thus far not produced a palliative treatment. However, the number of different pharmaceutical agents actually tested clinically or preclinically has been remarkably small, while the pharmacokinetic characteristics of the agents used has been notoriously poor. Additionally, clinicians are still trying to develop reliable means of measuring age-related cognitive disturbances, using tests that not only reflect meaningful activities of daily living, but are also sensitive to the modest improvements likely to be expected with current drugs. Thus, success in this area likely depends on the continued progress in multiple, inter-dependent areas of study.

At the same time, a number of laboratories have continued to try to develop model systems using animal species to gain insight or provide direction for additional chemical studies. One approach that has begun to gain increased attention involves the use of aged non-human primates. Because of their relatively close phylogenetic relationship to humans, as well as numerous similarities in aged-associated changes in the CNS and behavior, many investigators have argued that aged non-human primates provide potential advantages in terms of predicting drug responses in aged or demented humans, or at least identifying promising directions for human trials (reviewed in Dean and Bartus, 1988).

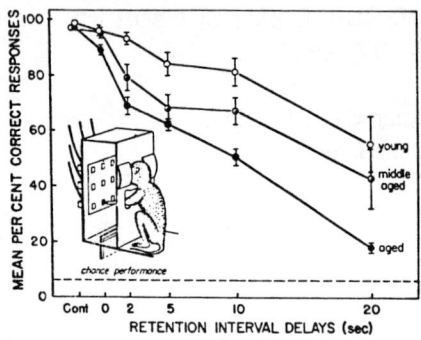 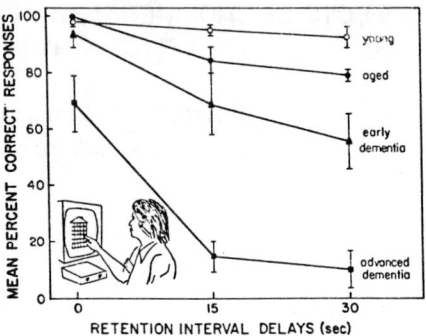

Figure 1. (Left panel) Differences in performance of young, middle-aged, and aged Cebus monkeys on an automated, delayed response procedure. Note progressively greater differences between age groups as duration of retention interval is increased. (Adapted from Bartus et al., Neurobiology of Aging 1:145-152, 1980).

(Right panel) Differences in visuospatial recall of young normal, elderly, mildly-to-moderately demented, and moderately-to-severely demented humans. Subjects were asked to remember which room of a 25-room house presented on a video monitor, had a light on in the window. (Adapted from Flicker et al., Neurobiology of Aging 5:275-283, 1984).

Several years ago we developed an automated test procedure that can be used to measure a recent memory impairment that is qualitatively similar to one that occurs in aged and demented humans (Bartus and Johnson, 1976; Bartus, 1978). Not only is this impairment the most severe and consistent cognitive defect we have seen in aged monkeys, but it also is conceptually and operationally similar to one of the most consistent and severe cognitive defects seen in non-demented aging people (Bartus, 1979; Dean and Bartus, 1988; Flicker et al., 1984). Additionally, it is one of the earliest (and therefore possibly one of the most fundamental) cognitive problems seen in the earliest phases of Alzheimer's disease (Fig. 1). Pharmacological studies conducted with young and aged monkeys over the past several years have added credence to the potential predictive values of this model (reviewed in Dean and Bartus, 1988). For example, when the centrally-acting anticholinergic scopolamine is given to young monkeys, a specific impairment in recent memory is produced that bears close resemblance to that which occurs naturally in aged monkeys and humans (depicted in Fig. 1) (Bartus and Johnson 1976; Bartus 1980). Thus, these effects, as well as those from numerous other drugs, parallel the effects on memory in human subjects (see Bartus, et al 1987b, for review). Finally, drug studies with aged monkeys intending to improve their performance on memory tasks, has produced a pharmacological profile which is generally consistent with the majority of the data from human subjects. For example, general cerebral vasodilators, CNS stimulants, neuroleptics, etc. have proven to be clearly ineffective in clinical trials while similarly negative data have been reported in tests with aged monkeys (Bartus, et al 1987). In contrast, tests with certain nootropics and neuropeptides continue to generate mixed and highly controversial effects in humans, while studies in aged monkeys have shown that many of these drugs can improve performance on recent memory tasks, but the effects are far from robust, and need to be carefully teased out on an individual basis (Bartus, et al 1983; Bartus, et al 1982).

Finally, tests with drugs which have reasonably selective effects on central neurotransmitter systems have shown that classic cholinomimetics can produce reliable improvement in aged patients under carefully controlled conditions (see Bartus et al 1987a;b, for review). In contrast, neither cholinergic precursors nor drugs affecting numerous other (non cholinergic) neurotransmitter systems have generated support as impressive or consistent. In aged monkeys, a very similar pharmacological profile has been obtained (see Bartus et al 1987a;b, for review). Thus, a developing logical and empirical basis exists for using aged monkeys to evaluate and help guide alternative treatment approaches that might be attempted in humans.

STUDY I : COMPARISON OF THA, 3,4 DAP & PHYSOSTIGMINE

Recently, interest in anticholinesterase therapy as a palliative treatment for AD has intensified. This most recent interest follows a report that significant improvement in the cognitive symptoms of AD patients was achieved in a double blind trial using an anticholinesterase, tetrahydroaminoacridine (THA) (Summers, et al 1986). Despite the interest, a number of questions and criticisms have been raised about this study and its conduct (Levy 1987; Parasol, et al 1987; Herrmann, et al 1987; Small, et al 1987; Tariot and Caine 1987), and numerous geriatric centers are performing follow up trials in an effort to replicate the initial findings. In addition to the questions regarding the reliability and validity of the initial claims, at least two interesting empirical suggestions have emerged from the recent discussions and interpretation of the positive claims. The first is that THA may be more efficacious than others tested to date in Alzheimer's patients. The second idea is that THA may possess important pharmacological properties which are separate from, and independent of its inhibition of acetylcholinesterase, making it a better therapeutic agent. This later suggestion arises from the fact that the chemical structure of THA is quite similar to a portion of diaminopyridine (DAP) (Summers, et al 1986). Among the biochemical properties of DAP is the ability to block potassium channels which, in turn, enhances calcium influx into presynaptic terminals, thus increasing release of certain neurotransmitters, including acetylcholine (Soni and Kam 1982). On this basis, it has been suggested that if THA, can indeed provide superior memory-enhancing effects in AD patients, all or part of this effect might be due to a similar acetylcholine-releasing activity (Summers, et al 1986). Although both of these suggestions of THA are interesting and potentially important in terms of directing future chemical synthesis, little direct evidence for either exists. In fact, very few investigators have even studied the effects of THA, DAP and other anticholinesterases under the same conditions. To help provide more information toward this end, the first study reported here directly compared the effects of THA, 3,4 DAP and physostigmine, evaluating their ability to improve the performance of memory-impaired aged monkeys (Bartus and Dean, 1988).

Methods

Ten aged Cebus monkeys were tested in the AGED, using an indirect delayed response paradigm, the same as was used to assess recent memory (Bartus and Johnson 1976; Bartus, et al 1978; Bartus, et al 1980). All monkeys were highly test-sophisticated, having been run in that paradigm for tens of thousands of trials prior to this study. Moreover, while all could perform in the apparatus quite well, all had confirmed deficits in recent memory ability.

All dosing was done by oral administration, 60 minutes prior to the initiation of testing, using a non-traumatic procedure whereby the drug was camouflaged in maple syrup and voluntarily consumed by the monkeys. Each test session lasted approximately 45 minutes. Over the course of this experiment, all monkeys were given several doses of each drug (as well as placebo) with a maximum of two doses given per week and a minimum of two days between doses. The doses were selected on the basis of pilot studies to bracket the most efficacious dose range for each drug. Finally, non-dose test days were used to calculate baseline confidence limits (p<.05) in order to make the assessment of the relative effects of the three drugs and placebo more easy and definitive. If a drug or placebo score exceeded the monkey's baseline confidence limit of performance, that same dose was repeated to determine the reliability of the effect.

TABLE 1: OVERALL EVALUATION

	#DISCRETE DOSES	#TIMES > C.I.	# REPLICATED	P VALUE
PLACEBO	40	3 (7%)	1 (2%)	......
PHYSOSTIGMINE	40	14 (35%)	11 (27%)	P < 0.005
THA	36	12 (33%)	6 (17%)	P < 0.05
3,4 DAP	36	9 (25%)	4 (11%)	P < 0.10
CLONIDINE: ACUTE	19	0 (0%)	......	P > 0.5
SUBCHRONIC	29	2 (7%)	0 (0%)	P > 0.5

Results

The data from an elaborate comparative study such as this can be evaluated a number of different ways. As an initial comparison, all the scores under the three drugs condition, as well as placebo, were compared directly to the range of baseline (non drug) scores from each individual monkey to determine the extent to which each monkey's performance exceeded what would be expected by chance. Table 1 shows that while only 3 of 40 placebo scores exceeded the baseline confidence limits, performance under physostigmine exceeded baseline performance 35% (14 of 40) of the time across the dose range tested, while under THA and 3,4 DAP that occurred 33% and 25% of the time, respectively. More importantly, when each individual monkey's best dose was selected and replicated, most of those under physostigmine were confirmed (11 of 14; p<.005, chi square test, compared to placebo; see Fig. 2), while only 6 of 12 were replicated with THA (p<.05) and 4 of 9 with 3, 4 DAP (p<.10). Thus, in terms of the entire group of 10 monkeys, physostigmine produced greater numbers of, and more consistent test scores which exceeded the baseline performance, than did either THA or 3,4 DAP.

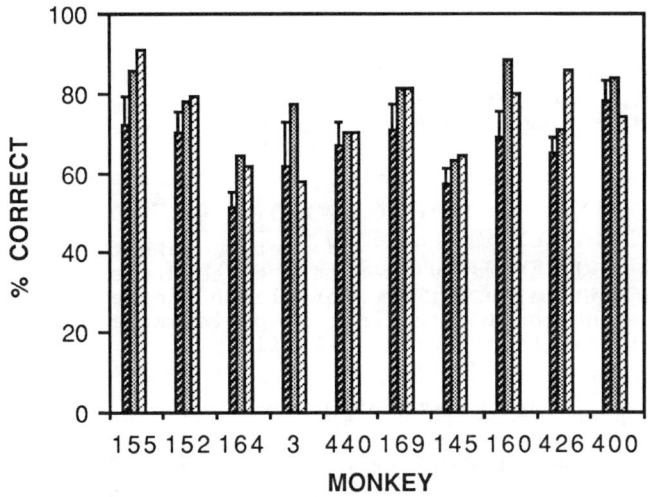

Figure 2. Effects of physostigmine, depicting the 'best dose' effects per each monkey tested. Each monkey is identified by number along abscissa and its performance depicted by the group of three bars. The first bar (darkest) is the mean baseline performance level, with the vertical line depicting the p<.01 confidence limit for the range of normal variability. The middle bar is the performance under that monkey's best dose of physostigmine, while the third bar depicts the monkey's performance when that dose was retested.

Another way to examine the data would be to look at the effect of each drug, as they affect each individual monkey. This analysis is shown in Table 2, which indicates that physostigmine improved performance (i.e.-score exceeded baseline confidence limit) in 9 of the 10 monkeys on at least 1 dose each (for a total of 14 doses), and that in 7 of the 9 monkeys, their best dose again produced significant improvement when retested (p<.01, t-test compared to placebo). Finally, the amount of improvement achieved under the 7 replicated doses was 21% over baseline.

TABLE 2: OVERALL SUMMARY

PHYSOSTIGMINE: 9/10 MONKEYS: INITIALLY IMPROVED (AT 14 DOSES)
7/10 MONKEYS: REPLICATED
70% MONKEYS "RESPONDERS" (P<0.01)
21% IMPROVEMENT (BEST DOSE)

THA: 6/10 MONKEYS: INITIALLY IMPROVED (AT 12 DOSES)
4/6 MONKEYS: REPLICATED
40% MONKEYS "RESPONDERS" (P<0.10)
18.5% IMPROVEMENT (BEST DOSE)

3,4 DAP: 7/10 MONKEYS: INITIALLY IMPROVED (AT 9 DOSES)
4/7 MONKEYS: REPLICATED
40 % MONKEYS "RESPONDERS" (P<0.10)
17.8% IMPROVEMENT (BEST DOSE)

SUBCHRONIC CLONIDINE: 2/8 MONKEYS: INITIALLY IMPROVED (AT 2 DOSES)
0/2 MONKEYS: REPLICATED
0% MONKEYS "RESPONDERS" (P>0.50)
0% IMPROVEMENT (BEST DOSE)

A similar analysis of THA revealed that 6 of the 10 monkeys exhibited improvement over baseline on at least 1 dose each, but that in only 4 monkeys were these 'best dose' effects replicated (p<.10). The mean percent improvement for these 4 monkeys on their best dose was 18.5%.

Somewhat similar effects were achieved with 3, 4 DAP. 7 of the 10 monkeys initially exhibited improvement on at least 1 dose each, but only 4 of these effects were replicated on the second test (p<.10). The mean improvement of these four monkeys was 17.8%.

Thus, as in the prior analysis, this analysis of the data also showed that under these test conditions, physostigmine was more effective and more reliable than either THA or 3,4 DAP.

Discussion

By testing a fairly large group of aged monkeys on a range of doses in a paradigm intended to measure recent memory, it was possible to make direct comparisons of the effects of THA, 3,4 DAP and physostigmine (Bartus and Dean 1988). This comparison revealed that physostigmine seemed to be the more consistently efficacious agent of the three drugs tests. However, a number of caveats exist which should temper attempts to over-generalize these data without the benefit of additional tests. One difficulty faced in any comparative pharmacological study involves trying to standardize parameters while simultaneously optimizing idiosyncratic conditions for each drug. For example, by carefully controlling the time between dosing and test, we most likely minimized certain pharmacokinetic advantages that one drug might have over another. Although the minimal 1 hour absorption period should have been sufficient to permit adequate brain levels to be achieved at certain doses within the range tested, the negative effects of the relatively short half life of physostigmine are clearly reduced by this paradigm.

Moreover, preliminary studies by others suggest that peak plasma levels of THA may not be achieved until after 1 to 2 hours after P.O. administration (Fitten, et al 1988). Although optimal drug levels should have been achieved during most of the test sessions, especially with the range of doses tested, only with additional tests can one be certain. Another caveat involves the fact that the effects of single doses may not predict chronic dosing effects. Additionally, the paradigm employed in the present experiment intentionally minimized the effects of numerous nonmnemonic variables (such as attention, motivation, etc.)which might be positively affected by certain drugs, but not detected in the present situation. Similarly, possible effects on cognitive variables distinct from recent memory might not be detected in the present study. Finally, it is still impossible to be certain how well data from aged monkey generalize to aged or demented humans.

Despite the concerns and cautions mentioned above, it is still possible to make a number of conclusions or suggestions from the present data. First, the results of this study lends no support to the idea that THA represents a significantly superior treatment for age-related memory disturbances, nor that THA has major advantages regarding unique specific mechanism(s) of action. Moreover, if the recent clinical results with THA prove to be accurate, it seems more likely that they might be due to a relatively 'enlightened' clinical protocol. That is, Summers, et al (1986) describes a protocol which included unusual efforts to carefully select a homogeneous group of patients, monitor blood levels, and select individual best doses as part of a preliminary screening strategy. These are all features of clinical protocols that are likely to maximize one's ability to detect subtle or individually variant treatment effects (Bartus, et al 1983; Bartus, et al 1987). Given this, it is conceivable that other drugs might also produce effects similar to those reported for THA, especially if the attempts to replicate the reported data indeed prove successful and similar attention is given maximizing the sensitivity of the clinical protocol. Indeed, the possibility that the reported effects might primarily be due to the protocol employed rather than the unique advantages of THA, per se, might deserve greater attention.

STUDY II: EFFECTS OF CLONIDINE

Although substantial evidence has accumulated that deficiencies in cholinergic transmission contribute to the cognitive defects seen in advanced age and dementia, it is commonly recognized that other neurotransmitter systems must also be involved (Bartus, et al 1982). One such neurotransmitter implicated in the memory loss of aging and dementia is noradrenalin (see reviews by Arnsten and Goldman-Rakic 1985a, Kubanis and Zornetzer 1981). Some of the more impressive evidence supporting this idea indicates that clonidine, an alpha-2 adrenergic receptor agonist, significantly improved performance of aged Rhesus monkeys (Arnsten and Goldman-Rakic, 1985b). However, an earlier preliminary report failed to demonstrate positive effects of clonidine in aged monkeys, (Bartus, et al 1983) and thus, evidence or its potential beneficial effects are not completely consistent. For this reason we attempted to further extend the work on clonidine (Bartus and Dean 1988b).

Effects of acute clonidine

Four monkeys were each given a range of 4 to 5 different doses of clonidine (.0025 to .04mg/kg; ip, 30 min) as described for the previous experiments. However, none of the monkeys performed better under any dose of clonidine, as compared to their range of control scores (see Table 1). In fact, significant impairments occurred in two of the monkeys in a dose-related fashion.

Effects of subchronic clonidine

Clonidine is known to have sedative effects at certain doses, especially before tolerance to these effects has a chance to occur. Because these sedative effects might mask otherwise positive effects of clonidine on performance of a memory task, we designed an additional test of clonidine's potential efficacy in treating age-related memory loss. Once again, several monkeys were tested over a range of doses. However, in this experiment each monkey was given four consecutive days of each dose to increase the likelihood of maximizing clondine's potential therapeutic effects, while simultaneously minimizing its contraindicated side effects. The results of this experiment are shown in Fig. 3 which illustrates little evidence of improvement. Two of the eight monkeys initially exhibited performance which apparently exceeded their baseline levels of performance on a single dose, each. However, in each case, the improvement was extremely subtle and when the dose was repeated a second time (see Fig. 3, connecting bracket) no replication of the trend was observed (Tables 1 and 2). Moreover, once again, some of the monkeys actually performed more poorly under clonidine (Bartus and Dean, 1988b).

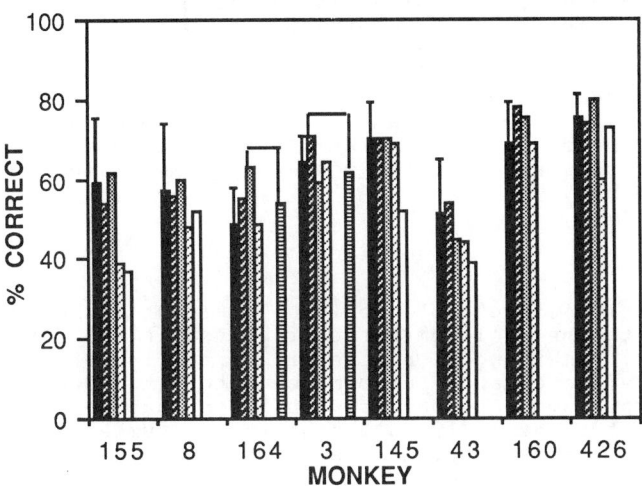

Figure 3. Effects of several separate doses (.01 to .06 mg/kg of clonidine, administered subchronically (four consecutive days) administration of clonidine. Although two of eight monkeys exhibited extremely modest apparent improvement (at a single dose each), in neither case was this effect replicated (separate bar connected by bracket). Once again, performance deficits were seen in some monkeys at the higher doses; compare lack of improvement here to performance under physostigmine in Figure 2.

Effects of clonidine combined with muscarinic agonists

It is commonly recognized that the memory loss observed with advanced age and dementia is likely the result of multiple factors involving more than a single neurotransmitter system. In fact, it has been suggested that in order to achieve substantial reductions in age-related memory loss, it may be necessary to improve simultaneously the function of two or more independent neurotransmitter systems (Bartus, et al 1982). Thus, as a final test of clonidine's potential efficacy in improving age-related memory deficiencies, we tested the effects of combining clonidine with two conventional muscarinic cholinergic agonists, arecoline and oxotremorine.

Monkeys from the same colony were selected for these tests and run on the same apparatus and paradigm described previously. The first study involved testing either .025 or .05 mg/kg of clonidine in combination with .05 mg/kg of arecoline (the lower dose of the active range, based on prior studies in our laboratory) (Bartus, et al 1980). These tests failed to demonstrate any improvement (p>.5); moreover, in certain cases, the monkey's performance fell below baseline levels.

A second study combined subchronic doses of clonidine (three consecutive days), with simultaneous administration of oxotremorine, oxotremorine-M and oxotremorine-2 (oxotremorine analogs with three and two extra methyl groups on the side chain, respectively) (Bartus, et al 1986). A range of doses of each oxotremorine compound was given (.00125 to .005 mg/kg), while the dose of clonidine was held constant (.04 mg/kg). Once again, however, none of the fifteen clonidine/oxotremorine combination doses produced improvement over the monkeys' normal baseline levels of performance.

Discussion

The negative effects of clonidine contrast markedly with those obtained with the cholinomimetics reported in the previous sections. Indeed, combining clonidine with arecoline or oxotremorine failed to enhance performance within the dose range where the muscarinic agonists had previously enhanced performance by themselves (Bartus, et al 1980a). Although a number interpretations for these observations exist, one likely explanation may involve the fact that, in addition to serving as an alpha-2 adrenergic agonist, clonidine also has anticholinergic properties (Buccafusco, et al 1988). Thus, although these data do not provide support for an adrenergic approach to treating age-related impairments, they also cannot be interpreted as evidence clearly inconsistent with the other idea either. It remains possible that other adrenergic agents, possessing a more advantageous pharmacological profile, might ultimately prove beneficial.

The question of why clonidine was reported by another laboratory (Arnsten and Goldman-Rakic 1985b) to produce significant improvement in memory may not be as easy to determine. Although the paradigm used in our laboratory is not identical to that used in the other study, both paradigms are intended to measure recent spatial memory. Also, although one used Rhesus monkeys and the other Cebus, all evidence to date suggests that the effects in one species generalizes to the other quite well (reviewed in Dean and Bartus 1988). Certainly, there is no _a priori_ reason to believe that clonidine represents a unique exception to this rule. In fact, given the complete lack of even a hint of positive activity in the present study, it seems likely that some _fundamental_ difference between the two studies must exits. One possibility may lie with the apparatus and procedure used in the other study. By using a non-automated, two choice, experimenter-paced task, we believe we minimized the opportunity for a number of unintentional experimenter biases and non-mnemonic variables to confound measures of recent memory, as well as the effects of clonidine. Possibilities that are less well controlled in other test situations include experimenter enhancement of cue distinctiveness, as well as the monkeys' overt posturing, visual attention to initiation of the trial and motivation and/or ability to keep pace with the task. While potential drug effects on some of these factors may represent interesting pharmacological responses, they have little direct relationship to memory and are less likely to be important to the cognitive problems of elderly and demented humans. Only through additional research, under carefully controlled conditions will we be able to resolve the differences in clonidine's effects that seem to exit.

REFERENCES

Arnsten, A.F.T., and P.S. Goldman-Rakic 1985a, Catecholamines and cognitive decline in aged nonhuman primates, in: Annuals of the New York Academy of Sciences, Vol. 444: Memory Dysfunction: An Integration of Animal and Human Research from Preclinical and Clinical Perspectives (D.S. Olton, E. Gamzu, and S. Corkin, eds.), New York Academy of Sciences, New York, pp. 218-241.

Arnsten, A.F.T., and P.S. Goldman-Rakic 1985b a_2-Adrenergic mechanisms in prefrontal cortex associated with cognitive decline in aged nonhuman primates, Science, 230:1273-1275.

Bartus, R.T. 1980, Cholinergic drug effects on memory and cognition in Animals. Aging in the 1980s: Psychological Issues (L.W. Poon, ed.), American Psychological Assoc., pp 163-179, Washington, D.C.

Bartus, R.T. and H.R. Johnson 1976, Short-term memory in the Rhesus monkey: disruption from the anti-cholinergic scopolamine. Pharmacology Biochem. & Behavior 5; 39-46.

Bartus, R.T., R.L. Dean and C. Flicker 1987, Cholinergic psychopharmacology: an integration of human and animal research on memory. Psychopharmacology: The Third Generation of Progress, (H.Y. Meltzer, ed.) pp 219-232 Raven Press, NY

Bartus, R.T. 1979, Effects of aging on visual memory, sensory processing, and discrimination learning in a nonhuman primate. Sensory, Systems and Communications in the Elderly, Aging 10 (Ordy, J.M. and K. Brizzee, eds) pp 85-113, Raven Press, NY

Bartus, R.T. and R.L. Dean 1988, Tetrahydroaminoacridine, 3,4 diaminopyridine and physostigmine: direct comparison of effects on memory in aged primates. Neurobiology of Aging 9:4 in press.

Bartus, R.T. and R.L. Dean 1988, Lack of positive effects of clonidine on memory tests in aged Cebus monkeys. Neurobiology of Aging 9:4 in press.

Bartus, R.T., D. Fleming and H.R. Johnson 1978, Aging in the rhesus monkey: debilitating effects on short-term memory, J. Gerontol. 33:858-871.

Bartus, R.T., R.L. Dean and B. Beer 1980a, Memory deficits in aged cebus monkeys and facilitation with central cholinomimetics, Neurobiology of Aging 1:145-152.

Bartus, R.T., R.L. Dean and B. Beer 1982a, Neuropeptide effects on memory in aged monkeys, Neurobiology of Aging 3:61-68.

Bartus, R.T., R.L. Dean, B. Beer and A. Lippa 1982b, The cholinergic hypothesis of geriatric memory dysfunction, Science 217:408-417.

Bartus, R.T., R.L. Dean and B. Beer 1983a, An evaluation of drugs for improving memory in aged monkeys: Implications for clinical trials in humans, Psychopharmacol. Bull. 19:168-184.

Bartus, R.T., R.L. Dean III and B. Beer 1984, Cholinergic precursor therapy for geriatric cognition: Its past, its present, and a question of its future, in: Aging, Vol. 26: Nutrition and Gerontology (J.M. Ordy, D. Harman, and R. Alfin-Slater, eds.), Raven Press, New York, pp.191-225.

Bartus, R.T., R.L. Dean and S.K. Fisher 1986, cholinergic treatment for age-related memory disturbances: Dead or barely coming of age?, in Treatment Development Strategies for Alzheimer's Disease (T. Crook, R. Bartus, S. Ferris, S Gershon, eds.) M. Powley Assoc., Madison, CT, 00 421-450.

Bartus, R.T., T.H. Crook, and R.L. Dean 1987. Current progress in treating age-related memory problems: A perspective from animal preclinical and human clinical research, in: Geriatric Clinical Pharmacology (W.G. Wood, and R. String, eds.) Raven Press, New York, pp 71-94.

Buccafusco, J.J., J.H. Graham and R.S. Aronstam 1988. Behavioral effects of toxic doses of soman, an organophospate cholinesterase inhibitor, in the rat: protection afforded by clonidine. Pharmacology, Biochemistry and Behavior, 29:309-313.

Dean, R.L. and R. T. Bartus 1988. Behavioral models of aging in non-human primates. In: Handbook of Psychopharmacology,Vol. 20: Behavioral Pharmacology, L.L. Iverson, S.D. Iverson, S.H. Snyder (Eds.), Plenum Press, NY pp. 325-392.

Fitten, L.J., K. Perryman, K. Tachiki and A. Kling 1988. Oral tacrine administration in middle-aged monkeys: Effects on discrimination learning. Neurobiology of Aging 9:221-224.

Flicker, C., R. Bartus, T. Crook, and S.H. Ferris 1984. Effects of aging and dementia upon recent visuospatial memory, Neurobiology of Aging 5:275-283.

Herrmann, N., J. Sadavoy, and A. Steingart 1987. Oral Tetrahydroaminoacridine in the treatment of senile dementia, Alzheimer's type, The N, Engl. J. of Med.316:1603-1604.

Kubanis, P. and S.F. Zornetzer 1981. Age-related behavioral and neurobiological changes: A review with an emphasis on memory, Behav. Neural Biol. 31:115-172.

Levy, R. 1987. Tetrahydroaminoacidine and Alzheimer's disease. Lancet 1:322.

Parasol, F.J., D.S. Baskin, Swihart, and S.H. Appel 1987. Oral Tetrahydroaminoacridine in the treatment of senile dementia, Alzheimer's type, The N, Engl. J. of Med.316:1603.

Soni, N. and P. Kam. 4-Aminopyridine - a review. Anaesth. Intensive Care 10:120-126.

Small, G.W., J.E. Spar and D.A. Plotkin 1987. Oral Tetrahydroaminoacridine in the treatment of senile dementia, Alzheimer's type, The N. Engl. J. of Med.316:1604.

Summers, W.K., L.V. Majovski, G.M. Marsh, K. Tachiki and A. Kling 1986. Oral tetrahydroaminoacridine in long-term treatment of senile dementia, Alzheimer type, The N, Engl. J. of Med. 315:1241-1245.

Tariot, P.N. and E.D. Caine 1987. Oral Tetrahydroaminoacridine in the treatment of senile dementia, Alzheimer's type, The N, Engl. J. of Med.316:1604-1605.

ACTIONS OF THA, 3,4-DIAMINOPYRIDINE, PHYSOSTIGMINE AND GALANTHAMINE ON NEURONAL K$^+$ CURRENTS AT A CHOLINERGIC NERVE TERMINAL

Alan L. Harvey and Edward G. Rowan
Dept. of Physiology & Pharmacology, Univ. of Strathclyde, Glasgow G1 1XW, U.K.

INTRODUCTION

Since the discovery that Alzheimer's type senile dementia is associated with a loss of acetylcholine-containing neurons in the brain, there have been several attempts to treat the disease by enhancing the effectiveness of the remaining functional cholinergic neurons. Strategies that have been tried include the administration of precursors in attempts to boost levels of acetylcholine available for release, the use of cholinoceptor agonists to cause direct activation of target neurons, anticholinesterase therapy in order to prolong the effects of endogenously released acetylcholine, and drugs that act presynaptically to facilitate the release of acetylcholine. Clinical trials with single agents have tended to demonstrate little or no functional improvement, but combinations of drugs that have different mechanisms of action may have synergistic effects, with consequently greater therapeutic benefits (Jorm 1986). The reports (Summers et al. 1981; 1986) of positive clinical effects with tetrahydroaminoacridine (THA) are especially interesting because THA has both anticholinesterase actions and K$^+$ channel blocking properties (which might lead to facilitation of transmitter release). Hence, this one compound combines two of the potentially useful mechanisms of action, although their relative importance has not yet been determined.

One difficulty underlying the study of the relative effects of drugs of interest for dementias is that of having a test system that allows simultaneous assessment of effects on presynaptic ion channels, on acetylcholine release, and acetylcholinesterase activity. Although such experiments cannot be carried out readily on cholinergic synapses in the brain, they can be performed on a well-characterised peripheral cholinergic synapse, the neuromuscular junction. Action potentials at motor nerve terminals can be recorded with extracellular electrodes so that effects on Na$^+$ and K$^+$ currents and on neuronal excitability can be determined; acetylcholine release can be quantified by measuring the quantal content of evoked endplate potentials (e.p.p.s) with classical intracellular recording techniques; and effects on acetylcholinesterase activity can be revealed by prolongation of the decay phase of e.p.p.s or spontaneously occurring miniature e.p.p.s (m.e.p.p.s).

In this chapter, we report a preliminary comparison of four agents of potential relevance to the drug treatment of

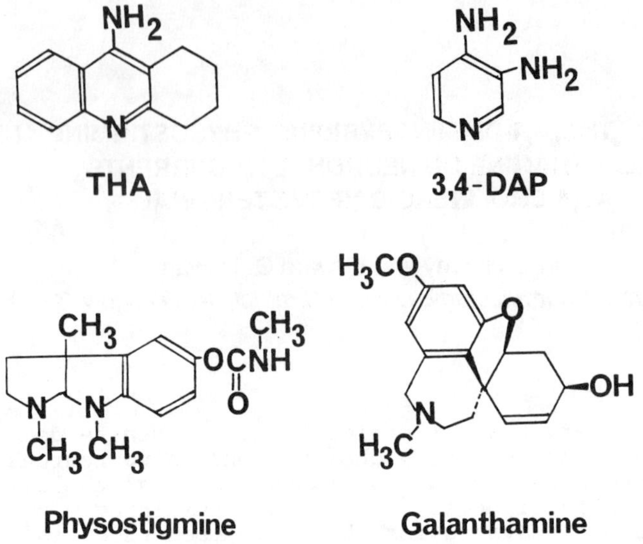

THA

3,4-DAP

Physostigmine

Galanthamine

Figure 1. Structures of the compounds studied. THA is tetrahydroaminoacridine; 3,4-DAP is 3,4-diaminopyridine.

Alzheimer's dementia on presynaptic action potentials and acetylcholinesterase activity at a mouse neuromuscular junction. The compounds studied were THA, 3,4-diaminopyridine (as a blocker of voltage-dependent K^+ channels of nerves), physostigmine, and galanthamine (another centrally active anticholinesterase drug; Paskov 1986); their structures are shown in Figure 1.

METHODS

Experiments were performed on the left triangularis sterni nerve-muscle preparation (McArdle et al. 1981) isolated from 17-22g male mice. The complete dissection of the muscle with its three nerves was performed under continuous perfusion with physiological salt solution (aerated with oxygen containing 5% carbon dioxide) of the following composition (mM): NaCl, 118.4; KCl, 4.7; $CaCl_2$, 2.5; $MgSO_4$, 1.2; KH_2PO_4, 1.2; glucose, 11.1; and $NaHCO_3$, 25, to buffer at pH 7.3.

The preparation was pinned thoracic side downwards to the base of a 2-3 ml tissue bath and perfused at a rate of 5-10 ml/min with the physiological solution described above, to which α-bungarotoxin (120 nM) was added to abolish postsynaptic activity. Experiments were performed at 18-22°C. The intercostal nerves were stimulated via a suction electrode, every 2 sec with pulses of 50 μsec duration and supramaximal voltage.

Presynaptic waveforms were recorded by a glass microelectrode (filled with 2M NaCl, resistance 5-15 megohms) placed inside the perineural sheath (near endplate areas) of one of the branches of an intercostal nerve (see Mallart 1985; Penner and Dreyer 1986; Anderson and Harvey 1988). The potential difference between the silver/silver chloride reference electrode in the bath and the recording electrode was measured by a high impedance unity gain electrometer (W-P Instruments, model M-701), displayed on a dual beam storage oscilloscope and simultaneously stored on FM tape (Racal 4DS). Usually 20-25 waveforms were recorded at each time period. As the shape of the waveform recorded was very dependent on the electrode position, waveforms were monitored continuously from the same site before and throughout application of drugs.

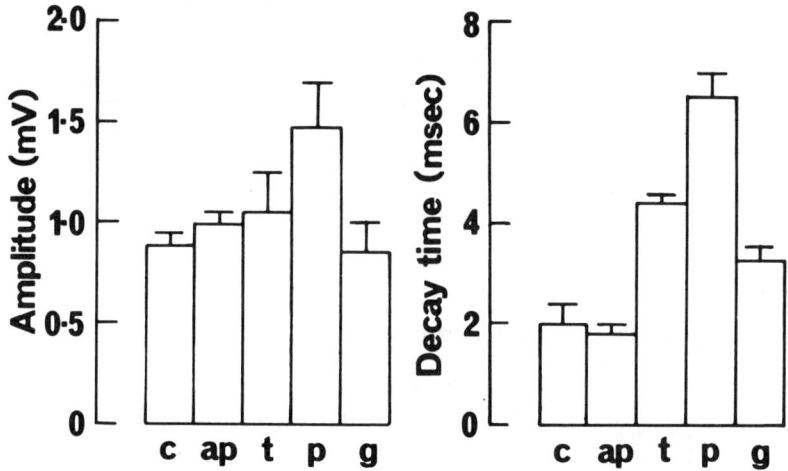

Figure 2. Effects of the compounds on the amplitude and decay time of miniature endplate potentials, c, control; ap, 3,4-diaminopyridine (600 µM), t, THA (1 µM); p, physostigmine (15 µM); g, galanthamine (3 µM). Each column represents the mean of averages obtained from 6 different muscle fibres; the vertical bar indicates the standard error of the mean.

M.e.p.p.s were recorded using conventional intracellular recording techniques from endplates in nonparalysed preparations. Rise times of m.e.p.p.s was 1 msec or less. At least six endplates in each preparation were sampled before and after addition of the test drugs.

ANTICHOLINESTERASE EFFECTS

The anticholinesterase effects of the four compounds at the neuromuscular were assessed by measuring the changes induced in the amplitude and time course of m.e.p.p.s (Figure 2). 3,4-Diaminopyridine at 600 µM had no effect on m.e.p.p.s, whereas 1 µM THA approximately doubled the decay phase of m.e.p.p.s but had little effect on their amplitude. Galanthamine (3 µM) also had little effect on m.e.p.p. amplitude, but it caused a significant prolongation of m.e.p.p.s. Physostigmine at 15 µM produced a 50% increase in the average amplitude of m.e.p.p.s and a three-fold increase in their time to decay.

EFFECTS ON ACTION POTENTIALS OF NERVE TERMINALS

3,4-Diaminopyridine, at concentrations of 50 µM and above, reduced the second negative component of perineural waveforms. At concentrations above 100 µM, 3,4-diaminopyridine abolished the second negative component, and revealed a small positive component and a late negative component (Figure 3). The positive component is associated with inward Ca^{2+} currents at the motor nerve terminals, while the late negative components represents a Ca^{2+}-activated K^+ current at the terminals. The maximum effect of 3,4-diaminopyridine occurred with 600 µM.

THA also reduced the second negative component of perineural waveforms without affecting the first negative component (Figure 4). At 10-100 µM, there was a concentration-dependent decrease in the second negative component with little effect on the time course of the signal. At 200 µM, THA further reduced this component, but also caused a marked broadening of the perineural signal (Figure 4). With 400 µM THA, the signal became very prolonged and the first negative component was also reduced in amplitude.

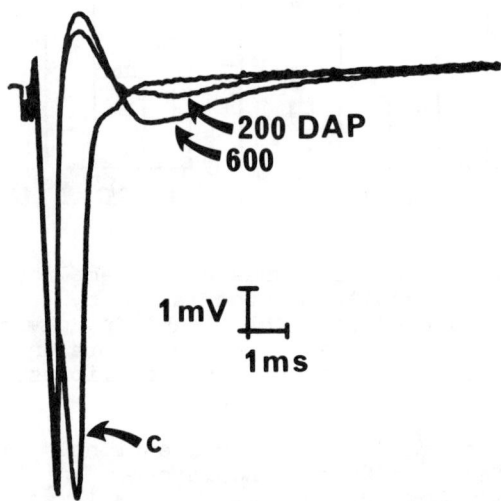

Figure 3. Effect of 200 and 600 µM 3,4-diaminopyridine (DAP) on perineural waveforms. The control record is indicated by the arrow labelled c. Note that 3,4-diaminopyridine abolished the second negative downward component of the control waveform and revealed a positive component that was followed by a small, late negative component. Each record is the average of 20 responses, all from the same site.

The anticholinesterases, galanthamine and physostigmine, were also tested for effects on perineural waveforms. Galanthamine (3-300 µM) had little consistent effect on any component of the signal (Figure 5). Physostigmine at 1.5 and 15 µM produced a small, but reproducible decrease in the second negative component (Figure 6). At considerably higher concentrations (75-300 µM), physostigmine produced a progressive decrease in the amplitude of the second negative component, without altering other parts of the waveform (Figure 6).

DISCUSSION

All four drugs augment cholinergic transmission at the neuromuscular junction and elsewhere. 3,4-Diaminopyridine facilitates acetylcholine release by blocking presynaptic K^+ channels, hence prolonging the depolarization that opens the voltage-sensitive Ca^{2+} channels associated with excitation-secretion coupling. At the concentration that produced a maximal effect on presynaptic action potentials, 3,4-diaminopyridine had no appreciable anticholinesterase activity. THA was also found to be able to block presynaptic K^+ channels, and it would, therefore, be predicted to augment acetylcholine release. However, the concentrations of THA that had pronounced effects on presynaptic action potentials were at least 10 times higher than those that blocked acetylcholinesterase. Moreover, the highest concentrations that were tested revealed that THA was not simply acting to block

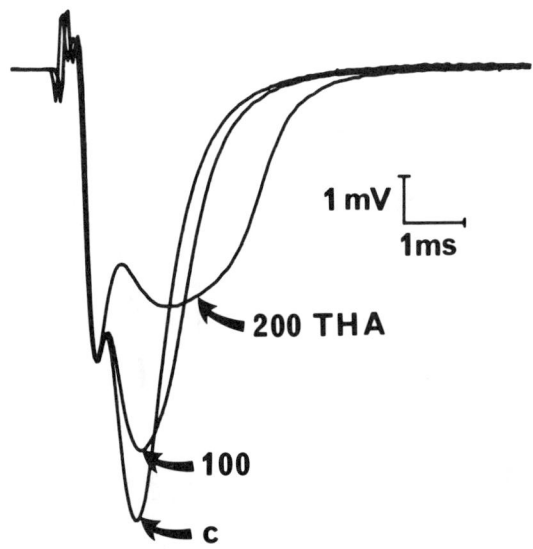

Figure 4. Effect of THA (100 and 200 μM) on perineural waveforms. Note that there is no change in the first negative deflection, and that the second negative deflection is reduced and prolonged.

presynaptic K^+ channels, but that it had additional actions that caused a marked prolongation of the perineural signal. Such an effect may be a consequence of the ability of THA to slow inactivation of Na^+ channels, as has been reported elsewhere. (Schauf & Sattin, 1987). Nevertheless, it seems feasible that THA at clinically relevant concentrations may be able to facilitate neurotransmitter release in addition to having anticholinesterase actions.

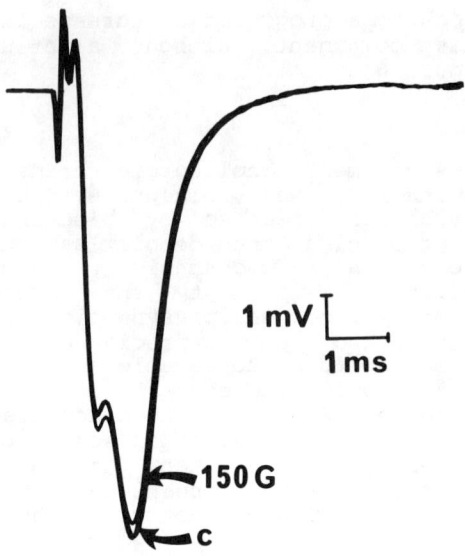

Figure 5. Perineural waveforms before (c) and in the presence of 150 μM galanthamine (G).

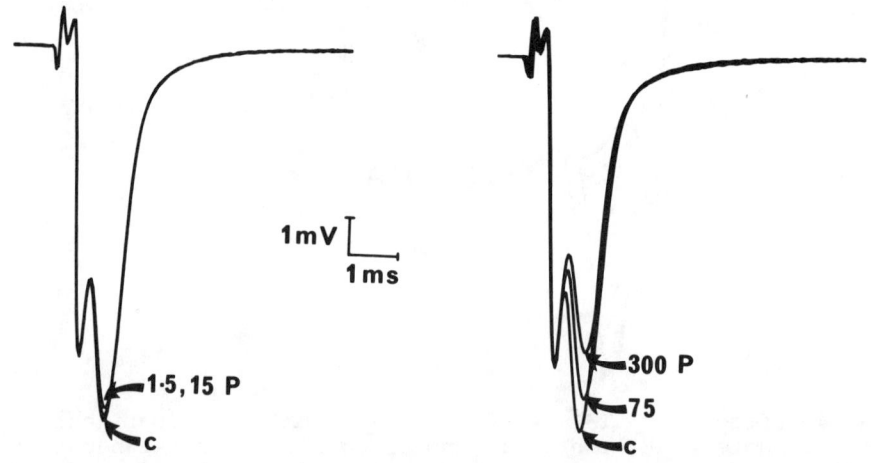

Figure 6. Effects of physostigmine (P) on perineural waveforms. Effects of 1.5 and 15 μM are shown in the left panel, and effects of 75 and 300 μM are shown on the right. All records were obtained from the same recording site.

Physostigmine was also found to affect presynaptic action potentials in a manner consistent with block of K^+ channels. However, the concentrations that were required to elicit this effect were greatly in excess of those needed to inhibit acetylcholinesterase. It is unlikely, therefore, that a channel blocking action would contribute to any clinical effects of physostigmine. The other centrally acting anticholinesterase, galanthamine, had no detectable effects on presynaptic action potentials.

REFERENCES

Anderson, A. J. and Harvey, A.L. 1988. Effects of the potassium channel blocking dendrotoxins on acetylcholine release and motor nerve terminal activity. Brit. Journal Pharmac. 93: 215-221.

Jorm, A.F. 1986. Effects of cholinergic enhancement therapies on memory function in Alzheimer's disease: a meta-analysis of the literature. Austral. N.Z. Journal Psychiat. 20: 237-240.

Mallart, A. 1985. Electric current flow inside perineural sheaths of mouse motor nerves. Journal Physiol. 368: 565-575.

McArdle, J.J., Angaut-Petit, D., Mallart, A., Bournaud, R., Faille, L. and Brigant, J.L. 1981. Advantages of the triangularis sterni muscle of the mouse for investigations of synaptic phenomena. Journal Neurosci. Meth. 4: 109-116.

Paskov, D.S. 1986. Galanthamine. In Handbook of Experimental Pharmacology, Vol. 79, ed. D.A. Kharkevich, 653-677. Berlin: Springer-Verlag.

Penner, R. and Dreyer, F. 1986. Two different presynaptic calcium currents in mouse motor nerve terminals. Pflugers Arch. 406: 190-197.

Schauf, C.L. and Sattin, A. 1987. Tetrahydroaminoacridine blocks potassium channels and inhibits sodium inactivation in Myxicola. Journal Pharmac. Exper. Ther. 243: 609-615.

Summers, W.K., Viesselman, J.O., Marsh, G.M. and Candelora, K. 1981. Use of THA in treatment of Alzheimer-like dementia: pilot study in twelve patients. Biol. Psychiat. 16: 145-153.

Summers, W.K., Majovski, L.V., Marsh, C.M., Tachiki, K. and Kling, A. 1986. Oral tetrahydroaminoacridine in long-term treatment of senile dementia, Alzheimer type. New Engl. J. Med. 315: 1241-1245.

EFFECTS OF CHOLINERGIC DRUGS USED IN ALZHEIMER THERAPY AT THE MAMMALIAN NEUROMUSCULAR JUNCTION

Ronald J. Bradley, Mark T. Edge, Stephan G. Moran, and Arthur M. Freeman
*The Neuropsychiatry Research Program & Dept. of Psychiatry,
School of Medicine, Univ. of Alabama, Birmingham, AL*

INTRODUCTION
Summers et al. (1986) have reported that 9-amino-1,2,3,4-tetrahydroacridine (THA) is effective in the treatment of some patients with severe Alzheimer's disease (AD). Low concentrations of THA have been shown to inhibit preparations of the enzyme acetylcholinesterase (Heilbronn 1961; Kaul 1962). In addition it has been proposed that THA might increase neurotransmitter release by blocking potassium channels in excitable cells because of its structural relationship to 4-aminopyridine (Summers et al. 1986). In fact, THA is known to excite pyramidal cells of the guinea pig hippocampal slice preparation by blocking a potassium conductance but the concentration required is 30 times greater than the serum levels of THA found in patients (Stevens and Cotman 1987). It has been reported that the IC_{50} for block of the potassium mediated A-current in cultured hippocampal neurons is 30 µM (Rogawski 1987). Park et al. 1986 found that patients who were improved after treatment with THA showed a range of THA serum concentrations between 5 ng/ml (20 nM) and 70 ng/ml (287 nM). Higher concentrations up to 176 ng/ml were associated with toxic side effects including nausea. Very high concentrations of THA well beyond the clinical range have also been reported to block potassium currents in guinea pig ventricular myocytes (Osterrieder 1987) and block sodium and potassium channels in Myxicola giant axons (Schauf and Sattin 1987). It has been reported that there is a marked reduction in nicotinic acetylcholine receptors (AChRs) in the cerebral cortical tissues in AD but it is not yet possible to say whether these AChRs are of pre-synaptic or post-synaptic origin (Kellar et al. 1987). It has not been demonstrated that THA affects this system but the most parsimonious theory of THA action in AD is that it inhibits acetylcholinesterase (AChE) in the brain and thereby raises the probability of synaptic transmission. This concept is supported by our finding that clinical concentrations of THA reverse curare-induced block at the neuromuscular junction.

PHARMACOLOGICAL EFFECTS OF DRUGS USED IN AD THERAPY
The effects of THA at a well defined nicotinic cholinergic synapse have not been directly measured. Therefore it is uncertain whether therapeutic concentrations of THA have any effect on cholinergic mechanisms. In order to answer such questions, a range of THA concentrations were studied at the rat neuromuscular junction where synaptic transmission was measured in the diaphragm after stimulation of the phrenic nerve. The AChR at the post-synaptic membrane of the neuromuscular junction is the best understood receptor in any neuronal tissue (Peper et al. 1982). Recent investigations of AChRs of CNS or ganglionic origin have shown that they have sequence homology with the muscle AChR and

also operate by opening a fast ion channel upon activation by ACh (Boulter et al. 1987). It is not yet possible to directly study the behavior of neuronal AChRs at a defined synapse in the CNS therefore we chose to investigate the effects of the drugs used in AD at the rat neuromuscular junction. The phrenic nerve was supramaximally stimulated at frequencies up to 100 Hz for 1 second and the consecutive muscle compound action potentials (CAPs) and the contraction of the rat diaphragm strip were simultaneously measured using a special in vitro muscle bath as previously described (Pagala 1983; Bradley et al. 1986a). The action potential is generated on a muscle fiber when the membrane depolarization induced by the nerve-released ACh reaches a threshold amplitude. Therefore the CAP reflects the average number of synapses where transmission has been successful. Failure in transmission due to a synaptic mode-of-action can occur under the following circumstances:

a) When too little ACh is released by the nerve. This failure can be achieved by lowering the calcium concentration, increasing the magnesium concentration which blocks the calcium uptake (Del Castillo and Engbaek 1954; Bradley 1986a) or by using an aminoglycoside antibiotic such as neomycin which is thought to block calcium uptake (Fiekers 1983a&b; Bradley 1986b).

b) When AChR function is altered so that channel conductance or open time is shortened. This effect could be caused by drugs which block the AChR channel after it opens. Another possibility is that some drug or experimental procedure could accelerate AChR desensitization within a nerve stimulation train thus causing failure (Bradley and Edge 1987a&b; Bradley et al. 1987).

c) When the number or density of AChRs on the post-synaptic membrane is reduced. This can be achieved by using a competitive antagonist for the AChR such as alpha-bungarotoxin (Bradley et al. 1987). In the disease Myasthenia Gravis an autoimmune attack at the neuromuscular junction results in a reduction in AChR density and a fade in tetanic tension (Satyamurti et al. 1975).

d) Some investigators believe that there are presynaptic AChRs which can be activated by ACh during a nerve stimulation train and thus increase or decrease the release of ACh. It is proposed that a drug which blocks these pre-synaptic AChRs could cause synaptic failure by reducing the amount of ACh which is released by the nerve (Bowman et al. 1988). There is however evidence that pre-synaptic AChRs are not present at the mammalian neuromuscular junction (See Bradley et al. 1987).

e) When too much ACh reaches the post-synaptic AChRs. This can occur when the AChE is inhibited by for example THA or physostigmine. The reason for this failure in transmission is that the membrane may be depolarized due to the excess ACh present in the synapse after inhibition of the AChE (Thesleff 1955) or the excess ACh may cause desensitization of the AChRs and thus block transmission (Magleby and Pallotta 1981). When the pre-synaptic potassium channels are blocked by 4-aminopyridine there is an increase in ACh release and at very high drug concentrations a block in transmission can be observed (Thesleff 1980).

In figure 1 the CAP and muscle twitch are shown for a single supramaximal stimulus of the nerve. The CAP represents the sum of all action potentials from the approximately 500 muscle fibers which are contained in the strip of diaphragm used for our experiments. It is important to point out that the individual APs are summed in the temporal sense in that an individual AP will be much shorter by comparison. The amplitude of the CAP is highly reproducible and will remain at this value for up to 6 hours at 30º C. The response at 30 Hz and 100 Hz nerve stimulation for 1 s is shown in figure 1. The amplitude of the 100 consecutive CAPs was evenly maintained throughout the 1 s period and showed no fade or facilitation of the response. Similarly the tetanic contraction of

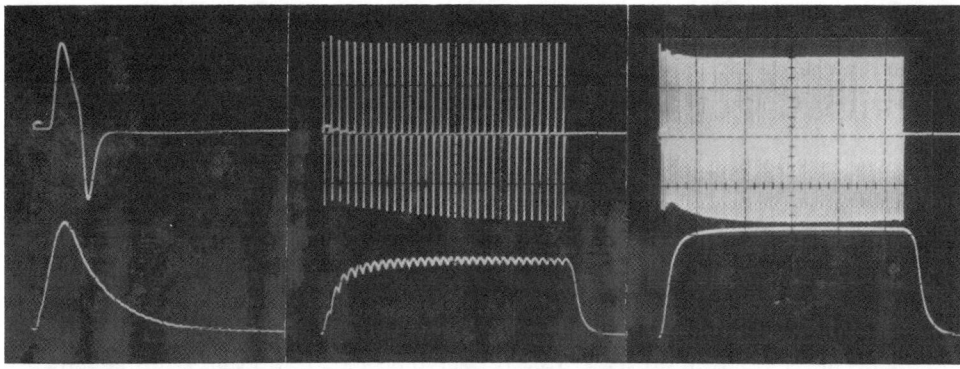

Figure 1. Compound Action Potential and Contraction of Rat Diaphragm Strip. The normal compound action potential (top) and tension (bottom) of rat diaphragm after phrenic nerve stimulation with a single supramaximal pulse and after a 1 sec train at 30 Hz and 100 Hz. The bath solution was composed of 135 mM NaCl, 5 mM KCl, 2 mM $CaCl_2$, 1 mM $MgCl_2$, 1 mM Na_2HPO_4, 15 mM $NaHCO_3$, 11 mM glucose, pH 7.4 and bubbled with 95% O_2 and 5% CO_2. The temperature was controlled at 30°C. **(Left)** Single nerve stimulation. Error Bars: (top) 10 mV/5 msec, (bottom) 200 mV/50 msec. **(Center)** 1 sec nerve stimulation train at 30 Hz. Error Bars: (top) 10 mV/200 msec, (bottom) 1 V/200 msec. **(Right)** 1 sec nerve stimulation train at 100 Hz. Error Bars: (top) 10mv/200 msec, (bottom) 1 V/200 msec.

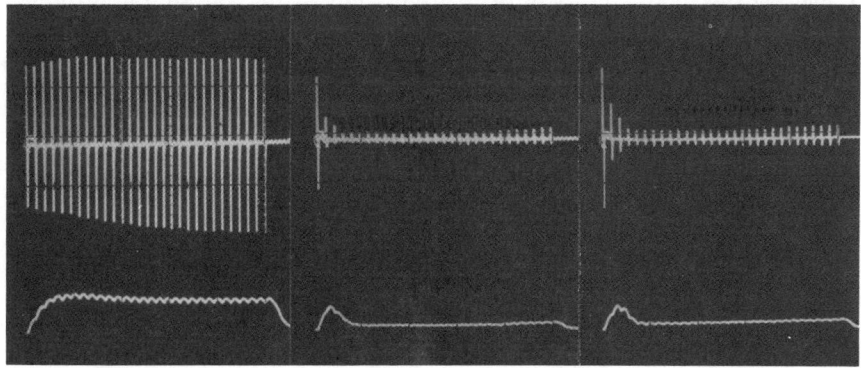

Figure 2. Effects of 12.5 µM THA at the Rat Neuromuscular Junction. The phrenic nerve was stimulated for 1 sec at 30 Hz as in figure 1. **(Left)** Control stimulation. **(Center)** After incubation with 12.5 µM THA for 60 min. **(Right)** After 180 min incubation with THA. Error Bars: (top) 10 mV/200 msec, (bottom) 1 V/200 msec.

the muscle was fused and like the CAP, the amplitude was evenly maintained throughout the 1 s period. In general, fade is never observed in a normal muscle up to 100 Hz nerve stimulation but occasionally there is a slight initial increase in the CAP amplitude during the stimulation train. The reason for this facilitation is unknown but may reflect an initial increase in the EPP amplitude at the initiation of nerve stimulation. In figure 2 the effect of a high concentration of THA (12.5 µM) is demonstrated for a nerve stimulation train at 30 Hz. The control stimulation as in figure 1 shows no fade in the CAP or tetanic contraction during the stimulation train. The first CAP was reduced in amplitude compared to the control value and the second CAP was further reduced to near zero. Thereafter the CAPs showed a slight and gradual increase in amplitude throughout the stimulation train. This dramatic decrement between the first and second CAP was frequency dependent and was more pronounced at higher frequencies of nerve stimulation. In figure 2C another stimulation train is shown after the THA had been present for 2 hours. The effect was less severe in that all the CAPs in the train were larger than in figure 2B. There are three phenomena which require explanation in these recordings. 1) The decrease in amplitude in the first CAP appears to be a result of baseline membrane depolarization due to excess ACh after the inhibition of AChE. Some muscle fibers are depolarized after THA treatment so that they do not respond with an AP. After some time these fibers repolarize to normal membrane potential and can again produce an AP. Therefore in figure 3C the first CAP is increased in amplitude towards the control value. Thesleff (1955) measured membrane potential in individual muscle fibers after treatment with the AChE inhibitor neostigmine and found that the fiber was initially depolarized but then returned to normal even though the neostigmine was still present. AChE inhibitors have direct effects on the AChR and it has been shown in iontophoretic experiments that pyridostigmine can both activate and desensitize the muscle AChR (Bradley et al. 1986b). Therefore the recovery observed in figure 2C during incubation with THA is probably due to an increase in the number of desensitized receptors which will reduce membrane depolarization. 2) The dramatic reduction in CAP between the first and second CAPs is not caused by any class of drug other than AChE inhibitors. It appears to be due to the fact that the EPP is made larger and longer by the inhibition of AChE. Most fibers cannot respond after the first CAP because they are still depolarized from the first EPP. Therefore the effect is frequency dependent in that these fibers can respond at lower frequencies of stimulation if sufficient time is left between consecutive EPPs. The proof for this theory is that any procedure which reduces the amplitude and time constant of the EPP will reverse such effects of AChE inhibitors (Bradley 1986a&b; Bradley et al. 1986a; Bradley and Edge 1987a&b; Bradley et al. 1987) 3) The CAP amplitude increases gradually throughout the 1s train after the initial decrement. This could reflect a build-up of AChR desensitization throughout the 1 s train caused by the accumulation of ACh which is not metabolized by AChE. Another possibility is that the increased ACh concentration displaces the THA from the AChE thus reversing the inhibition. However, a similar recovery of CAP amplitude within the stimulation train occurs in the case of DFP which is an irreversible organophosphate AChE inhibitor (Bradley 1986b) therefore the latter theory seems unlikely.

In figure 3 the effects of 200 nM curare are shown for nerve stimulation at 60 Hz. The decrement induced by curare was quite different from that produced by AChE inhibitors such as THA. In the case of curare there was a gradual development of the fade with little or no decrement in the second CAP. In figure 1A the first four CAPs are almost normal in amplitude and then the fade began again and reached a minimum at 150 ms. Thereafter the CAP increased slightly in amplitude and reached a maximum after another 100 ms but then decreased

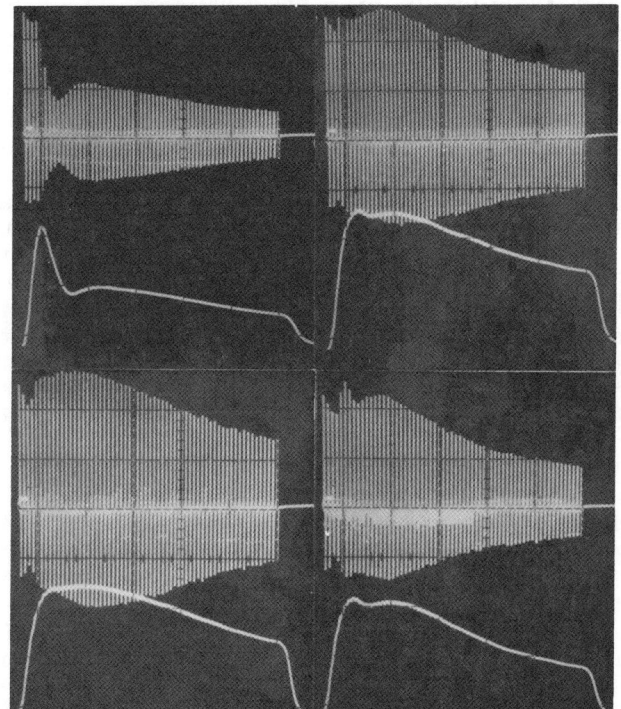

Figure 3. Effects of THA on the Fade Produced by 200 nM Curare.
The phrenic nerve was stimulated for 1 sec at 60 Hz. **(Upper Left)** After incubation with 200 nM curare for 120 min. **(Upper Right)** After addition of 300 nM THA for 30 min with the curare still present. **(Lower Left)** After addition of 900 nM for 30 min. **(Lower Right)** After addition of 1.5 µM THA for 30 min. Error Bars: (top) 10 mV/200 msec, (bottom) 1 V/200 msec

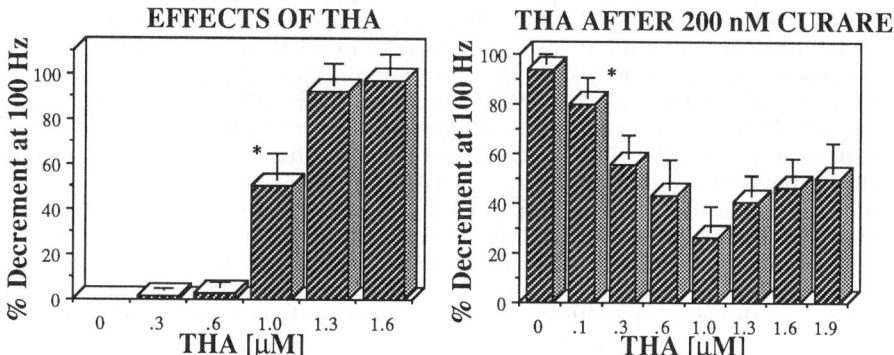

Figure 4. Comparison of THA Effects on Normal NeuromuscularTransmission and After Fade Induced by 200 nM Curare.
The phrenic nerve was stimulated for 1 sec at 100 Hz as in figure 1. The % decrement was calculated as 100x(1st CAP - 10th CAP)/1st CAP. **(Left)** Effects of THA on normal neuromuscular transmission. **(Right)** Effects of THA after incubation with 200 nM curare for 120 min. Each concentration of THA was added for 30 min and then increased to the next dose.

again until the end of the stimulation period. The reason for this intermediate increase in CAP fade is not clear but it may reflect an initial increase in EPP amplitude occurring during the overwhelming fade response due to curare. Another possibility is that there are multiple synaptic effects of curare with different rate constants which are manifested as the above complex pattern of fade. In fact all curare-like non-depolarizing blockers show this same phenomenon of fade with a superimposed increase in CAP. Very low concentrations of THA partially reversed the effects of curare. In figure 3B 300 nM THA has been added for 30 min with the curare still present and the initial decrement has been almost completely reversed. Increasing the THA concentration to 900 nM improved the reversal even more but 1.5 µM had less effect and the reversal was less complete. It is apparent that the effects of THA in inhibiting the AChE overwhelm the response at 1.5 µM producing a smaller reversal. The fact that a less optimal reversal is produced when the THA concentration is too high suggests that there is excess membrane depolarization or perhaps excess desensitization of AChRs due to the buildup of ACh. An examination of the shape of the fade produced when 1.5 µM THA is added after the curare shows very complex kinetics which indicate that the first CAP is reduced in amplitude and there is an initial fade which may be characteristic of the AChE inhibitor effect. In summary, it was found that low concentrations of THA which did not affect normal neuromuscular transmission were capable of reversing the synaptic failure that was caused by curare treatment. These experiments are summarized in figure 4 showing that a concentration of THA as low as 100 nM can reverse the effects of curare but have no effect on normal transmission. It appears that the increase in the normal EPP amplitude and time constant produced by these concentrations of THA have no effect on normal transmission up to 100 Hz but this increase can raise the EPP above threshold values after curare has blocked the AChRs.

The use of cholinergic agonists and precursors has been pursued in the treatment of AD. Sterz et al. (1986) have reported that choline has no positive effects on cholinergic transmission at any concentration but higher concentrations (1 mM) actually block normal transmission. In fact a fade response similar to the effects of curare was observed. Choline treatment was not found to reverse curare induced fade but rather 1 mM choline increased the fade which was produced by curare. In figure 5 the effects of 3 µM carbachol are shown when it is administered after 150 nM curare. The characteristic effects of curare at both 30 and 100 Hz nerve stimulation were only accelerated by the carbachol treatment. The carbachol concentration was increased in steps of 500 nM up to 3 µM. Carbachol had no significant effect below a concentration of 2 µM but above 2 µM carbachol the curare-induced fade was seen to increase. Under no circumstances did any concentration of carbachol reverse the effects of curare. The carbachol appears to increase desensitization during the nerve stimulation train and thereby accelerates the fade response. This possibility of reversal by carbachol might be suggested by the pre-synaptic theory of curare action which would predict that the fade is due to blockade of pre-synaptic AChRs by curare and this fade might be reversed by carbachol which would displace the curare (Bowman et al. 1988). Although the fade response was not reversed in the above experiments it could be argued that the carbachol had desensitized the putative pre-synaptic AChRs.

A decrement in neuromuscular transmission can also be produced by lowering the output of ACh from the pre-synaptic terminal which results in failure because the EPP amplitude does not reach threshold. The type of decrement produced by lowering the ACh output is quite different to the fade produced by curare-like

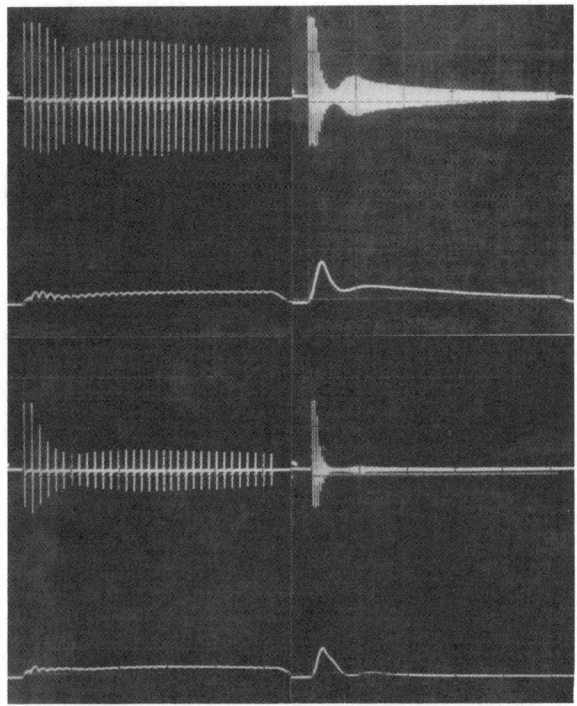

Figure 5. Effects of Carbachol Administered After Curare.
The phrenic nerve was stimulated for 1 sec at 30 and 100 Hz. **(Upper Left)** After incubation with 150 nM curare for 120 min and stimulation at 30 Hz. **(Upper Right)** After incubation with 150 nM curare for 120 min and stimulation at 100 Hz. **(Lower Left)** After addition of 3 µM carbachol for 30 min with the curare still present. **(Lower Right)** After addition of 3 µM carbachol for 30 min. <u>Error Bars:</u> (top) 10 mV/200 msec, (bottom) 1 V/200 msec.

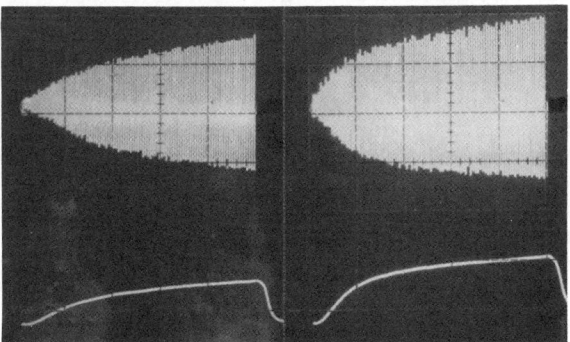

Figure 6. Effects of THA After Neuromuscular Transmission Has Been Blocked by Lowering the Calcium Concentration.
The phrenic nerve was stimulated for 1 sec at 100 Hz as in figure 1. **(Left)** Control stimulation. The buffer was as in figure 1 but the calcium was reduced to 150 µM and there was no magnesium present. **(Right)** After 30 min incubation with 300 nM THA. <u>Error Bars:</u> (top) 2 mV/200 msec, (bottom) 100 mV/200 msec.

drugs. After reducing the calcium concentration there is only an even reduction in the CAP or contraction and at very low calcium concentrations (< 100 µM) there is a facilitation in the response as shown in figure 6A. This gradual increase in the response may be due to the buildup of calcium in the pre-synaptic terminal with repetitive nerve stimulation. A similar use-dependent facilitation in neuromuscular transmission is seen in the Eaton-Lambert Syndrome where there is an autoimmune attack directed at the pre-synaptic terminal of the human neuromuscular junction which blocks calcium uptake. As shown in figure 6B, the administration of 300 nM THA can partially reverse this decremental response presumably by inhibiting the AChE and increasing the amount of ACh which reaches the AChRs. Our preliminary studies show that this reversal of the pre-synaptic deficit may be less complete than in the case of post-synaptic block. It is likely that after calcium reduction there is a limit to the amount of ACh which is available at the post-synaptic membrane even after AChE inhibition so there is a limitation to the number of fibers which can be rescued.

Table 1 is a summary of the effects of the drugs used in Alzheimer therapy on the muscle preparation at 60 Hz nerve stimulation. The average values are given for the concentrations required to reverse the curare-induced fade by a significant amount as defined in the table legend. In addition the concentration is given which affects normal transmission. Under these conditions an average of 117 nM THA was required to reverse the curare-induced block by 10% but approximately a ten-fold higher concentration was required to affect normal transmission. These results are significant because the concentrations of THA which reversed the block were well within the range of concentrations which are measured in the sera of AD patients (< 287 nM). In the case of physostigmine it has been reported that a single oral dose of 2 mg produces a brief maximum serum concentration of 1ng/mL or 3.6 nM (Sharpless and Thal 1985). In the above experiments this dose of physostigmine produced little effect in reversing block and a 10% change was observed at an average concentration of 6.1 nM. A close derivative of THA was tested, HP 029 (1,2,3,4-tetrahydro-9-amino-1-ol maleate), and was found to be less potent than THA but nonetheless at higher concentrations it was just as effective as THA in reversing block. As yet the efficacy of this drug in Alzheimer therapy is unknown but it will be interesting to compare the serum concentrations in patients during current clinical trials with the dose-response data obtained at the neuromuscular junction. A high concentration of 4-aminopyridine (7.8 µM) was required to reverse the curare-induced block but serum concentrations in the range of only 1 or 2 µM are probably attained in humans after a single oral or intravenous dose of 20 mg (Uges et al. 1982). In general 4-aminopyridine was not as effective as the other drugs tested with regard to the degree of reversal observed throughout a range of concentrations. As clinically relevant concentrations were inactive in the above experiments there is reason to expect that 4-aminopyridine might not be useful in AD therapy. Our unpublished studies show that 4-aminopyridine may inhibit AChE and block the AChR channel as well as blocking potassium channels. The block of normal transmission which occurred at an average concentration of 141 µM is probably due in part to AChE inhibition and AChR channel block although significant potassium channel block occurs at this concentration.

At the concentrations of THA used in the above experiments there was no significant change in the shape or temporal characteristics of the compound muscle action potential or the compound nerve action potential. It is therefore likely that the low concentrations of THA which are required to reverse the transmission block produced by curare do not significantly affect the pre-synaptic potassium channels but rather inhibit the AChE. This would to some extent reverse the effects of curare on the EPP and therefore cause some fibers to cross the depolarization threshold again.

Table 1. Effects of Drugs Used in Alzheimer Therapy on Normal Neuromuscular Transmission and Reversal of Fade Produced by 200 nM Curare.

The average concentration (± standard error) of drug is shown which produces a 10% reversal in the curare-induced fade at 60Hz nerve stimulation. The values were extrapolated from dose-response curves (N=4). The decrement is defined as 100x(1st CAP - 10th CAP)/1st CAP. In the case of normal transmission the drug concentration is given which produces a 50% decrement in the 3rd CAP at 60 Hz nerve stimulation.

	REVERSAL OF CURARE EFFECT	**EFFECT ON NORMAL TRANSMISSION**
THA	117 ± 14.6 nM	1.08 ± 0.37 µM
PHYSOSTIGMINE	6.1 ± 1.43 nM	54.7 ± 17.1 nM
HP 029	327 ± 67.2 nM	4.3 ± 0.84 µM
4-AMINOPYRIDINE	7.8 ± 1.84 µM	141 ± 33.2 µM

CONCLUSIONS

The drugs which are used in Alzheimer therapy were tested at the rat neuromuscular junction. These drugs are known to inhibit AChE in the case of THA, HP 029 which is close derivative of THA and physostigmine. On the other hand, 4-aminopyridine may increase ACh release by blocking pre-synaptic potassium channels. When transmission was blocked by reducing the release of ACh or by treatment with curare, it was found that THA could reverse the block at concentrations which are well within the range found in the sera of AD patients treated with THA (< 287 nM). The THA derivative HP 029 was less potent than THA but at higher concentrations it was as effective as THA in reversing block. The concentration of physostigmine required to reverse the block was higher than the maximum concentration which is found in serum after a single 2 mg oral dose. For the above three drugs a ten-fold higher concentration was required in order to block normal neuromuscular transmission. In the case of 4-aminopyridine the concentration required to reverse block was also higher than has been reported in human sera. However, the effects of 4-aminopyridine are complex and may involve AChR channel block as well as AChE inhibition. It is possible that the reversal of curare-induced fade reported above for 4-aminopyridine may involve AChE inhibition as well as potassium channel block. The low concentrations of THA, HP 029 or physostigmine which were found to reverse transmission block did not affect the shape of the compound nerve action potential or the compound muscle action potential. It is therefore likely that low concentrations of these drugs do not affect potassium channels but rather inhibit the AChE at the synapse so that additional ACh is available to increase depolarization. The small increase in ACh concentration reaching the AChRs after treatment with therapeutic concentrations of THA is not sufficient to interfere with normal synaptic transmission. The most parsimonious theory of THA action in AD is that it inhibits AChE in the brain and thereby raises the probability of synaptic transmission. This concept is supported by our finding that clinical concentrations of THA reverse curare-induced block at the neuromuscular junction. The other drugs tested were not so effective as THA in reversing cholinergic block at therapeutic concentrations. The agonists choline or carbachol do not reverse curare-induced block but intensify this block, therefore the concept of AD therapy with agonists is not supported by studies at the mammalian neuromuscular junction.

ACKNOWLEDGMENTS
This work was supported in part by U.S. Army Medical Research and Development Command, Contract No. DAMD 17-84-C-4182, USPHS grant MH39115 and the Alabama Chapter of the Myasthenia Gravis Foundation. We wish to thank Dr. Michael F. Murphy of Hoechst-Roussel Pharmaceuticals Inc. for the generous gift of HP 029.

REFERENCES
Boulter, J., J. Connolly, E. Deneris, D. Goldman, S. Heinemann, and J. Patrick. 1987. Functional expression of two neuronal nicotinic acetylcholine receptors from cDNA clones identifies a gene family. Proc. Natl. Acad. Sci. USA. 84: 7763-7767.
Bowman, W.C., I.G. Marshall, A.J. Gibb, and A.J. Harborne. 1988. Feedback control of transmitter release at the neuromuscular junction. Trends Pharmacol. Sci. 9: 16-20.
Bradley, R.J. 1986a. Calcium or magnesium concentration affects the severity of organophosphate-induced neuromuscular block. Eur. J. Pharmac. 127: 275-278.
Bradley, R.J. 1986b. Reversal of organophosphate-induced muscle block by neomycin. Brain Res. 381: 397-400.
Bradley, R.J., M.K.D. Pagala, and M.T. Edge. 1986a. Curare can reverse the failure in muscle contraction caused by an AChE inhibitor. Brain Res. 377: 194-198.
Bradley, R.J., R. Sterz, and K. Peper. 1986b. Agonist and inhibitory effects of pyridostigmine at the neuromuscular junction. Brain Res. 376: 199-203.
Bradley, R.J. and M.T. Edge. 1987a. In vitro reversal of soman effects on the rat diaphragm. In Proceedings of the Sixth Medical Chemical Defense Bioscience Review. US Army Medical Research and Development Command p. 575.
Bradley, R.J., and M.T. Edge. 1987b. Forskolin counteracts the effects of the organophosphate soman at the neuromuscular junction. Brain Res. 425: 401-406.
Bradley, R.J., M.K. Pagala, and M.T. Edge. 1987. Multiple effects of α-toxins on the nicotinic acetylcholine receptor. FEBS Lett. 224: 277-282.
Del Castillo, J., and L. Engbaek. 1954. The nature of the neuromuscular block produced by magnesium. J. Physiol (London) 124: 370-392.
Fiekers, J.F. 1983a. Effects of the aminoglycoside antibiotics, streptomycin and neomycin, on neuromuscular transmission. I. Presynaptic considerations. J. Pharmacol. Exp. Ther. 225: 487-495.
Fiekers, J.F. 1983b. Effects of the aminoglycoside antibiotics, streptomycin and neomycin, on neuromuscular transmission. II. Postsynaptic considerations. J. Pharmacol. Exp. Ther. 225: 496-502.
Heilbronn,E. 1961. Inhibition of cholinesterase by tetrahydroaminoacridine. Acta Chem. Scand. 15: 1386-1393.
Kaul, P.N. 1962. Enzyme inhibiting action of tetrahydroaminoacridine and its structural fragments. J. Pharm. Pharmacol. 14: 243-248.
Kellar, K.J., P.J. Whitehouse, A.M. Martino-Barrows, K. Marcus, and D.L. Price. 1987. Muscarinic and nicotinic cholinergic binding sites in Alzheimer's disease cerebral cortex. Brain Res. 436: 62-68.
Magleby, K.L., and B.S. Pallotta. A study of desensitization of acetylcholine receptors using nerve-released transmitter in the frog. J. Physiol. (London)

316: 225-250.

Osterrieder, W. 1987. 9-Amino-1,2,3,4-tetrahydroacridine (THA) is a potent blocker of cardiac potassium channels. Br. J. Pharmac. 92: 521-525.

Pagala, M.K.D. 1983. An in vitro fluid electropde technique to record compound muscle action potential along with nerve evoked tension. J. Electrophysiol. Tech. 10: 111-118.

Park, T.H., K.H. Tachiki, and W.K. Summers. 1986. Isolation and the fluorometric, high-performance liquid chromatographic determination of tacrine. Anal. Biochem. 159: 358-362.

Peper, K., R.J. Bradley, F. Dreyer. 1982. The acetylcholine receptor at the neuromuscular junction. Physiol. Rev. 62: 1271-1340.

Rogawski, M.A. 1987. Tetrahydroaminoacridine blocks voltage-dependent ion channels in hippocampal neurons. Eur. J. Pharmac. 142: 169-172.

Satyamurti, S., D.B. Drachman, and F. Slone. 1975. Blockade of acetylcholine receptors: A model of myasthenia gravis. Science 187: 955-957.

Schauf, C.L., and A. Sattin. 1987. Tetrahydroaminoacridine blocks potassium channels and inhibits sodium inactivation in Myxicola. J. Pharmac. Exp. Therap. 243: 609-613.

Sharpless, N.S., and L.J. Thal. 1985. Plasma physostigmine concentrations after oral administration. Lancet i: 1351.

Sterz, R., K. Peper, J. Simon, J.P. Ebert, M. Edge, M. Pagala, and R.J. Bradley. 1986. Agonist and blocking effects of choline at the neuromuscular junction. Brain Res 385: 99-114.

Stevens, D.R., C.W. Cotman. 1987 Excitatory actions of tetrahydro-9-aminoacridine (THA) on hippocampal pyramidal neurons. Neurosci. Lett. 79: 301-305.

Summers, W.K., J.O. Viesselman, G.M. Marsh, K. Tachiki, A. Kling. 1986. Oral tetrahydroaminoacridine in long-term treatment of senile dementia, Alzheimer type. N. Eng. J. Med. 315: 1241-1245.

Thesleff, S. 1955. The mode of neuromuscular block caused by acetylcholine, nicotine, decamethonium and succinylcholine. Acta Physiol. Scand. 34: 218-231.

Thesleff, S. 1980. Aminopyridines and synaptic transmission. Neuroscience 5: 1413-1419.

Uges, D.R.A., Y.J. Sohn, B. Greijdanus, A.H.J. Scaf, and S. Agoston. 1982 4-Aminopyridine kinetics. Clin. Pharmacol. Ther. 31: 587-593.

THA - MEMORY & COGNITIVE ENHANCEMENT IN MOUSE, MONKEY AND MAN

Jaime Fitten, Kent Perryman, and Patricia Gross
Sepulveda VA Medical Center, Sepulveda, CA and
U.C.L.A. School of Medicine, Los Angeles, CA

INTRODUCTION

Since the appearance of the original reports describing central cholinergic defects in brains of Alzheimer patients over 10 years ago (Bowen 1976; Davies 1976; Perry 1977), cholinergic enhancement strategies have been at the forefront of efforts to pharmacologically palliate the cognitive symptoms of the disease (Bartus 1985). Yet today, relatively few cholinergic agents suitable for clinical use are available. One of the least frequently used is the anticholinesterase THA (1,2,3,4-tetrahydro-9-amino-acridine; tacrine). It has been recently suggested as an agent with substantial therapeutic potential (Summers 1981; Summers 1986). However, only three exploratory clinical studies have employed THA as a cognitive activator in demented patients. One study reports modest facilitation of memory in higher functioning patients after brief THA therapy (Kaye 1972). The other two report more dramatic improvements in both acute and longer term applications (Summers 1981; Summers 1986). Nevertheless, the clinically important questions of short and long-term efficacy and of chronic toxicity remain without conclusive answers. The animal literature has made only a small contribution to our present knowledge of THA's memory enhancing capacity. Only a surprisingly few animal studies exploring THA's effects on learning and memory have appeared in the literature in spite of the fact that its anticholinesterase activity has been known for over 30 years (Shaw 1955).

ANIMAL STUDIES

Three recent investigations using different species have reported THA's beneficial effect on memory and learning. Flood and co-workers (1985) demonstrated a memory-enhancing effect with short-term parenterally administered THA in mice by using a T-maze shock-avoidance paradigm. The results were comparable to those obtained in the same experimental paradigm with the anticholinesterase Edrophonium and the direct agonists arecoline and oxotremorine. Bartus and colleagues (1983) used acute parenteral THA in aged non-human primates (C. appella) and reported improved performance on an indirect, delayed matching to sample task. In this study they also described memory-enhancing effects of physostigmine, arecoline, and oxotremorine in the same subjects.

We recently reported the chronic effects of oral THA on discrimination learning in late middle-aged macaque monkeys (Fitten 1988). This study was carried out as a two-phase experiment. Initially one of five subjects served as his own control in a multiple dose, placebo-controlled crossover trial intended to establish a dose-response curve and an effective dose range based on THA serum concentrations. Subsequently, the other four monkeys were given the previously established "best" THA dose while learning up to four color pair discriminations. They also learned up to four other comparable color-pair discriminations while on placebo. Two of the subjects were given THA first, then placebo; the other two received placebo first then THA. Results demonstrate that the dose-response curve from the first phase had characteristically the shape of an "inverted" U (doses < 1 mg/kg/day were behaviorally ineffective, while doses > 2 mg/kg/day produced classic acute cholinergic toxicity, e.g. diarrhea, salivation, tremor, etc.). In phase two no order effects were noted. Combined scores for THA tests were compared to placebo scores. The difference was significant at $p<0.01$ with all four THA treated monkeys requiring fewer trials to reach learning criterion. Low chronic toxicity was suggested by the survival time of the monkeys which has now reached over two years without observable health impairment.

Few THA toxicology studies have been published to date. THA belongs to the monoamine acridine class of compounds which can produce hepatotoxicity and bonemarrow suppression (for review see Summers 1980). Of particular clinical interest are studies exploring the possible consequences of long-term oral THA ingestion. One recent study examined THA's ability to induce chromosome aberrations in mouse bone-marrow after repeated intraperitoneal and oral applications. Cytogenetic analysis revealed an increased percentage of chromatid and chromosome breaks in THA treated mouse cells versus controls. The authors suggest further long-term animal studies before commencing clinical trials (Sram 1984).

We have recently concluded a study of long-term, memory-enhancing doses of oral THA in mice (Fitten 1987). In this study, we evaluated THA's potential for chronic toxicity and assessed its ability to sustain the retention-enhancing capacity previously shown when given over brief periods of time (Flood 1985). Previously effective short-term memory-enhancing doses were given eight-week old mice for either four or six months. Two groups of twenty each received vehicle only (placebo) for four or six months while two other equally-sized groups received THA plus vehicle in the same manner. At the end of four or six months, memory retention was assessed by a T-maze footshock avoidance paradigm. Toxicity was evaluated through 1) behavioral measures (e.g. tight rope, otorod activity wheel, etc.), 2) biochemical test (liver-specific serum ornithine transcarbamylase activity; OTC), and by 3) histological studies of liver tissue, and 4) gross examination of major organs after sacrifice. A sustained, memory-activating effect of THA was suggested by improved memory retention scores ($p=.001$) when compared to the vehicle-only groups. A lack of significant chronic toxicity was indicated by 1) no impairment on tests of behavioral toxicity, 2) a significant but mild elevation of OTC in the THA treated groups yet not present in all THA mice. Over half of the latter had OTC actvity levels indistinguishable from control values. Finally, no gross visceral organ pathology was detected in THA subjects and a blinded histological examination of THA treated and control liver tissues revealed no abnormalities in either one.

It is difficult to extrapolate results from healthy animal models, even if they are aging or primates, directly on to humans with AD. These models, nonetheless, are relevant to the cognitive problems experienced by elderly and by demented individuals. They can make important contributions by 1) providing a rationale for clinical trials, 2) helping define drug toxicities, and 3) verifying clinical observations. Thus, further behavioral and toxicologic animal studies are necessary to enrich the presently scant THA literature. With the newly generated clinical interest in THA, these studies should be soon forthcoming.

HUMAN STUDIES

Presently three studies comprise the entire clinical literature on THA as a cognitive activator. A detailed review of these studies will not be attempted here. Only major points regarding these investigations will be discussed. The first two of these reports (Summers 1981; Kaye 1982) are pilot studies and were the first to stimulate interest in the cognitive tests and clinical staging. Encouraging results were reported, with six out of 12 patients showing memory improvement and nine out of twelve improving in clinical staging. Four out of those nine did so dramatically.

Kaye and colleagues in contrast, used oral THA and lecithin separately and in combinations which included placebo. Each of four combinations was given randomly to ten patients with Primary Degenerative Dementia (PPD) in divided doses several hours before testing. Slightly more than a day separated each of the testing sessions. Memory test scores were the only outcome measure. Interestingly, only the lecithin plus THA combination demonstrated improved serial word learning. This occurred only in higher functioning patients. Thus Kaye et al. report less clinically promising results than Summers.

While methodologically different, these studies do share certain limitations. The first relates to the use of somewhat vague diagnostic criteria for AD. As a result there were probably several non-AD subjects included in each study, In both studies, the length of treatment was extremely short and both were open trials. Outcome measures were either very restricted (e.g. memory only) or were measured by infrequently used instruments. Dose ranges were quite narrow and fixed. Yet in spite of these limitations, both studies established THA as a potentially useful agent in the treatment of AD.

It was not until 1986 that the third study was published (Summers 1986). In this investigation more patients were involved, lecithin was given along with THA to prevent possible neuronal membrane phospholipid breakdown after sustained inhibition of cholinesterase, long-term therapy was provided, and a randomized, double-blind crossover design was described. The study had other useful features such as a dose-titration phase to determine individual optimal doses, as well as an improved method of patient selection. Results indicated dramatic improvement in cognition and functional capacity in most patients. Initial enthusiasm was dampened however, by fairly widespread concerns over the study methods. These have ben reported elsewhere (Pirozzolo 1987, Hermann 1987, Small 1987, Kopelman 1987, Tariot 1987, Weintraub 1988) and will not be discussed here. Consequently, the short and long-term clinical efficacy and chronic toxicity of THA still remain in doubt.

A multicenter replication study under the combine sponsorship of the NIA , ADRDA and Warner-Lambert Laboratories was recently launched to attempt to answer the many lingering questions about THA's clinical usefulness in AD. Early on, however, the appearance of significant liver enzyme abnormalities in about 25% of the patients receiving THA caused the study to be temporarily suspended. At this writing the study has just restarted. Two important questions remain, however, about whether this large-scale study, will adequately replicate Summers' recent investigation. The first centers on the use of lecithin with THA. Ulus and Wurtman (1988) have recently reiterated a rationale for the combined use of these agents in treating AD. In sum, combined therapy would lessen the probability of neuronal membrane phospholipid depletion during prolonged cholinesterase inhibition because sufficient dietary lecithin would provide an adequate supply of extracellular choline during increased acetylcholine synthesis and thus prvent "autocannibalism" of choline-containing membrane phospholipids. Consequently, long-term therapy using cholinesterase inhibitors alone might possibly hasten cholinergic cellular breakdown, countering whatever initial benefits might be derived from the anticholinesterase effect. While still speculative, this recommendation does appear to be supported by in vitro results (Ulus 1988). The multicenter study, nonetheless, does not include lecithin in its protocol.

A second concern revolves around the THA doses to be given patients in the revised multicenter study. Adjusted maximal doses could be half the original ones in order to significantly lessen the risk of hepatotoxicity for large numbers of patients. Such doses, if so limited, would be very substantially below the daily amounts of THA given by Summers to his patients. In a smaller replicative study, larger adjusted doses probably given. Thus the multicenter study, though it will undoubtedly provide much valuable information about THA in the treatment of AD, may still not be able to furnish all of the needed data, particularly if results fail to be as robust as Summers' (Summers 1986).

Our clinical experience with THA began in 1985. We used non-human primate THA data and two pilot AD patients meeting NINCDS-ADRDA criteria for probable AD to estimate the range of potentially clinically effective THA doses and to determine the time-course of THA serum concentration following an oral dose. After repeated trials over several months we concluded that a potentially effective THA dose could fall between 1 mg/kg/day and approximately 3 mg/kg/day when given in divided doses. Peak THA serum concentrations generally occurred between two and three hours after oral dose. THA was virtually undetectable in serum after 24 hours.

One such sequence of tests on a patient is shown in Table 1. It should be noted that the first two scores are probably spuriously low because the patient was still adapting to the inpatient environment. Subsequently, after adaptation, the patient did show as much as a 3 to 4 point variability in Mini Mental State Exam (MMSE) scores in repeated testings while not receiving THA. These scores were in the low and mid-teen range. This patient was also later able to tolerate doses in excess of 2 mg/kg/day before developing mild cholinergic side effects.

TABLE 1. THA SEQUENCE IN AN ALZHEIMER'S DISEASE PATIENT

Period	MMSE Score	Serum THA Conc. ng/ml
Baseline	8	0
THA 25 mg BID	9	< 3
Washout	12	0
THA 50 mg ng BID	11	3.1
Placebo	12	0
Washout	N/A	N/A
THA 50 mg BID + Lecithin	14	22.7
THA 75 mg BID	16	30
Placebo	11	0
THA 75 mg BID	18	34

Periods of treatment are one week each except for the final three periods which are two days each. Serum THA concentration was determined at the time of testing for all but the first four periods. Mental status testing occurred approximately two hours after the morning THA dose. MMSE is Mini Mental Status Exam by Folstein et al.

After the pilot studies, we began and recently completed an inpatient randomized, double-blind, placebo-controlled crossover study of oral THA treatment in ten patients meeting NINCDS-ADRDA criteria for AD. We are currently in the process of data analysis. However, we note that one patient had minor hepatic enzyme elevation, while three others developed significant liver ewnzyme abnormalities at the end of the blinded phase. In two patients, these values remained elevated for several weeks after the end of the study. None became symptomatic. Thus, we are left with the possibility that THA could be hepatotoxic for a substantial portion of the AD population. Hopefully, toxicity will prove to be minor and transient if it occurs. If dramatic cognitive results can be confirmed, they may perhaps balance the toxic risk suggested by the multicenter study findings as well as our own. Thus we are obliged to wait longer before a clearer picture of the true therapeutic potential of THA begins to emerge.

REFERENCES

Bartus, R.T., R.L. Dean, and B. Beer. 1983. An evaluation of drugs for improving memory in aged monkeys: implications for clinical trials in humans. Psychopharmacol. Bull. 19: 168-184.

Bartus, R.T., R.L. Dean, M.H. Pontecorvo, and C. Flicker. 1985. The cholinergic hypothesis: a historical overview, current perspectives and future directions. In Memory dysfunction: an integration of animal and human research from preclinical and clinical perspectives, ed. D.S. Olton, E. Gamzu, and S. Corkin, 332-358. New York Academy of Sciences, New York.

Bowen, D.M., C.B. Smith, P. White, and A.N. Dawson. 1976. Neurotransmitter-related enzymes and indices of hypoxia in senile dementia and other abiotrophies. Brain: 99: 459.

Davies, P. and A.J.F. Maloney. 1976. Selective loss of central cholinergic neurons in Alzheimer's disease. Lancet 2: 1403.

Fitten, L.J., K. Perryman, K. Tachiki, A. Kling. 1988. Oral tacrine administration in middle-aged monkeys: effects on discrimination learning. Neurobiol. of Aging 9: 221-224.

Flood, J.F., G.E. Smith, and A. Cherkin. 1985. Memory enhancement: supra-additive effect of subcutaneous cholinergic drug combinations in mice. Psychopharmacology 86: 61-67.

Kaye, W.H., N. Sitaram, H. Weingartner, M.H. Ebert, S. Smallberg, and J.C. Gillin. 1982. Modest facilitation of memory with combined lecithin and anticholinesterase treatment. Bio. Psychiatry 17: 275-279.

Perry, E.K., R.H. Perry, G. Blessed, and B.E. Tomlinson. 1977. Necropsy evidence of central cholinergic deficits in senile dementia. Lancet 1:189.

Pirozzolo, F.J., N. Hermann, G.W. Small, M. Kopelman, P.N. Tariot, et al. 1987. Oral tetrahydroaminoacridine in the treatment of senile dementia Alzheimer-type (Letters) N. Engl. J. Med., 316:1603-1605.

Shaw, F.H. and G.A. Bentley. 1953. Pharmacology of some new anticholinesterases. Aust. J. Exp. Biol. Med. Sci. 31: 573.

Sram, R.J., J. Kocisova, and I. Kodytkova. 1984. Chromosomal aberrations in mouse bone-marrow induced by tacrin. Activ. Nerv. Sup. (Praha) 26:84-86.

Summers, W.K., K.R. Kaufman, E. Altman, and J.M. Fischer. 1980. THA- a review of the literature and its use in treatment of five overdose patients. Clin. Toxicol. 16: 269-281.

Summers, W.K., L.V. Majovski, G.M. Marsh, K. Tachiki, and A. Kling. 1986. Oral tetrahydroaminoacridine in long-term treatment of senile dementia, Alzheimer type. N. Engl. J. Med. 315: 1241-1245.

Summers, W.K., J.O. Viesselman, G.M. Marsh, and K. Candelora. 1981. Use of in treatment of Alzheimer-like dementia: a pilot study in twelve patients. Biol. Psychiatry 16: 145-153.

Weintraub, M. and R. Standish. 1988. Tetrahydroaminoacridine: a possible treatment for senile dementia of the Alzheimer type. Hosp. Formul. 23: 31-35.

Ulus, I.H. and R.J. Wurtman. 1988. Prevention by choline of the depletion of membrane phosphatidylcholine by a cholinesterase inhibitor. (Letter) N. Engl. J. Med. 318: 191.

TACRINE: Levels and Effects on Biogenic Amines and Their Metabolites in Specific Areas of the Rat Brain

Ken H. Tachiki, Karen Spidell, Lisa Samuels, Ronald Ritzmann, Alan Steinberge, Robert L. Lloyd, William K. Summers, and Arthur Kling

Psychiatry Service, VA Medical Center, Sepulveda, CA;
Dept. of Psychiatry & Biobehavioral Sciences, U.C.L.A., Olive View;
U.C.L.A. Medical Center and VA Medical Center, Brentwood, CA

INTRODUCTION

Tacrine (Shaw 1953; Bartus et al. 1982; Bajgar et al. 1979; Huger et al. 1986; tetrahydro-9-aminoacridine, THA) and physostigmine are generally recognized as potent inhibitors of the enzyme acetylcholinesterase (AChE) (Shaw et al. 1953; Bartus et al. 1982; Bajgar et al. 1979). Physostigmine is reported to be both a more potent AChE inhibitor than THA (Bajgar et al. 1979; Huger et al. 1986) and more effective in enhancing memory and learning behaviors in animal studies (Bartus et al. 1988). THA has also been reported to enhance learning and memory in animals (Bartus et al. 1988; Flood et al. 1985). However, in contradistinction to the limited clinical efficacy of physostigmine, THA has recently been reported to be effective in the treatment of Alzheimer's disease (Summers et al. 1986; Ohman et al. 1988; Gauthier et al. 1988), a disorder often attributed in part to a deficit in cholinergic activity (Bartus et al. 1982). This suggests that there may be a difference between these drugs in their respective pharmacokinetics such that THA gets to the site of action in the CNS more readily than physostigmine and/or that THA may be acting in the CNS in ways other than as an inhibitor of AChE.

Recent reports show that, in vivo, the maximum inhibition of brain AChE, and consequent behavioral effects of physostigmine, occur within 1 hour after administration while maximal effects of THA occur between 2 to 3 hours after i.p. injection (Park et al. 1986; Sherman 1988). This difference in pharmacokinetics favors physostigmine getting to the site of action in the CNS more readily than THA. Both the higher potency of inhibition and more rapid pharmacokinetics of physostigmine seem antithetical to the observed greater efficacy of THA in the treatment of Alzheimer's disease. Investigation of possible alternative mechanisms of action of THA may help resolve this incongruity.

Suggestive of an alternative mechanism of action, THA was recently reported to affect sodium and potassium channels (Stevens et al. 1987; Schauf et al. 1987; Harvey et al. 1988) whereas physostigmine had an effect only on potassium channels (Harvey and Rowan 1988). Also, contrary to reports that physostigmine shortens alcohol induced sleep-time (AIST) (Rashti et al. 1988), a behavior associated with CNS changes in dopaminergic activity (Kiianmaa et al. 1984), THA was found to prolong AIST (Rashti et al. 1988).

To further explore the possibility of alternate mechanisms of action of THA, dopamine (DA), serotonin (5-HT), norepinephrine (NE) and their various metabolites (HVA, DOPAC, 5-HIAA) were assessed in rat brain regions following THA administration by i.p. injection. Concurrently, we employed a highly sensitive fluorometric method to assay THA in the same samples of brain tissue.

Methods

Experimental procedures: THA (3mg/kg, i.p.) was administered to Wistar rats (180-210 gm). Two hours post injection, animals were sacrificed, and the following brain regions isolated: cortex, striatum, hippocampus, pons, thalamus, hypothalamus, medulla and cerebellum (Konig and Klippel 1974). Tissue was immediately frozen on dry ice and stored at -70C. All tissue was analyzed within seven days of dissection.

Biochemical Assays: For assay of brain samples, the tissue samples were weighed and homogenized in 5 volume (w/v) of 0.5 M HCl. An aliquot of homogenate was removed for protein assay. The remaining homogenate was centrifuged at 15,000 xg for 10 min in an Eppendorf microcentrifuge. The supernatant extracts were utilized for quantification of THA and of biogenic amines and metabolites. These samples were analyzed directly via high performance liquid chromatography (HPLC, Spectra Physics) without any preliminary sample purification. For assay of THA, an IBM fluorometric detector, 3u cyano analytical column (ALLTECH; 100 mm length). The mobile phase employed consisted of 5% methanol, 7% acetonitrile, and 87% of a triethylamine/phosphate buffer (10) at pH 5.5 (v/v/v). For assay of brain DA, 5-HT, NE and their various metabolites (HVA, DOPAC, 5-HIAA), a 50 ul aliquot of the homogenate supernatant containing dihydroxybenzylamine (DHBA) as an internal standard was analyzed using HPLC (Spectra Physics) equipped with a BAS electrochemical detector set at +1 volt. Chromatographic separation was achieved with a 3u, C-18 reverse phase column (Axxiom, 100 mm length) at a temperature of 45°C. The mobile phase was a sodium phosphate (10.5 gm/L) buffer containing sodium octyl sulfate (50 mg/L), disodium EDTA (377 mg/L), 8% methanol (v/v) at a pH of 3.8. The minimum level of detection was about 1 ng injected.

Results

Table 1 gives the levels of THA in brain regions two hours after administration. The values represent the median percentage of THA contained in all brain regions tested. THA levels in the various brain areas ranged from 104 to 1142 ng per gram wet weight of tissue. The highest percentages of THA were found in lateral and medial cortex, and in the striatum. These areas are involved in cognitive and motor functions. The areas with the lowest percentages were the brain stem and thalamus.

Table 1 also shows the changes in brain amines and metabolites, relative to values obtained for control animals injected with saline, in these same brain regions. These data are expressed as percentages of control values. Moderate decreases in levels of NE were noted only in the thalamus and the frontal cortex. DA levels were doubled in the striatum, medulla and lateral cortex, while decreases were observed in the thalamus, pons and cerebellum. The DA metabolites DOPAC and HVA showed similar changes as DA with increases in the striatum, medulla and lateral cortex; HVA was decreased in the pons, while both these metabolites were decreased in the thalamus. The metabolites were not quantified in the cerebellum. Increases in 5-HT and 5-HIAA were noted in the thalamus and the medulla, while a decrease in 5-HT was found in the striatum. Levels of 5-HOAA were not quantified for the striatum.

Discussion

These results indicate that THA is not uniformly distributed throughout the brain. However, even in brain regions where the highest levels of THA are

TABLE 1

PERCENT CHANGES IN BRAIN METABOLITE LEVELS
AFTER THA ADMINISTRATION

BRAIN AREA	NE	DA	DOPAC	HAVA	5-HT	5-HIAA	THA*
Cerebellum	99	75	--	--	102	91	8.6
Hippocampus	93	104	131	214	72	--	10.3
Hypothalamus	101	123	138	158	108	107	9.5
Lateral Cortex	92	192	183	136	113	--	13.2
Medial Cortex	90	125	135	120	99	--	14.1
Frontal Cortex	85	102	99	117	113	--	9.2
Medulla	99	229	142	158	140	135	7.3
Pons	87	65	95	75	95	128	7.7
Striatum	95	188	211	214	72	--	12.6
Thalamus	77	38	52	42	179	130	7.5

-- Not Determined

* The THA levels are expressed as median percentage of total distribution in those areas tested.

found, these levels are still well below the reported IC50 level for the inhibition of AChE (19). The results also demonstrate that these doses of THA produce significant changes in biogenic amines and their metabolites. It is interesting to note that the changes in the respective metabolites tend to parallel the changes in the parent amine compound. The decrease in DA and its metabolites suggests a decrease in functional activity of the dopaminergic neurons. The increases in DA and metabolites could reflect either a decrease or an increase in the functional activity of DA neurons. However, given that the monoamine oxidase for DA is most probably localized in cells other than the DA neuron, the results indicate an increase in dopaminergic activity in the striatum, medulla and lateral cortex.

While the changes in amines do not parallel the peak levels of THA, it is possible that there is a difference in sensitivity and/or site of action in different brain regions. The absence of correlation between concentration and effect argues against a nonspecific effect (e.g., ion channel blockade) which is associated with a non-uniform distribution of THA in the brain. Rather, these data suggest that THA may be highly specific in affecting a select population of dopaminergic neurons within a specific brain structure, e.g., the ventral tegmental projections to the lateral cortex and not those to the frontal cortex. Alternatively, THA may be exerting differential effects at different pre-synaptic endings of the same dopaminergic cell or cell population.

While it is tempting to speculate that THA is producing its primary effect on memory through alterations in amines, the preponderance of evidence suggests that memory is primarily mediated by cholinergic mechanisms. To gain insight to this perplexing problem, further work is needed to assess the effects of THA on other components of the cholinergic system. In addition, the effects of other anticholinesterase compounds on biogenic amines and their metabolites needs to be investigated.

Relatively pure anticholinesterase drugs can only effect the remains of a severely decimated cholinergic system. These preliminary data suggest that THA acts through other systems which may be relatively intact in Alzheimer's disease. This may explain the observed therapeutic benefits of THA.

Acknowledgement: This work was supported by funds from the Research Services of the Sepulveda and Brentwood Veterans Administration Medical Centers.

REFERENCES

Bajgar J; Fusek J; Patocka J; Hrdina V. 1979. In vivo kinetics of blood cholinesterase inhibition by 9-amino-1,2,3,4-tetrahydroacridine, its 7-methoxy derivative and physostigmine in rats. Bohemoslov. Physiol. 28:31-34.

Bartus RT; Dean RL. 1988. Direct comparison of physostigmine, THA and 3,4-DAP: Differential effects on improving memory in aged monkeys. Intl. Symp. Adv. in Alzheimer Therapy: Cholinesterase Inhibitors Abst., 13.

Bartus RT; Dean RL; Beer B; Lipa AS. 1982. The cholinergic hypothesis of geriatric memory dysfunction. Science, 217(30):408-417.

Flood JF; Smith GE; Cherkin A. 1985. A memory enhancement; supra-additive effect of subcutaneous cholinergic drug combinations in mice. Psychopharmacol. 86:61-67.

Gauthier S; Masson H; Gauthier L; Bouchard R; Baily R; Becker R; Collier B; Gayton D; Kennedy J; Kissel C; Lamontagne A; Nair NPV; Ratner J; St-Martin M; Morin J; Suissa S; Tesfaye Y; Vida S. 1988. Tetrahydroaminoacridine and lecithin in Alzheimer's disease, Intl. Symp. Adv. in Alzheimer Therapy: Cholinesterase Inhibitors Abst., 19.

Harvey AL; Rowan EG. 1988. Actions of THA, 3,4-Diaminopyridine, physostigmine and Galanthamine on neuronal K+ currents at a cholinergic nerve terminal, Intl. Symp. Adv. in Alzheimer Therapy: Cholinesterase Inhibitors Abst., 14.

Huger FP; Robertello G; Petko W. 1986. Molecular forms of actylcholinesterase: Brain regional differences in sensitivity to physostigmine. Neurosci. Abs. 12:888.

Kiianmaa K; Tabakoff B. 1984. Catecholaminergic correlates of genetic differences in ethanol sensitivity. In Catacholamines: Neuropharmacology and Central Nervous System-Theoretical Aspects. Alan R. Liss, Inc., NY.

Konig J; Klippel R. 1974. The rat brain: A sterotaxic atlas. Robert E. Krieger Publishing Co., Huntington, NY.

Ohman G; Nyback H. 1988. Preliminary experiences with THA for the amelioration of symptons of Alzheimer's disease, Intl. Symp. Adv. in Alzheimer Therapy: Cholinesterase Inhibitors Abst., 18.

Park TH; Tachiki KH; Summers Wk; Kling D, Fitten J; Perryman K; Spidell K and Kling AS. 1986. Isolation and the fluorometric, high-performance liquid chromatographic determination of tacrine, Anal. Biochem. 159:358-362.

Rashti A; Childres S; Tachiki K; Melchior C; Steinberg A; Ritzmann RF. 1988. Alterations in response to ethanol by tacrine and physostigmine, Neursci. Abs..

Schauf CL; Sattin A. 1987. Tetrahydroaminoacridine blocks potassium channels and inhibits sodium activation in myxicola, JPET 243:609-613.

Shaw, FH, Bentley, GA. 1953. The pharmacology of some new anticholinesterases, Austral. J. Exp. Biol. 31:573-576.

Sherman KA. 1988. Blood cholinesterase inhibition as a guide to the efficacy of putative therapies for Alzheimer's, Intl. Symp., Adv. in Alzheimer Therapy: Cholinesterase Inhibitors Abst., 6.
Summers WK; Majovski LV; Marsh GM; Tachiki K and Kling A. 1986. Oral tetrahydroaminoacridine in long-term treatment of senile dementia, alzheimer type, New Eng. J. Med., 315:1241-1245.
Stevens DR; Cotman CW. 1987. Excitatory actions of tetrahydro-9-aminoacridine (THA) on hippocampal pyramidal neurons, Neurosci. Let. 79:301-305.

Part IV

Clinical Experience with Cholinesterase Inhibitors

Part C

EFFICACY & SIDE EFFECTS OF THA IN ALZHEIMER'S DISEASE PATIENTS

Vinod Kumar
Dept. of Psychiatry, S.I.U. School of Medicine, P.O. Box 19230, Springfield, IL

INTRODUCTION

Alzheimer's disease (AD) is a progressive chronic degenerative brain disorder manifesting with several deficits in cognitive functions, behavior, activities of daily living, and several other mental functions. It seems to be well established that several neurotransmitters are affected in the brains of the AD patients (Corkin et al. 1982, Whitehouse 1985, Roth and Wischik 1985). The most severe and consistently reported deficit seems to be of acetylcholine and associated enzymes choline actyltransferase (CAT) and acetylcholinesterase (Davies and Maloney 1977, Perry et al. 1977). There have been several attempts to correct the cholinergic deficits by precursor loading, use of cholinesterase inhibitors, muscarines or nicotinic agonist and the combination of precursor loading and the cholinesterase inhibitors. Unfortunately most of the attempts have been marginally successful and without any meaningful clinical improvement (see review by Becker and Giacobini 1988). Since the publication of a report by Summers et al. (1986), there has been a great deal of interest in the use of Tacrine (Tetrahydroaminocrindine), a long acting reversible cholinesterase inhibitor, for the treatment of Alzheimer's disease.

This report suggested that THA produced significant improvement in orientation, global assessment, name learning test, and Alzheimer deficit scale. After the publication of this report several researchers all around the world have started their own studies and some of the results were presented at International Symposium "Advances in Alzheimer Therapy: Cholinesterase Inhibitors" held at Springfield, Illinois in 1988. Findings from most of these studies are being published in this volume, but in the present chapter I will try to assess the overall evidences concerning the efficacy and side effects of this new drug and possible future strategies.

EVIDENCE FOR EFFICACY of THA

Before the publication of this book, there were only three published studies on the efficacy and side effects of THA in Alzheimer's disease patients, involving no more than a total of 39 patients in all three studies (Summer et al. 1981,1986 and Kaye et al. 1982). First of these three studies was done by Summers and colleagues (1981), who gave 0.5 mg to 1.5 mg/kg THA, intravenously to 12 Alzheimer disease patients in an open trial and reported that 6/12 (50%) showed significant improvement in their memory and 9/12 (75%) improved in their clinical staging. A year later Kaye et al. (1982) gave 30 mg of THA orally and 60 gm of lecithin for a day in four combinations using a placebo. All four treatment combinations were separated by at least 56 hours. They observed no significant improvement in memory functions using a 60 gm total dose of lecithin or a 30 mg total dose of tacrine. However, they found that lecithin in combination with THA facilitated some cognitive functions in the less impaired patients. Third study (Summers et al. 1986), surprised most of the medical and scientific community when the report was published in the New England Journal of Medicine. Summers and colleagues reported a dramatic improvement in 17 AD patients. This study involved three phases I, II, III, and 17 patients completed phase I (7-10 days), 14 patients phase II (3 week) and 12 patients were in phase III (3-36 months). All patients were given lecithin 10 +gm/day with THA (25-200 mg/day). The clinical evaluation was done by global assessment, orientation test, name learning test, and Alzheimer deficit scale. Overall improvement in all patients was significant in all three phases on all four tests when treatment phase scores were compared with placebo or predrug period. However there was some indication that more severely impaired patients showed a lesser degree of improvement on various clinical measures.

All three studies described above indicate the need to evaluate the efficacy of THA and if we concentrate on Dr. Summer's finding (1981, 1986) then we have already met that need. But after publication of Summers report in the New England Journal of Medicine, several concerns were raised about the methodological problems associated with this study (Pirozzolo 1987, Small 1987, Triot 1987) which might have contributed to dramatic results. There was an urgent need to replicate the above study. Besides these three published studies there are now 5 more studies in progress looking at the efficacy of THA and preliminary results from four of these studies were presented at the International Conference on "Advances in Alzheimer's Disease Cholinesterase Inhibitors." The detailed description of these studies has been given in this volume in various chapters but I would like to briefly comment on the question of efficacy in these new studies.

Canadian study (Gauthier et al. in this volume). This study involved 51 patients but has reported results on 19 patients, who have completed six weeks of placebo controlled dose titration phase. Patients were given 7.2 gm lecithin/day with 25-150 mg/THA/day. The precise information is not available to establish the double blindedness of this phase of the study. They observed a statistically significant improvement in functions (ADL), mini mental scores, verbal word fluency test and selection reminding task (Buschke 1974). However these results are encouraging, but the effects were small and side-effects were seen in 80% of subjects including changes in liver enzymes in 34% of the patients. A double-blind phase of the study still remains to be done. This study supports the positive finding with THA reported by Summers and his colleagues, but did not demonstrate the consistently dramatic effects reported by Summers et al. (1986).

Swedish Study (Nyback et al. in this volume). This group has reported the results of an open study involving 10 AD patients. Patients were given THA without lecithin in the doses of 25 to 200 mg/day. Five of the 10 patients (50%) showed a measurable improvement in their cognitive function (verbal ability and memory) and in their behavior. Results from this study are encouraging, but again it is not clear that the size of the effects are commensurate with those reported by Summers et al. (1986).

British Study (Levy R. 1988). The detailed data were not presented at conference but this study included six A.D. patients and three of them (50%) showed some improvement in cognitive function and assessment of daily living. Unfortunately all three improved patients also had raised liver enzymes. Double blind phase of the study is in progress.

U.S. Study. (Fitten et al. 1988 and in this volume). This group had studied 10 patients in a double blind placebo controlled crossover design. Patients were given THA (25-250 mg/day) in combination with lecithin for a week. This group reported no significant difference in placebo vs drug treatment phase performance of 10 patients on various neuropsychological tests. Supplementary data analysis recognized only one patient who could be termed as a mild responder.

NIA Sponsored Multi-Centre THA Study. The efficacy data are not available on this study, but it was stated that the outcome measures were summarized when the temporary halt was applied to the study because of elevated liver enzymes. The data from the titration phase, which included a blinded placebo treatment, were sufficient to justify continued evaluation of tacrine despite the obvious side effect profile. However, this statement should not be considered as evidence of efficacy since the patients were only exposed to one week at each treatment level. This study is in progress at present and the efficacy data should be available in the foreseeable future.

To date the data are available on seven studies on the use of THA in Alzheimer's disease and five of them report measurable improvement and two studies (Kaye et al. 1982, Fitten et al. 1988) have not found a significant improvement in their patients. One could say that both of these studies used THA for a very short duration of time, Kaye et al. (1982) for one day and Fitten et al. (1988) for one week only. However these two nearly negative studies indicate a need to evaluate the efficacy of the THA in a well designed study. Among the several ongoing studies, the U.S. Multi-center Study is the largest study which is well organized and plans to enroll the highest number of subjects. We expect to study 300 patients over a two year period. There were no data presented from the US study other than the statement that the summary of the data gathered prior to suspension of the study justified the continuation of the study in spite of observed side effects. It was interesting to note that during this conference a number of questions were raised about the absence of concomitant use of lecithin in the US multi-centre study. In the present circumstances one has to wait to see the effectiveness of lecithin in combination with THA in other ongoing studies in England and Canada. It is not surprising but important to note that all these ongoing studies are using different cognitive and behavioral measurements to quantify the change or improvement. There is a fair chance for various studies getting different results from THA therapy.

SIDE EFFECTS

In spite of somewhat encouraging results showing the efficacy of THA, the frequency and type of side effects are going to be one of the major stumbling blocks in the wide spread use of THA. Summers et al. (1981)

reported that 33% of patients experienced nausea and diaphoresis and 16% emesis by iv administration of 0.5 - 1.5 mg/kg THA. However all side effects appeared at 1 mg/kg or more of THA. Kaye et al. (1982) did not report any side effects but they used very small doses of THA (30 mg/day). Summers et al. (1986) reported a total of 18 side effects in 17 patients in phase I, nausea in 6 patients, diaphoresis in 3, belching in 2, emesis in 3, abdominal discomfort in 3, diarrhea in 2, excessive micturition in 1 patient. However all these side effects subsided by the reduction of the dose of THA or the administration of glycopyrrolate.

As is evident from the studies described in this volume the frequency of side effects seems to be under reported previously. New reports suggest that from 50% to 80% patients on THA suffer from various side effects (Canadian Study 80%, British 50%, Swedish 70%). US multi-centre study has not reported the details of the side effect other than some reports suggesting 20% of the patients showed elevated liver enzymes (ADRDA News Letter 1987). Among various side effects, the elevation in liver enzymes seems to be a difficult problem to deal with during these drug trials: Canadian study reported that 34% of the patients had elevated liver enzymes and British study reported that 50% of their patients had elevated liver enzymes. One has to be prudent enough while reading these percentages because the number of patients is very small. There are some suggestions that elevation of liver enzymes is more prevalent with higher doses of THA, but Gauthier et al. (in this volume) reported that some patients had raised liver enzymes even with 25 mg of THA in combination with lecithin. These conflicting reports suggest that, in fact, we still do not know the exact mechanism and the cause or causes of the raised liver enzymes. However, there is good news that the elevated liver enzymes come back to normal after stopping the drug but it takes several weeks (6-8 weeks) before the enzyme levels reach the predrug level.

FUTURE OF THA

There appears to be enough evidence to suggest that THA should continue to be evaluated for its efficacy and side effects in Alzheimer disease patients. It is not going to be 'The Drug' to treat all symptoms of Alzheimer's disease in all Alzheimer's patients as sometimes portrayed in public media, but it may be the drug to help some Alzheimer disease patients in some of these symptoms for a limited period of time. There seems to be enough evidence to suggest that Alzheimer's disease patients have multiple neurotransmitters deficits and will require multiple neurotransmitter replacements to produce maximum improvement in their cognition and behavior. Someone has to be the first to try to treat Alzheimer's disease patients with a combination of drugs effecting several neurotransmitters at the same time. One may produce more side effects than with a single drug but it is worth taking the risk.

REFERENCES

ADRDA Newsletter 1988. THA Update, Vol.8 No. 1, Spring.
Becker, R.E., E. Giacobini. 1988. Mechanisms of cholinesterase
    inhibitions in senile dementia of the Alzheimer type: clinical,
    pharmacological and therapeutic aspects.
    Drug Development Research 12:163-95.

Corkin, S. 1982. Some relationships between global amnesias and the memory impairments in Alzheimer's Disease. In Alzheimer's Disease: A Report of Progress. Aging vol. 19, ed. S. Corkin. New York:Raven Press.

Davies, P., A.J.F. Maloney. 1976. Selective loss of central cholinergic neurons in Alzheimer's disease. Lancet 2 1403.

Fitten, L.J., K.M. Perryman, P. Gross, A. Steinberg. (1988). Chronic oral THA administration in mice, monkeys and man. Presented at International Symposium "Advances in Alzheimer Therapy: Cholinesterase Inhibitors".

Kaye, W.H., N. Stiaram, H. Weingartner, M.H. Ebert, S. Smallberg, and J.C. Gillin. 1982. Modest facilitation of memory in dementia with combined lecithin and anticholinesterase treatment. Biol. Psychiatry. 17:275-280.

Levy, R. 1988. An open trial of high dose lecithin and THA in Alzheimer's disease. Listed abstract International Symposium "Advances in Alzheimer Therapy: Cholinesterase Inhibitors" held at Springfield, Illinois March 19-20,.

Perry, E.K., B.E., Tomlinson, G. Blessed, K. Bergman, P.H. Gibson, R.H. Perry. 1978. Correlation of Cholinergic abnormalities with senile plaques and mental test scores in senile dementia. Brit Med Journal 2:1457-9.

Pirozzolo, F.J., G.W. Small, P.N. Tariot. 1987. Oral tetrahydroaminoacridine in the treatment of senile dementia Alzheimer-type (Letters) N. Engl. J. Med. Sci 31:573.

Roth, M. and C.M. Wischik. 1985. The heterogeneity of Alzheimer's disease and its implication for scientific investigations of the disorder. In Recent Advances In Psychogeriatrics ed. A.T. Churchill, 71-92. New York: Livington Inc.

Summers, W.K., J.O. Viesselman, G.M. Marsh, K. Candelora. 1981. Use of THA in treatment of Alzheimer-like dementia: pilot study in twelve patients Biol Psychiatry 16:145-53.

Summers, W.K., L.V. Majowski, G.M. Marsh, K. Tachiki, and A. Kling. 1986. Oral tetrahydroaminoacridine in long-term treatment of senile dementia, Alzheimer's type. N. Engl. J. Med. 315:1241-1245.

Whitehouse, P.J. 1986. Development of neurotransmitter-specific therapeutic approaches in Alzheimer's disease. In Treatment Development Strategies for Alzheimer's Disease. Ed. J. Crook, R. Bartus, S. Ferris, S. Gershon, 483-498. Connecticut: Mark Pawley Associates, Inc.

PRELIMINARY EXPERIENCES AND RESULTS WITH THA FOR THE AMELIORATION OF SYMPTOMS OF ALZHEIMER'S DISEASE

H. Nybäck[1], H. Nyman[1], G. Ohman[3], I. Nordgren[2], and B. Lindström[4]
[1]Dept. of Psychiatry and Psychology, [2]Dept. of Toxicology,
[3]Hospital Pharmacy, Karolinska Institute and Hospital, Stockholm, Sweden;
[4]National Board of Health and Welfare, Dept. of Drugs, Uppsala, Sweden

The positive results of a clinical trial of oral tetrahydroaminoacridine (THA) in Alzheimer's dementia published in November 1986 (Summers et al. 1986) nave evoked many expectations and demands from both patients and relatives as well as from physicians of different disciplines. Whether the results by Summers and coll. will be replicated by other research groups and whether their findings will constitue "a triumph for the scientific method" (Davis and Mohs 1986) is, however, an open question. In any case, the cholinergic approach toward a remedy for the victims of Alzheimer's disease seems scientific rational and as long as we have relevant hypotheses about the neuro-chemical and patho-physiological basis of intellectual decline in dementia, there will be hope for future drug developments. Many cell groups and many transmitter systems are affected in the course of a dementing process but only the defects of the cholinergic pathways correlate significantly with the degree of mental impairment (Perry et al. 1978, Palmer and Bowen 1988). Also in animal experiments have the critical importance of cholinergic systems for memory and learning been shown (Deutsch 1973).

We have started a double-blind cross-over trial of THA in patients with Alzheimer's dementia by an open dose-finding study in ten patients. The drug was manufactured by Nobel Chemicals (Sollentuna, Sweden) and marketed by Nobel Medica (Sundbyberg, Sweden). Capsules of 25 and 50 mg were made and administered to patients in successively increasing daily doses from 25 mg up to 200 mg or until side effects such as nausea or vomiting occurred. The patients were kept at the highest tolerated doses for one to three weeks before plasma and CSF sampling and evaluation of side effects and therapeutic respones. Table 1 shows the sex of the patients, their age and their degree of dementia according to the Global Deterioration Scale of Reisberg and coll. (1982). The time on THA and the maximal drug dose is also seen.

The treatment response was evaluated using the Wechsler Adult Intelligence Scale (WAIS), the Wechsler Memory Scale, the Mini-Mental State Examination (Folstein et al. 1975), clincial rating scales (Adolfsson et al. 1981) and through interviews with the next of kin. Adverse effects were recorded through physical examination and laboratory tests of blood and urine. Effects of THA on brain aminergic systems were investigated by analysis of acetylcholine and monoamine metabolites in the CSF using mass fragmentography (Karlen et al. 1974, Swan et al. 1976). Levels of THA in plasma were determined using high-performance liquid chromatography (Lindström et al. to be published).

Table 1. Demographic Data

PATIENTS (sex)	AGE (years)	DEMENTIA STAGE (GDS)	WEEKS ON THA	MAX DOSE (mg/day)
1. Female	71	6 (severe)	5	120
2. Female	57	6 (severe)	3	170
3. Female	62	3 (mile)	1	120
4. Female	60	5 (moderately severe)	3	150
5. Male	63	4 (moderate)	3	150
6. Male	63	2 (very mild)	3	200
7. Male	62	4 (moderate)	3	150
8. Female	61	4 (moderate)	3	80
9. Female	57	5 (moderate severe)	3	150
10. Female	80	3 (mild)	2	150
Mean	63.9	4.2	2.9	144

Table 2. Effects of the THA Treatment.

PAT	ADVERSE EFFECTS	CLINCIAL RESPONSE
1.F	Muscular rigidity	Doubtful
2.F	Skin rash, itching	No
3.F	Nausea, vomiting	No
4.F	Vomiting	Improved verbal ability and less apraxia
5.M	No	Improved verbal ability
6.M	Nasal secretion	Improved verbal memory
7.M	Nausea	Improved memory
8.M	Nausea, vomiting, elevated liver enzymes	Improved memory and orientation
9.F	No	No
10.F	No	No

RESULTS

During the increase in the THA doses side effects were recorded in seven of the ten patients (Table 2). The side effects were mainly related to central or peripheral cholinergic activation. By decreasing the daily dose by 25 or 50 mg, the medication could be continued in all but one patient who had elevated liver enzymes, even at the low dose of 25 mg t.i.d.

Improvements in the clinical ratings and the neuropsychological tests, particularly in measures of verbal ability and memory, were seen in five of the ten patients. Reports from close relatives, who saw the patients during their visits at home over the week-ends, were also positive. In some instances, we got the impression that the spontaneous behavior of the patients was more improved than their test performance which could be confounded by frustration and anxiety. As an example of a relative's report, an extract from a letter may be quoted (a 29-year old daughter about her 60 year old mother, both Medical Doctors):

> "On THA she had more initiative and was less anxious to get lost. She did not loose words as frequently as before. She reacted emotionally more adequate and there were more emotions in her. She was able to remember and effectuate longer sequences of instructions. She gained weight and seemed physically and mentally improved. I know that this is my personal reaction, maybe expectations from my knowing of the THA treatment, but on the other hand, I am close to my mother and I have followed the progression of her disease in detail."

The levels of homovanillic acid (HVA), 3-methoxy-4-hydroxyphenyl glycol (MHPG) and 5-hydroxyindoleacetic acid (5-HIAA), metabolites of central dopaminergic, noradrenergic and serotonergic neurotransmission, were increased on THA as compared to pretreatment levels (Table 3.) This was particularly pronounced for HVA.

THA concentrations in venous blood before and after the morning dose (25-50 mg) varied considerably betwen patients. The highest concentration was found already at 90 minutes following drug intake (Figure 1). The mean plasma half-life was 3.0 hours (range 1.5-6 h). Patients with low plasma levels had less side effects and exhibited smaller changes in the concentrations of acetylcholin, HVA and 5-HIAA in the CSF than patients with high levels of THA.

CONCLUSIONS

The present study was undertaken as an introduction to a more extensive double-blind cross-over study of THA in Alzheimer's dementia. Despite the small number of subjects and the short treatment period, we found interesting clinical and biochemical results presented above. THA appears as a powerful psychotropic drug which may ameliorate critical symptoms of dementia such as afasia and amnesia. Side effects are, however, common, consisting of both cholinergic manifestations such as nausea and vomiting and toxic or allergic reactions such as liver cell damage. The resorption and the metabolism of the drug seem rapid, a more even plasma level and a greater tolerability may be achieved by a more frequent dosing than t.i.d.

With a sensitive method elevations of acetylcholine levels in the CSF can be detected as a reflexion of the inhibition of central cholinesterases Also, levels of HVA and 5-HIAA were significantly increased as an evidence of an interaction of the drug with other brain momoaminergic pathways. Balancing dopaminergic and cholinergic systems are known in the basal ganglia and dopamine receptor blocking drugs increase dopamine turnover and elevates HVA levels in the CSF (Sedvall et al. 1975). The present results with THA illustrate how the same effect can

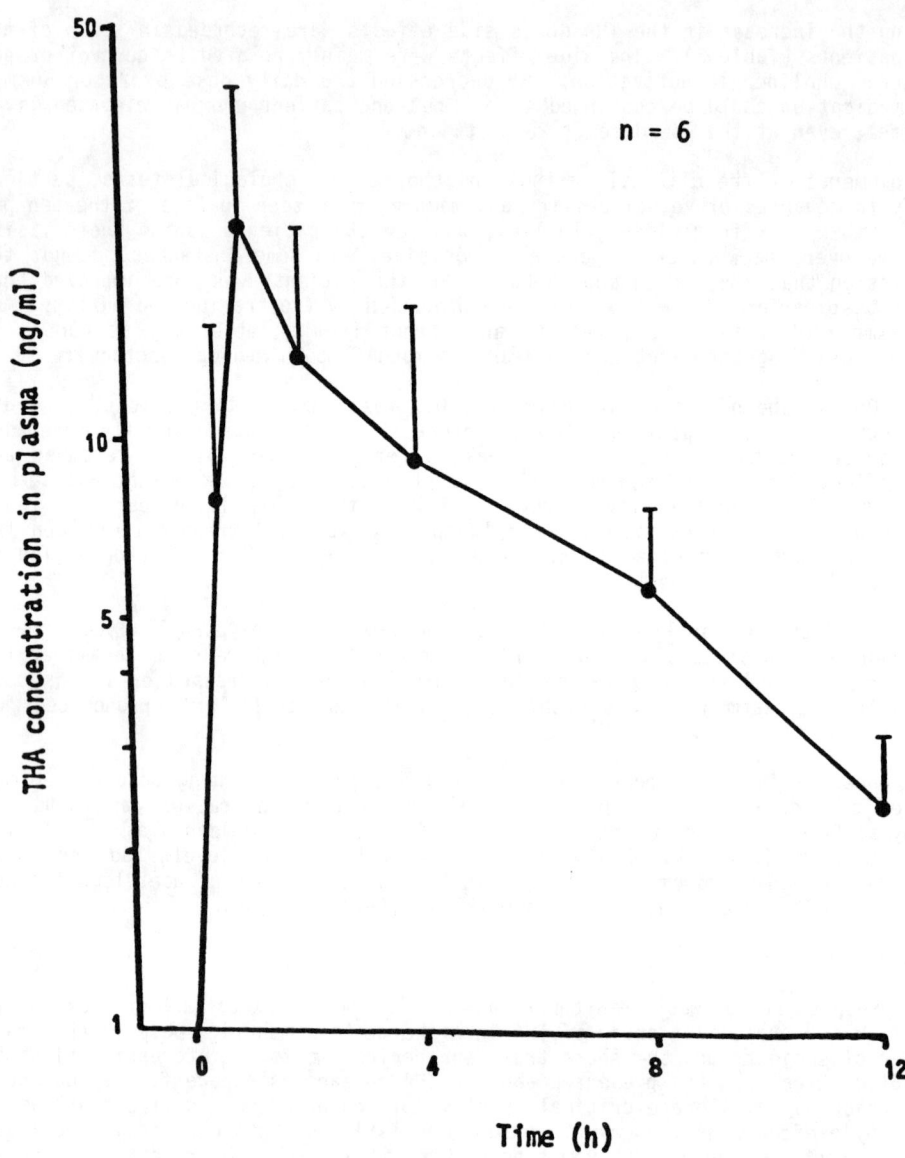

Figure 1. Drug level in the blood of patients following a morning dose of THA (50 mg).

Table 3. Levels of Monoamine Metabolites and Acetylcholine (pmol/ml) in the CSF of Patients during Treatment with THA.

PAT	HVA Before	HVA During	MHPG Before	MHPG During	5-HIAA Before	5-HIAA During	ACH Before	ACH During
1.	75	117	25	31	62	99	-	-
2.	107	129	30	29	48	41	1.4	1.9
3.	180	288	34	37	106	161	3.6	5.0
4.	145	158	30	37	82	99	0.6	3.8
5.	90	130	45	38	60	79	1.8	9.1
6.	288	383	45	46	108	154	-	-
Mean	148	201	35	36	78	106	1.9	5.0
± s.d.	±79	±109	±8.4	±6.0	±25	±46	±1.3	±3.0

be brought about by stimulation of cholinergic neurons which probably are located as interneurons in a feed-back regulation of the activity of dopamine neurons.

ACKNOWLEDGEMENTS

The skillful technical assistance of Mss Kjerstin Lind, Alexandra Tylec, Asa Franzen and Gun Jakobsson is gratefully acknowledged. The study was supported by grants from the Swedish Medical Research Council, the Bank of Sweden's Tercentennial Foundation, The Swedish Society of Medicien, Osterman's Foundation, Söderström-König's Foundation and the Karolinska Institute.

REFERENCES

Adolfsson, R., C.G. Gottfries, L. Nyström and B. Winblad. 1981. Prevalence of dementia disorders in institutionalized Swedish old people. Acta Psychiatr. Scand. 63: 225-244.
Davis, K.L. and R.C. Mohs. 1986. Cholinergic drugs in Alzheimer's disease. The New Engl. J. Med. 315: 1286-87.
Deutsch, J.A. 1973. The physiological basis of memory. New York Academic Press, 59-77.
Folstein, M.F., S.E. Folstein and P.R. McHugh. 1975. "Mini-Mental State". A practical method for grading the cognitive state of patients for the clinician. J Psychiatr. Res. 12: 189-198.
Karlen, B., G. Lundgren, I. Nordgren and B. Holmqvist. 1974. Ion pair extraction and gas phase analysis of acetylcholine and choline. In Handbook of chemical assay methods, ed. I. Hanin, 163-179, Raven Press.
Palmer, A.M. and D.M. Bowen. 1988. Neurotransmitter basis of the behavioural changes of Alzheimer's disease. Eighth European Winter Conference on Brain Research, Tignes, France, p 51.
Perry, E.K., B.E. Thomlinson, G. Blessed, K. Bergmann, P.H. Gibson and R.H. Perry. 1978. Correlation of cholinergic abnormalities with senile plaques and mental test scores in senile dementia. Brit. Med. J. 2: 1457-59.

Reisberg, B., S.H. Ferris, M.J. De Leon and T. Crook. 1982. The global deterioration scale for assessment of primary degenerative dementia. Am. J. Psychiatry 139: 1136-39.
Sedvall, G., B. Fyrö, H. Nybäck and F.A. Wiesel. 1975. Actions of dopaminergic antagonists in the striatum. Advances in Neurology 9: ed. D.B. Calne, T.N. Chase and A. Barbeau. 131-140, Raven Press.
Summers, W.K., L.V. Majovski, G.M. Marsh, K. Tachiki and A. Kling. 1986. Oral tetrahydroaminoacridine in long term treatment of senil dementia, Alzheimer type. New Engl. J. Med. 315: 12451-1245.
Swan, C.G., B. Sandgärde, F.A. Wiesel and G. Sedvall. 1976. Simultaneous determination of the three major monoamine metabolites in brain tissue and body fluids by mass fragmentography. Psychopharmacology 48: 147-152.

TETRAHYDROAMINOACRIDINE AND LECITHIN IN ALZHEIMER'S DISEASE

S. Gauthier, H. Masson, L. Gauthier, R. Bouchard, B. Collier, Y. Bacher, R. Bailey, R. Becker, H. Bergman, R. Charbonneau, D. Dastoor, D. Gayton, J. Kennedy, C. Kissel, M. Krieger, S. Kushnir, A. Lamontagne, M. St-Martin, J. Morin, N.P.V. Nair, L. Neirinck, J. Ratner, S. Suissa, Y. Tesfaye, S. Vida

McGill Centre for Studies in Aging, 1650 Cedar Ave., Montreal, Canada H3G 1A4

INTRODUCTION

Alzheimer's Disease (AD) is now one of the most devastating illness affecting our Senior Citizens and it adds greatly to the costs of our Health Care Systems. Any treatment that could improve some of its symptoms (memory loss, speech impairment and spatial disorientation) is to be given a fair trial.

Since the discovery of a significant deficit in the ability to synthesize acetylcholine in the cortex and hippocampus of individuals suffering from AD, many attempts have been made to supplement the brain levels of this transmitter. Thus choline and lecithin have been administered orally in an attempt to increase the availability of choline (as the natural precursor) in the brain. Although this approach is safe and can be considered as a form of dietary supplementation, it did not help patients affected with AD (Etienne 1981). An alternative line of treatment has been to replace acetylcholine with substitutes. In order to deliver such analogs to the brain while avoiding peripheral side-effects, infusion pumps have been implanted subcutaneously with catheters running from the abdomen to the cerebral ventricles. These systems have been tested using bethanechol chloride as a direct muscarinic agonist. Such protocols involved surgery under general anesthesia, and published results showed little therapeutic value (Gauthier 1986, Penn 1988; Harbaugh in preparation).

One other approach has been to inhibit acetylcholinesterase activity (AchE), and thus potentiate residual cholinergic function. Physostigmine is the best known of such agents. Various reports have documented a beneficial effect of physostigmine in AD, whether used orally or subcutaneously (Davis 1982; Thal 1983), although other studies failed to show a clinical beneficial effect (Caltagirone 1982).

Unfortunately peripheral side-effects, a short half-life and a narrow therapeutic window limit the usefulness of physostigmine.

Tetrahydroaminoacridine (THA), which has a pharmacological action similar to physostigmine but with much less toxicity, has been tested by Summers et al in 1981 intravenously and in 1986 orally, with encouraging results. The latter study involved an open titration period to the most effective dose, then a randomized double-blind cross-over period. Those who were responsive were allowed to continue for up to twelve months of treatment. Although the number of patients studied was small (N=17) a beneficial effect was seen in early and late stages of the disease. A modest amount of lecithin was ingested by all patients throughout the study and this may have contributed to its success, although the pharmacological level of interaction between the precursor and the AchE inhibitor is not yet clear. Another reason to combine the two agents is the potential protective effects of lecithin on phospholipid depletion induced by the AchE inhibitor (Ulus 1987). Finally, it is quite possible that THA exerts therapeutic effects by potentiating the action of non-cholinergic transmitters (Drukarch 1987).

The potential therapeutic value of THA with lecithin in the intermediate stages of AD was tested in Canadian patients and we report here the results of an eight weeks double blind study, which showed statistically significant improvements of self-care activities of daily living (ADL), the scores on the Folstein's Mini Mental State (MMS), the Selective Reminding Task (SRT) and on Verbal Word Fluency (VWF).

METHODOLOGY

Patients population

Fifty one patients (19 men, 32 women, mean age 68.5, range 53 to 85) with a clinical diagnosis of "probable" or "definite" AD (criterias of the NINCDS-ADRDA work group, McKhann 1984) participated freely in the study. An informed consent form was signed by the patients if they were competent and by their Curator if they were not. Patients were ambulatory and functional by themselves or with supervision by spouse, other family member or a dedicated paramedical employee. They were thus at a Stage 4 or 5 on the Global Deterioration Scale (Reisberg 1982). Patients with significant parkinsonian features (more than mild rigidity or akinesia), active heart disease (brady-arrhythmias), uncontrolled seizure disorder, active liver or hematological dysfunction had been excluded, as well as those with Modified Ischemic Scores (Hachinski 1975; Rosen 1979) of 4 or more. Investigators at nine sites were responsible for recruitment and supervision of their patients, with a common data base and procedure.

Laboratory, functional and neuropsychological assessments

Besides the standard diagnostic procedures used for AD such as Computer Assisted Tomography (CAT) and blood screen for reversible causes of dementia (NIH Consensus conference 1987), a complete blood count, liver screen and EKG were done immediately prior to entry into the study and serially thereafter. Patients were seen every two weeks at which time the following scales were administered by an investigator blind to the phase of treatment:

* Rapid Disability Rating Scale-2 (Linn 1982)   RDRS-2
* Geriatric Depression Scale (Brink 1982)   GDS
* Mini Mental State (Folstein 1975)   MMS
* Selective Reminding Task (Buschke 1974)   SRT
* Verbal Word Fluency (Spreen 1969)   VWF

Data was also collected on ability to draw geometric figures and on general behavior. A semi-structured diary was kept by the caregivers on functional behavior at home.

Drugs doses and administration

THA has been purchased from Pharmascience Inc., Montreal, Qué., Canada, as 25 and 50mg capsules. It was to be administered orally in 50mg increments over six weeks until a maximal daily dose of 200mg in three divided doses, as in Summers 1986 study, but because of GI intolerance and liver toxicity the titration was reduced to 25mg increments with a maximal daily dose of 100mg. Plasma samples have been collected for the assay of THA levels. Plasma cholinesterase (ChE) activity was measured before and on THA by the method of Ellman (1961). Lecithin capsules (1.2g, 12% phosphorylcholine content) were provided by Pharmascience Inc and ingested at meal time throughout the study, starting two weeks prior to THA in order to avoid confusion in possible side effects, as one capsule per meal for a week than two per meal (total daily dose of 7.2g). Choline blood levels were measured before and on lecithin by the method of Goldberg (1973). Patients and families were blind as to the content of the "THA capsules" during the entire study, as well as one investigator at each site responsible for the regular functional and cognitive assessments. Finally, glycopyrrolate (Robinul) was donated by A.H. Robins, Mississauga, Ont., and administered up to 2mg tid to patients suffering from GI side-effects; propantheline bromide (Probanthine) 15mg bid was used as an alternative in four cases.

Data analysis

All data were collected and mailed to a clinical monitor at the McGill Centre for Studies in Aging, who checked the completeness of the information and stored it on soft disk using a Data Base III program. Complications were reported to the Health Protection Branch of Health and Welfare Canada. The Wilcoxon signed-rank test was used to assess the significance of changes in plasma choline levels, plasma ChE activity, MMS, RDRS-2, VWF and SRT scores between the various weeks of treatment (Table 1 and 2).

RESULTS

Drop outs

All subjects complied with the protocol and only 6 (12%) dropped out because of GI side-effects. One was withdrawn because of unexplained fever. One had a thrombotic stroke on day 3 of the study, with an ischemic score of 4 but leukoaraiosis on CAT scan (Hachinski 1987). The study was stopped when liver enzymes elevation was found to occur in more than 30% of subjects. At that time 19, (10 men, 9 women, mean age 67.7, range 53-84) had completed the full 8 weeks of titration and the statistical analysis was done on their data.

Plasma levels

Choline levels did not change between baseline (mean 16.18±5.50 nmoles/ml) and after two weeks of lecithin (mean 16.05±4.96 nmoles/ml). ChE activity did not decrease between baseline (mean 1.86±0.52 umoles/min/ml) and after six weeks of THA (mean 1.91±0.44 umoles/min/ml). Plasma THA levels are not yet available.

Functional and cognitive changes

The MMS score improved significantly after four weeks of THA and that improvement was lost two weeks after cessation of THA (Table 1). The self-care ADL section of the RDRS-2 (items 1-8) showed a statistically significant improvement after 6 weeks of THA (Table 1). There was no change of mood as reflected by the GDS. Both the VWF and SRT improved in parallel and significantly after 4 weeks of THA; that improvement was lost 2 weeks after cessation of the drug (Table 2). The drawing that paralleled best the clinical improvement was of a clock drawn and set by the subjects at 11:10 (Figure 1). Half the patients improved in their ability to draw such a clock. Most families commented that the subjects returned to their baseline functional difficulties 2 to 4 weeks after cessation of THA.

Table 1. Summary of MMS and RDRS-2 Scores (N=19)

Week	Treatment	MMS	RDRS-2
0		17.8±4.1	12.4±3.0
	lecithin		
2		18.5±4.2	12.6±3.3
	lec+THA (mean dose 50mg/d)		
4		19.2±5.1	12.3±2.8
	lec+THA (mean dose 88mg/d)		
6		19.4±4.1*	11.9±2.9
	lec+THA (mean dose 77mg/d)		
8		19.4±5.3	11.2±2.6i
	lecithin		
10		17.9±4.6**	11.5±2.5

*   p = 0.025 with week 0 and 0.007 with week 10
** p = 0.026 with week 8   i   p = 0.0120 with week 0

Table 2. Summary of VWF and SRT Scores (N=19)

Week	Treatment	VWF	SRT
0		8.6±9.0	0.44±0.17
	lecithin		
2		8.9±7.7	0.37±0.16
	lec+THA (mean dose 50mg/d)		
4		11.2±9.4	0.45±0.15
	lec+THA (mean dose 88mg/d)		
6		10.6±7.9*	0.49±0.19i
	lec+THA (mean dose 77mg/d)		
8		11.2±8.2**	0.44±0.16ii
	lecithin		
10		8.8±8.3***	0.40±0.16iii

*   p = 0.0395 with week 0   i   p = 0.044 with week 0
**   p = 0.0415 with week 0   ii   p = 0.006 with week 2
*** p = 0.0252 with week 8   iii p = 0.007 with week 8

Side-effects and toxicity

80% of subjects experienced autonomic side-effects, including in order of frequency; nausea, vomiting, loss of appetite, loss of weight, headache, fatigue, dyspnea, restless sleep, nasal and lacrimal hypersecretion. Although a peripheral muscarinic blocker was used in 62% of subjects, side-effects were improved only by a reduction in the total daily dose of THA. Reversible elevations in liver enzymes levels were found in 17/50 subjects (34%) and ranged from a 50% elevation of alkaline phosphatase to a 20 fold increase in SGPT and a 11 fold increase in SGOT. There was no obvious pattern as to the age and sex of people affected, and the changes were seen at doses of THA as low as 25mg per day, and as rapidly as within a week. Enzymes levels normalized after 2 to 8 weeks off THA.

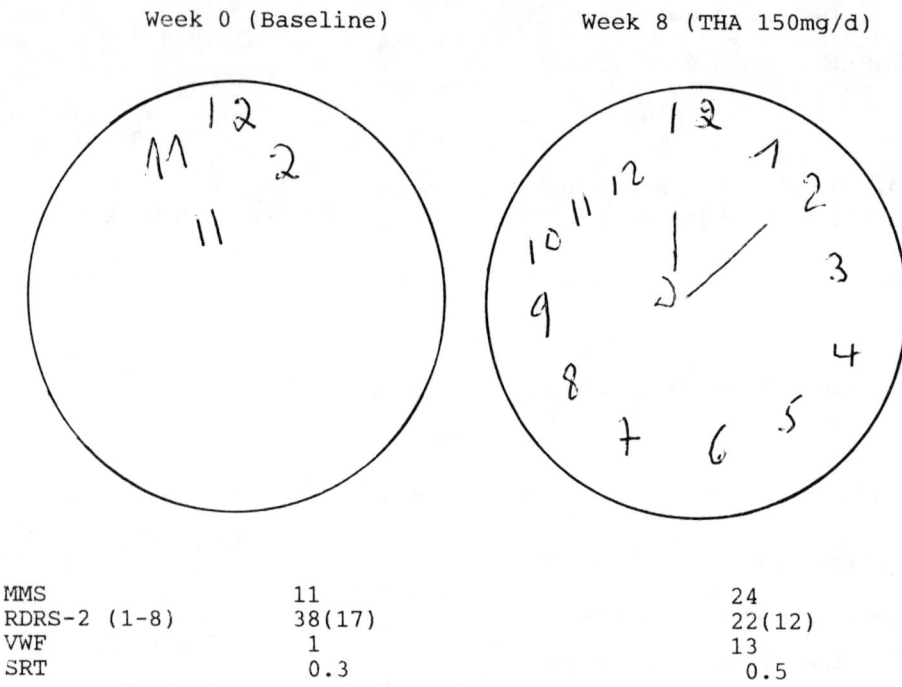

	Week 0 (Baseline)	Week 8 (THA 150mg/d)
MMS	11	24
RDRS-2 (1-8)	38(17)	22(12)
VWF	1	13
SRT	0.3	0.5

Figure 1. Performance of subject E.H., age 85, stage 5

DISCUSSION

The data available to date shows a small but measurable and statistically significant improvement in function as reflected by the self-care ADL portion of the RDRS-2. Decline in autonomy at work and at home have long been noted by clinicians to precede and then parallel measurable cognitive changes in the course of AD. Improvement during therapeutic drug trials in AD should thus be monitored at both functional and cognitive levels. Functional changes are best detected by caregivers and a standardized set of questions could be put to those by regular phone calls at home. The fact that a global non-specific cognitive scale such as the MMS could detect changes under THA treatment as well as selective tasks for memory (the SRT) and speech (the VWF) is of interest. These tasks had failed to improve on bethanechol intracerebral infusion whereas the MMS had increased significantly although slightly (Harbaugh in preparation). Whether one general scale, such as the modified MMS (Teng 1987), the Alzheimer Disease Rating Scale (Rosen 1984) or the Hierarchic Dementia Scale (Cole 1987), is preferable to the MMS remains to be answered.

If THA with lecithin is confirmed to be therapeutically effective for some stages of AD in future double-blind cross-over trials, some mechanisms of action other than potentiation of remaining cortical cholinergic activity is likely. Indeed, selective cholinergic interventions in the past have failed to help significantly. Furthermore, our data showed no increase of choline plasma levels and no reduction in plasma ChE activity with the current doses of lecithin and THA administered, which gave measurable clinical effects. Finally, in vitro animal pharmacological data have shown non-cholinergic effects of THA (Drukarch 1987). Such studies have also shown a down regulation of M_1 receptors with chronic THA intake (Flynn 1987).

The liver toxicity that was observed appears to be idiosyncratic and cytotoxic in pattern. Fortunately all changes were reversible. Future drug trials using THA will require careful serial monitoring of liver enzymes levels. Peripheral muscarinic blockers gave minimal help to relieve side-effects and can potentially add to the liver toxicity. They should thus not be used.

CONCLUSIONS

THA with lecithin over six weeks improved cognition and functional state in the intermediate stages of AD. Side-effects were seen in 80% of subjects and changes in liver enzymes occurred in 34%. For future trials using this combination we recommend the following:

* Smaller capsules (600mg) of enriched lecithin (75% PC) in order to double the plasma choline levels;
* slow titration of THA starting with 25mg at breakfast and increases of 25mg each other week till 100mg/day;
* lower doses of THA if side-effects occur rather than add peripheral muscarinic blockers;
* monitor plasma levels of liver enzymes regularly;
* monitor ADL changes at home by a structured questionnaire to the caregiver;
* a global cognitive scale performed by observers blinded to the treatment, after training;
* a wash out period of one month between the two phases of a double-blind cross-over trial, because of a long carry-over effect.

REFERENCES

Brink, T.L., J.A. Yesavage, O. Lum et al. 1982. Screening tests for geriatric depression. Clin. Geront. 1:37-43.
Buschke, H., P.A. Fuld. 1974. Evaluating storage, retention and retrieval in disordered memory and learning. Neurology 24:1019-1025.
Caltagirone, C., G. Gainotti, C. Masullo. 1982. Oral administration of chronic physostigmine does not improve cognitive or mnesic performances in Alzheimer's presenile dementia. J. Neuroscience 16:247-249.
Cole, M., D.P. Dastoor. 1987. A new hierarchic approach to the measurement of dementia. Psychosomatics 28:298-304.
Davis, K.L., R.C. Mohs. 1982. Enhancement of memory processes in AD with multiple-dose intravenous physostigmine. Am. J. Psychiatry 139:1421-1424.
Drukarch, B., K.S. Kits, E.G. Van der Meer et al. 1987. THA, an alleged drug for the treatment of AD, inhibits acetylcholinesterase activity and slow outward K+ current. Eur. J. Pharmacol. 141:153-157.
Ellman, G.L., D.K. Courtney, V. Andres et al. 1961. A new and rapid colourimetric determination of acetylcholinesterase activity. Biochem. Pharmacol. 7:88-95.
Etienne, P., D. Dastoor, S. Gauthier et al. 1981. Lack of effect of lecithin treatment for 3 months. Neurology 31:1552-1554.
Flynn, D.D., A. Suarez, A. Cordoves et al. 1987. Differential regulation of muscarinic receptor subtypes in the cerebral cortex by chronic THA. Society for Neuroscience Abstracts 13:726.
Folstein, M., S. Folstein, P.R. McHugh. 1975. Mini Mental State: a practical method for grading the cognitive state of patients for the clinician. J. Psychiatry Res. 12:189-198.
Gauthier, S., R. Leblanc, R. Quirion et al. 1986. Transmitter replacement therapy in AD using intracerebroventricular infusions of receptor agonists. Can. J. Neurol. Sco. 13:394-402.
Goldberg, A.M., R.E. McCaman. 1973. The determination of picomole amounts of acetylcholine in mammalian brain. J. Neurochem. 20:1-8.
Hachinski, V.C., L.D. Ilif, E. Zilkha et al. 1975. Cerebral blood flow in dementia. Arch. Neurol. 32:632-637.
Hachinski, V.C., P. Potter, H. Merskey. 1987. Leukoaraiosis. Arch. Neurol. 44:21-23.
Harbaugh, R.E. 1988 (in preparation). Intracerebroventricular bethanechol chloride infusion in AD: results of a collaborative, double-blind study.
Linn, M.W., B.S. Linn. 1982. The Rapid Disability Rating Scale-II. J. Amer. Ger. Soc. 30:378-382.
McKhann, G., D. Drachman, M. Folstein et al. 1984. Clinical diagnosis of AD. Neurology 34:939-944.
NIH Consensus Conference. 1987. Differential diagnosis of dementing diseases. JAMA 258:3411-3416.

Penn, R.D., E.M. Martin, R.S. Wilson et al. 1988. Intraventricular bethanechol infusion for AD. Neurology 38:219-222.

Reisberg, B., S.H. Ferris, M.J. de Leon et al. 1982. The Global Deterioration Scale for assessment of primary degenerative dementia. Am. J. Psychiatry 139:1136-1139.

Rosen, W.G., R.D. Terry, P.A. Fuld et al. 1979. Pathological verification of ischemic score in differentiation of dementias. Ann. Neurol. 7:486-488.

Rosen, W.G., R.C. Mohs, K.L. Davis. 1984. A new rating scale for AD. Am. J. Psychiatry 141:1356-1364.

Spreen, O., L.A. Benton. 1969. The neurosensory center comprehensive examination for aphasia. Victoria, B.C.: University of Victoria Neuropsychology Laboratory.

Summers, W.K., J.O. Viesselman, G.M. Marsh et al. 1981. Use of THA in treatment of AD-like dementia: pilot study in twelve patients. Biol. Psychiat. 16:145-153.

Summers, W.K., L.V. Majovski, G.M. Marsh et al. 1986. Oral THA in long-term treatment of senile dementia, Alzheimer type. NEJM. 315:1241-1245.

Teng, E.L., H.C. Chui. 1987. The modified Mini Mental State (3MS) examination. J. Clin. Psychiatry 48:314-318.

Thal, L.J., P.A. Fuld. 1983. Memory enhancement with oral physostigmine in AD. NEJM 308:720.

Ulus, I.H., R.J. Wurtman. 1987. Prevention by choline of the depletion of membrane phosphatidylcholine by a cholinesterase inhibitor. NEJM 318:191.

EFFECT OF THA ON ACETYLCHOLINE RELEASE AND CHOLINERGIC RECEPTORS IN ALZHEIMER BRAINS

Agneta Nordberg[1,2], Lena Nilsson[1], Abdu Adem[1,2], John Hardy[3], and Bengt Winblad[2]

[1]*Dept. of Pharmacology, Univ. of Uppsala, Box 591, S-751 24 Uppsala, Sweden;*
[2]*Dept. of Geriatric Medicine, Karolinska Institute, Huddinge Hospital, S-141 86 Huddinge, Sweden;* [3]*Dept. of Biochemistry, St. Mary's Hospital Medical School, London, UK*

INTRODUCTION

Alzheimer´s disease, senile dementia of Alzheimer type (AD/SDAT) is characterized by changes in several transmitter systems of the brain (Gottfries 1985, Hardy et al 1985). One of the most pronounced changes has been found in the cholinergic nervous system (for review see Perry 1986). Deficits in several presynaptic markers such as choline (Ch) uptake (Rylett et al 1983), choline acetyltransferase (ChAT) activity (Davies and Maloney 1976, Perry et al 1977, White et al 1977) acetylcholine (ACh) synthesis (Sims et al 1980), release (Nilsson et al 1986) and number of nicotinic receptors (Whitehouse et al 1986, Nordberg and Winblad 1986b) have been observed in AD/SDAT brains. The cholinergic muscarinic receptors which are mainly supposed to be postsynaptically located, seem to be preserved in AD/SDAT (Nordberg and Winblad 1986a). A loss in the muscarinic M_2 subtype can however not be excluded (Mash et al 1985), although data are somewhat conflicting (Caulfield et al 1983, Nordberg et al 1986).

Several strategies to increase the activity of the cholinergic system in AD/SDAT brains are possible, such as:

* increasing the uptake of Ch and synthesis of ACh (precursor administration)
* protecting ACh from being inactivated by acetylcholine-esterase (AChE) (administration of AChE inhibitors)
* direct action on the cholinergic receptors (muscarinic agonists).

As illustrated in Figure 1, drugs with different mechanisms of actions have been tried as therapeutic agents in AD/SDAT. The outcome of the clinical trials has been modest, except for those with AChE inhibitors, such as physostigmine and tetra-hydroaminoacridine (THA) (Mohs et al 1985, Hollander et al 1986, Summers et al 1986). A limiting factor for the effect of physostigmine has been its short half-life in plasma.

The aim of this presentation is to describe the underlying biochemical mechanisms for physostigmine and THA in control and AD/SDAT brain tissue. Due to the lack of a reliable animal model for AD/SDAT, human brain tissue with short postmortem delay has been used.

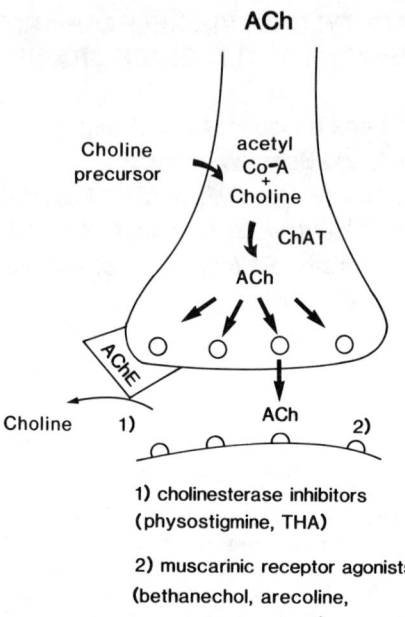

Figure 1 Drugs with different modes of action in the cholinergic synapse.

The findings indicate that the AChE inhibitors can restore the ACh release in AD/SDAT brain tissue. The mechanism of action of THA in the cholinergic nerve terminal involves inhibition of AChE, and interaction with muscarinic and nicotinic cholinergic receptors.

EFFECT OF THA ON BRAIN ACh RELEASE

To further investigate the underlying neurochemical mechanisms of action of AChE inhibitors in AD/SDAT brain tissue, an <u>in vitro</u> model for ACh release has been used (Nilsson et al 1986). Human brain tissue obtained with short postmortem delay (less than 14 hours) is incubated with labelled choline (^{3}H-Ch) and the calcium dependent potassium evoked release of ACh (^{3}H-ACh) is investigated under the influence of different drugs. Using this <u>in vitro</u> release system, a significantly lower amount of potassium evoked release of ACh is measured from AD/SDAT cortical slices compared to age-matched control tissue (Figure 3). This observation is in line with earlier findings of a decreased synthesis of ACh (^{14}C-ACh) in cortical biopsies from AD/SDAT patients (Sims et al 1980).

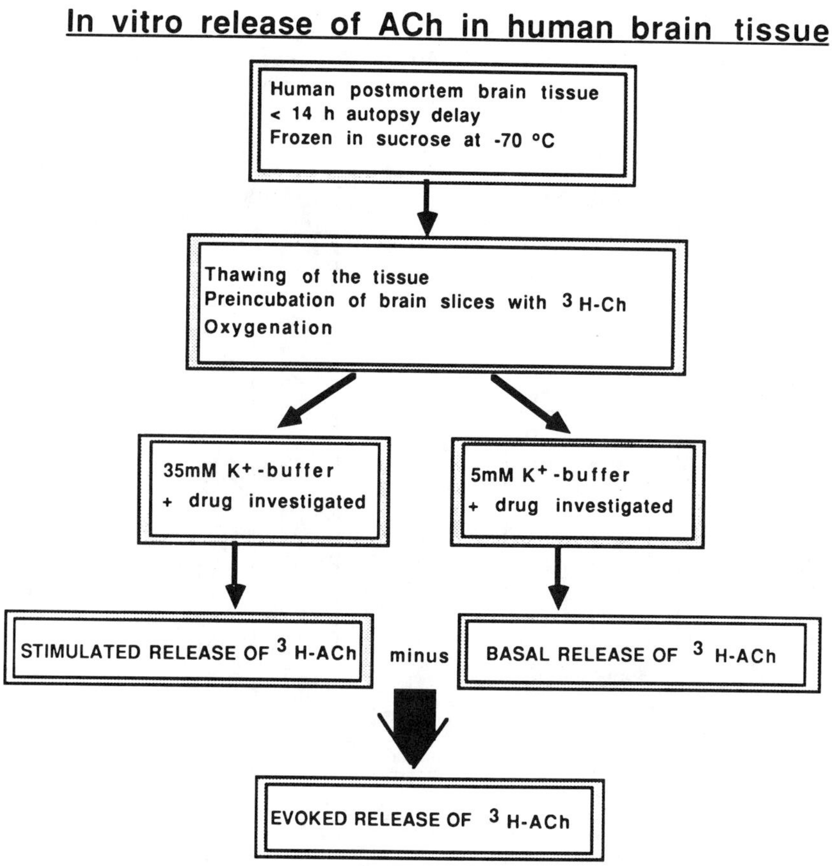

In vitro release of ACh in human brain tissue

```
┌─────────────────────────────────┐
│ Human postmortem brain tissue   │
│ < 14 h autopsy delay            │
│ Frozen in sucrose at -70 °C     │
└─────────────────────────────────┘
                │
                ▼
┌─────────────────────────────────────────┐
│ Thawing of the tissue                   │
│ Preincubation of brain slices with ³H-Ch│
│ Oxygenation                             │
└─────────────────────────────────────────┘
         │                │
         ▼                ▼
┌──────────────────┐  ┌──────────────────┐
│ 35mM K⁺-buffer   │  │ 5mM K⁺-buffer    │
│ + drug investigated│ │ + drug investigated│
└──────────────────┘  └──────────────────┘
         │                │
         ▼                ▼
┌───────────────────────┐  ┌───────────────────────┐
│STIMULATED RELEASE OF  │  │ BASAL RELEASE OF      │
│    ³H-ACh             │minus│    ³H-ACh          │
└───────────────────────┘  └───────────────────────┘
                │
                ▼
     ┌─────────────────────────┐
     │EVOKED RELEASE OF ³H-ACh │
     └─────────────────────────┘
```

Figure 2 Schematic presentation of an <u>in vitro</u> ACh release model using human postmortem brain tissue.

In the presence of the AChE inhibitor physostigmine or THA, the release of ACh from AD/SDAT brain slices is restored to control level (Figure 3). In cortical tissue from agematched controls a reduced release of ACh is observed in the presence of physostigmine or THA. Similar to the findings in human control brain tissue, animal experiments have also showed a reduced release of ACh in the presence of physostigmine (Szerb 1977, Marchi et al 1981). The reduction in ACh release seen in the presence of AChE inhibitors, can be explained by a negative feedback mechanism via presynaptic muscarinic autoreceptors. It is evident from our studies that the effect of AChE inhibitors on the <u>in vitro</u> ACh release in cortical control and AD/SDAT brain tissue is different.

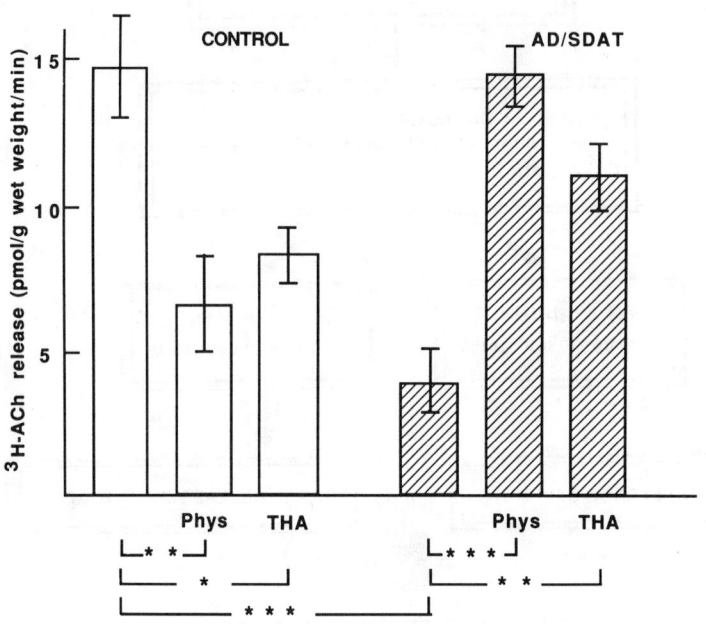

Figure 3 Effect of THA 10^{-4} and physostigmine 10^{-4}M on <u>in vitro</u> ^{3}H-ACh release from control and AD/SDAT brain slices (pmol/g wet weight/min). The values are the mean ± standard error of mean of duplicate or triplicate assays from 4-5 individuals.
* $p < 0.05$ ** $p < 0.01$ *** $p < 0.001$

EFFECT OF THA ON BRAIN CHOLINERGIC RECEPTORS

Recent observations indicate that the AChE inhibitors may not only act via inhibiton of the enzyme AChE but also

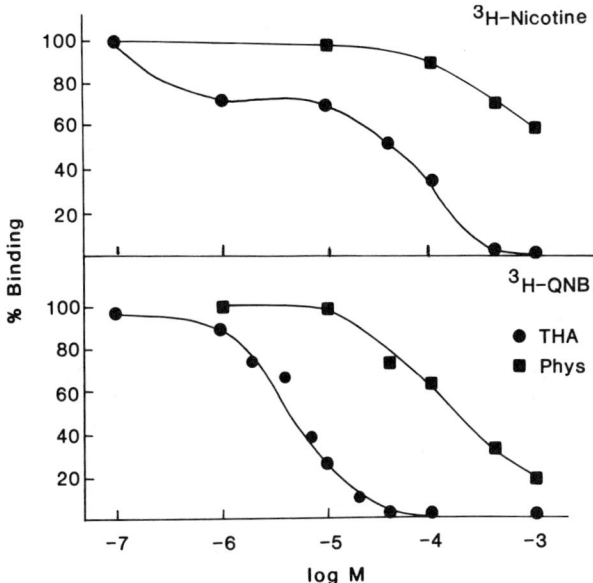

Figure 4 Displacement of ^{3}H-nicotine and ^{3}H-QNB receptor binding by THA (10^{-7}M-10^{-3}M) and physostigmine (10^{-7}M-10^{-3}M) to human control cortical tissue.

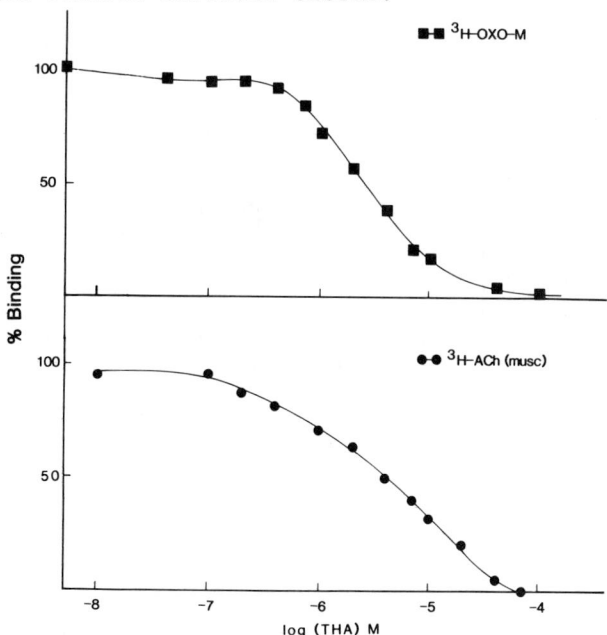

Figure 5 Displacement of muscarinic agonist (^{3}H-oxotremorine -M; ^{3}H-ACh) binding by THA (10^{-8}M-10^{-3}M) to human control cortical tissue.

via a direct effect on the cholinergic receptors. As shown in Figure 4, increasing concentrations of unlabelled THA compete with the muscarinic antagonist ^{3}H-QNB for the the binding to muscarinic binding sites in human control cortex. Similarly, THA also competes with the binding of ^{3}H-nicotine to nicotinic binding sites in human cortex. Physostigmine shows a lower binding affinity to both muscarinic and nicotinic cholinergic binding sites (Figure 4). The competition curves for physostigmine, are shifted to the right compared to the curves for THA. Moreover, THA also competes for the binding sites of different muscarinic agonists to human cortical tissue (Figure 5). Figure 5 illustrates the competition curves for THA and the muscarinic agonists ^{3}H-oxotremorine-M and ^{3}H-ACh (in the presence of cytisine). The results indicate that THA might interact with the muscarinic M_2 receptor subtype in human control cortex.

PUTATIVE MECHANISM OF ACTION OF THA

To further analyse the underlying mechanisms induced by THA in cortical control and AD/SDAT tissue, the inhibitory effect of different nicotinic and muscarinic receptor antag-

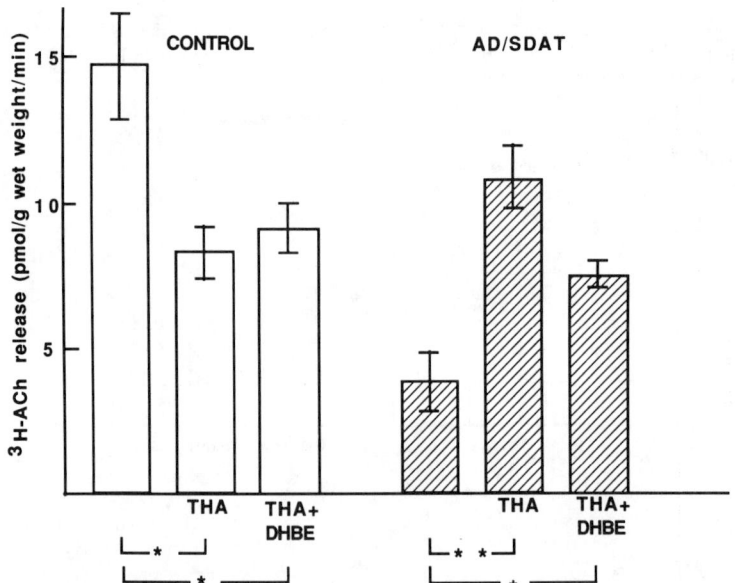

Figure 6 Effect of THA 10^{-4}M and THA 10^{-4}M plus dihydro-β-ery-troidine (DHBE) 10^{-6}M on in vitro release ^{3}H-ACh release from control and AD/SDAT brain slices (pmol/g wet weight/min). The values are the mean $\pm$ standard error of mean of duplicate or triplicate assay from 3-5 individuals.
* $p < 0.05$ ** $p < 0.01$

onists, on the THA induced ACh release has been investigated. In AD/SDAT cortical tissue the increased release of ACh by THA is partially blocked in the presence of the nicotinic receptor antagonist dihydro-β-erytroidine (DHBE) (Figure 6). A similar antagonistic effect was also observed for the nicotinic receptor antagonist tubocurarine concerning the facilitating effect of physostigmine on ACh release in AD/SDAT brain tissue (Nilsson et al 1987). The observation indicates, that some of the effects induced by THA in Alzheimer tissue may be mediated via nicotinic receptors. A reduced number of high affinity nicotinic receptors has been reported in Alzheimer cortical tissue (Whitehouse et al 1986, Nordberg and Winblad 1986), and the reduction in number of high affinity binding sites, can be explained by a shift in the proportion of high affinity nicotinic binding sites to low affinity nicotinic binding sites (Nordberg et al 1988). The shift in proportion of subpopulations of nicotinic receptors leads to a loss in receptor affinity. Nicotinic receptors that normally are easily desensitized may therefore respond to receptorstimulation. An induced release of ACh via nicotinic receptors (presynaptically located?) has been shown for nicotinic agonists, both in

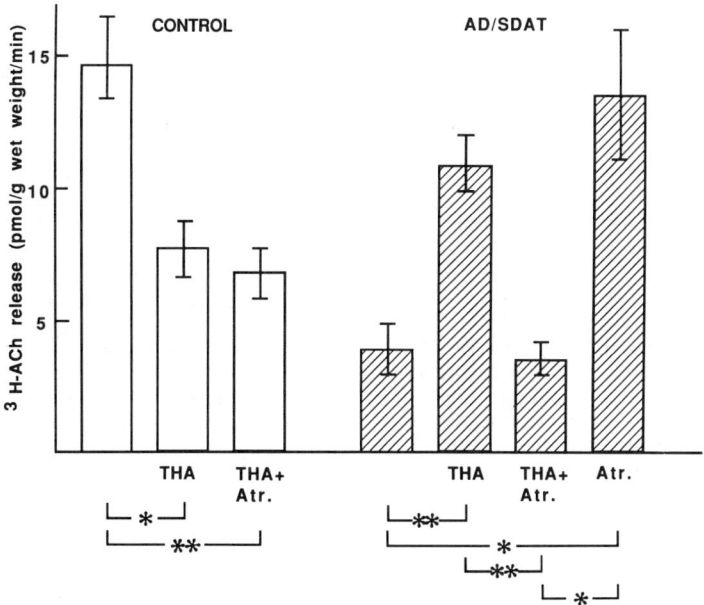

Figure 7 Effect of THA 10^{-4}M and THA 10^{-4}M plus atropine 10^{-5}M on in vitro release of ^{3}H-ACh from control and AD/SDAT brain slices (pmol/g wet weight/min). The values are the mean $\pm$ standard error of mean of duplicate or triplicate assay from 3-5 individuals.
* $p < 0.05$ ** $p < 0.01$

brain (Beani et al 1985) and peripheral nerve tissue (Wessler et al 1986). It is possible that THA in AD/SDAT tissue directly and/or indirectly via ACh might stimulate the nicotinic receptors to release ACh. In control cortical tissue, the nicotinic antagonists does not prevent the decreased release of ACh induced by THA, since the decreased ACh release in control tissue is induced via muscarinic autoreceptors.

The muscarinic antagonist atropine is known to increase the release of ACh both in rodent (Hadházy and Szerb 1977) and human brain tissue (Nordström et al 1982, Nilsson et al 1986), due to a blockade of the muscarinic negative feedback mechanism. The enhanced ACh release elicited by atropine in AD/SDAT tissue, is an indication of preserved muscarinic autoreceptors in AD/SDAT (Figure 7). When the AChE inhibitor THA plus atropine was added to the release medium the ACh release response was different in AD/SDAT and control cortical tissue (Figure 7). Atropine, in control human tissue, did not reverse the inhibitory effect of THA on ACh release. Atropine has earlier in rodent brain tissue been reported to restore the reduced release of ACh induced by physostigmine (Szerb and Somogyi 1973, Marchi et al 1981). The observation might indicate a difference in effect between THA and physostigmine. In AD/SDAT tissue on the other hand, atropine fully antagonized the increased release of ACh, induced by THA (Figure 7). The fact that THA and atropine separately can induce an increased release of ACh in AD/SDAT tissue, while combined, no effect on ACh release is observed, indicates different but interacting mechanism of action of THA and atropine. The muscarinic presynaptic autoreceptor controlling ACh release has been suggested to be identical with the muscarinic M_2 subtype (Raiteri et al 1984). Our receptor competition data support the assumption that THA might interact with the M_2 site in brain. Consequently, muscarinic receptors seem to be involved in THA:s mechanisms of action in AD/SDAT.

In conclusion, by using an <u>in vitro</u> ACh release model, it was found that physostigmine and THA can restore the release of ACh to control level in AD/SDAT brain tissue. The effect of THA seem to be multiple, since it not only acts as an AChE inhibitor but interacts with subtypes of muscarinic and nicotinic receptors in brain tissue.

Acknowledgements

This study was financially supported by grants from Swedish Medical Research Council, Swedish Tobacco Company, Stiftelsen för Gamla Tjänarinnor, Nobel Medica.

References

Beani, L., Bianchi, C., Nilsson, L., Nordberg, A., Romanelli, L., Sivilotti, L. 1985. The effect of nicotine and cytisine on ^{3}H-acetylcholine release from cortical slices of guinea pig brain. Naunyn Schmiedebergs Arch Pharmacol 331: 293-296

Caulfield, M.P., Straughan, D.W., Cross A.J., Crow T., Birdsall, N.J.M. 1982. Cortical muscarinic receptor subtypes and Alzheimer's disease. Lancet II: 1277

Davies, P., Maloney, A.J.F., 1976. Selective loss of central cholinergic neurons in Alzheimer's disease. Lancet II:1403

Gottfries, C.G. 1985. Alzheimer's disease and senile dementia: biochemical characteristics and aspects of treatment. Phychopharmacology 86: 245-252

Hadházy, P., Szerb, J.C. 1977. The effect of cholinergic drugs on ^{3}H-acetylcholine release from slices of rat hippocampus, striatum and cortex. Brain Res 123: 311-322

Hardy, J.A., Adolfsson, R., Alafuzoff, I., Bucht, S., Marcusson, J., Nyberg, P., Perdahl, E., Wester, P., Winblad, B. 1985. Transmitter deficits in Alzheimers disease. Neurochem Int 7: 545-563

Hollander, E., Mohs, R.C., Davies, K.L. 1986. Cholinergic approaches to the treatment of Alzheimer's disease. Br Med Bull 42: 97-100

Marchi, M., Paudice, P., Raiteri, M., 1981. Autoregulation of acetylcholine release in isolated hippocampal nerve endings. Eur J Pharmacol 73: 75-79

Mash, D.C., Flynn, D.D., Potter L.T. 1985. Loss of M2 muscarinic receptors in the cerebral cortex in Alzheimer's disease and experimental cholinergic denervation. Science 228: 1115-1117

Mohs, R.C., Davies, B.M., Mathé, A.A., Rosén, W.G., Johns, C.A., Greenwald B.S., Horvath T.B., Davies, K.L. 1985. Intravenous and oral physostigmine in Alzheimer's disease. Interdiscipl Top Gerontol 20: 140-152

Nilsson, L., Nordberg, A., Hardy, J., Wester, P., Winblad, B. 1986. Physostigmine restores ^{3}H-acetylcholine efflux from Alzheimer brain slices to normal level. J Neural Transm 67: 275-285

Nilsson, L., Adem, A., Hardy, J., Winblad, B., Nordberg, A. 1987. Do tetrahydroaminoacridine (THA) and physostigmine restore acetylcholine release in Alzheimer brains via nicotinic receptors? J Neural Transm 70: 357-368

Nordberg, A., Winblad, B. 1986a. Brain nicotinic and muscarinic receptors in normal aging and dementia. In: Fisher A., Hanin I., Lachman C (eds) Alzheimer´s and Parkinson´s disease: Strategies for research and development (Advances in behavioral biology, vol 29) Plenum Press, New York, pp 95-108

Nordberg, A., Adem, A., Hardy, J., Winblad, B. 1988. Changes in nicotinic receptor subtypes in temporal cortex of Alzheimer brains. Neurosci Lett in press.

Nordberg, A., Winblad, B. 1986b. Reduced number of ^{3}H-nicotine and ^{3}H-acetylcholine binding sites in the frontal cortex of Alzheimer brains. Neurosci Lett 72: 115-119

Nordberg, A., Alafuzoff, I., Winblad, B. 1986. Muscarinic receptor subtypes in hippocampus in Alzheimer´s disease and mixed dementia type. Neurosci Lett 70: 160-164

Nordström, Ö., Westlind, A., Undén, A., Meyersson, B., Sachs, C., Bartfai, T. 1982. Pre- and postsynaptic muscarinic receptors in surgical samples from human cerebral cortex. Brain Res 234: 287-297

Perry, E.K., Perry, R.H., Gibson, R.H., Blessed, G., Tomlinson, B.E., 1977. A cholinergic connection between normal aging and senile dementia in the human hippocampus. Neurosci. Lett. 6: 85-89

Perry, E.K. 1986. The cholinergic hypothesis: ten years on. Br Med Bull 42: 63-69

Raiteri, M., Leardi, R., Marchi, M. 1984. Heterogeneity of presynaptic muscarinic receptors regulating neurotransmitter release in rat brain. J. Pharm Exp Ther. 228: 209-214

Rylett, R.J., Bull, M.J., Colhoun, E.H. 1983. Evidence for high affinity choline transport in synaptosomers prepared from hippocampus and neocortex of patients with Alzheimer´s dicease. Brain Res 289: 169-175

Sims, N.R., Smith, C.C.T., Davison, A.N., Bowen, D.M., Flack, R.H.A., Snowden, J.S. 1980. Glucose metabolism and acetylcholine synthesis in relation to neuronal activity in Alzheimer´s disease. Lancet i: 333-336

Summers, W.K., Majovski, L.V., Marsh, G.M., Tachiki, K., Kling, A. 1986. Oral tetrahydroaminoacridine in long-term treatment of senil dementia, Alzheimer type. New Engl J Med 315: 1241-1245

Szerb, J.C., Somogyi, G.T. 1973. Depression of acetylcholine release from cortical slices by cholinesterase inhibition and by oxotremorine. Nature (New Biol) 241: 121-122

Wessler, I., Halank, M., Rasbach, J., Kilbinger, H. 1986. Presynaptic nicotine receptors mediating a positive feedback on transmitter release from the rat phrenic nerve. Naunyn Schmiedebergs Arch Pharmacol 334: 365-372

White, P., Goodhardt, M.J., Keet, J., Hiley, C.R., Carrasco, L.H., Williams, I.E.I., 1977. Neocortical cholinergic neurons in elderly people. Lancet I; 668-670

Whitehouse, P.J., Martino, A.M., Antuono, P.G., Lowenstein, P.R., Coyle, J.T., Price, D.L., Kellar, K.J. 1986. Nicotinic acetylcholine binding sites in Alzheimer´s disease. Brain Res 371: 146-151

TACRINE, BACK TO THE FUTURE?

W.K. Summers[1], L.V. Majovski[2], G.M. Marsh[3], K. Tachiki[4], and A. Kling[4]
[1] Solo Research, Arcadia, CA; [2] Huntington Medical Research Institute, Pasadena, CA; [3] Dept. of Biostatistics, Univ. of Pittsburgh, Pittsburgh, PA; [4] Sepulveda VA Hospital, Sepulveda, CA

The story of THA is curious and filled with serendipity. In 1907, Alois Alzheimer, M.D. described a progressive dementia in a 51-year-old woman. This illness, thought to be rare, came to bear his name. Today it is known that this illness is quite common and is the fifth largest cause of death in the United States.

In 1910, Benda first synthesized the diaminoacridine class of drugs in Germany. The intent was to develop an antihelmenic to treat intestinal worms. The chemical class had antiseptic effect and was used intravenously. By 1930, the diaminoacridines were found to have hepatic toxicity. Use of diaminoacridines was discontinued.

During the Second World War, Adrian Albert, Ph.D. and a group of Australian chemists synthesized monoamineacridine compounds, including Tacrine (Albert 1945). Dr. Albert's intent was to find a safe intravenous antibacterial agent that could be used as part of the war effort. This classified research was published after the Second World War. Toxicology studies of THA revealed no antibacterial effect. However, THA did have marked capacity to arouse the central nervous system. Coma induced by barbiturates, opiates and uremia were reversed by THA. It was used as a curiosity drug in Great Britain during the 1950's to prolong suxamethonium induced muscle relaxation in a surgical setting. Proposed mechanism of Tacrine's action was by means of inhibition of acetylcholinesterase.

The drug was used in the United States in the 1960's at the Missouri Institute of Psychiatry. Dr. Samuel Gershon and Dr. Turin Itil were working with phencyclidine-like agents (Itil 1966). The intent of this work was to discover the cause of schizophrenia. If these agents caused delirium, the delirium was reversed by THA. The schizophrenia puzzle was not solved by this work. However, THA was found to be substantially more effective and safer than physostigmine.

My relationship with tacrine began in 1975 when I was a senior psychiatric resident at Washington University (St. Louis). While on a consultation liaison rotation, I was presented with a case of "black patch delirium," otherwise known as "post cataract psychosis" (Summers 1979). After reviewing the case, I reasoned that this particular patient was suffering from an anticholinergic psychosis induced by atropinic eye drops. An intravenous infusion of physostigmine resulted in rapid lysis of the delirium. Regardless, the results were not fully pleasing. The patient, with intact cognitive function, threw up on me. Further, she had a sweaty, ashen appearance that was guaranteed to promote fear in house officers.

I searched the literature for a safer, effective drug and came up with the propsect of THA. I contacted Dr. Gershon. He was very encouraging and helpful. Without Dr. Gershon's help, my first IND in 1976 might not have been a possibility.

My first trial with THA was in overdose coma. Tacrine was often dramatic in its reversal of coma (Summers 1980). A second application was an attempt to reverse street PCP. This data, which was never published, failed to show improvement in 13 patients. My third application of THA was in 12 Alzheimer's patients where an intravenous injection of THA gave encouraging results.

In 1982, I was fortunate in meeting the late Dr. Arthur Cherkin. He had read my earlier papers on the use of THA, and introduced me to Dr. Arthur Kling. My association with UCLA commenced. Dr. Kling has supported this work through the gentle guiding of a senior scientist. He also initiated the primate and mice studies of THA. In 1983, Dr. Ken Tachiki joined the team and a sensitive serum assay for THA was developed. By the spring of 1984, the animal studies indicated that THA given orally was active and safe.

In June, 1984, the very first patient, SD-1, a 66-year-old female, was started on oral THA. The improvement was excelent. Six months later she continued to appear healthy and the improvement was sustained. Other patients were added cautiously. By December, 1985, Dr. Kling felt a paper was merited. This paper was published on November 13, 1986, in the New England Journal of Medicine.

The methodology used in this report is the same as that mentioned in the New England Journal of Medicine paper (Summers et al. 1986). Dosage used in these subjects was determined clinically. Namely, the presence of minor anticholinergic excess resulted in lowering the dose. Toxicity was noticed usually within 12-48 hours of starting the highest dose. In our patients, maintenance versus maximum dose is given in Table 1.

TABLE 1

THA Dose	Highest Dose Exposed To		Maintenance Dose	
200 mg/day	55.6%	(25)	15.5%	(7)
150 mg/day	35.6%	(16)	42.2%	(19)
125 mg/day	4.4%	(2)	6.7%	(3)
100 mg/day	4.4%	(2)	28.9%	(13)
<75 mg/day	-	(0)	6.7%	(3)
TOTALS	100%	(45)	100%	(45)

Legend: Number of subjects given in parentheses.

The acute adverse effects of THA are well-known. They are, as with physostigmine and other cholinergic enhancing drugs, in order of ascending importance: nausea, diaphoresis, excessive salivation, vomiting, diarrhea, hypermicturation, bradycardia, seizures and respiratory arrest.

The chronic adverse effects of THA are unknown. The purpose of this paper is to review our experience in patients on tacrine for more than three months with regard to liver toxicity.

Seventeen of 48 subjects in our study **FAILED TO COMPLETE THREE MONTHS** of THA treatment. Eight of 17 were diagnosed as non-Alzheimer's dementias. Five of these 8 were exposed to tacrine without benefit. Three Alzheimer's patients were not benefitted by tacrine. Three of 17 subjects were noncompliant to the protocol. Two of 17 left the study per family request. One was removed for medical reasons.

Thirty-one subjects of the 48 patients involved in the study were on tacrine for **MORE than three months.** Eleven have left the study for a variety of reasons. Four of the 11 were removed for noncompliance. Two of the 11 were dropped per family request. Two of 11 expired and one had a post-operative cardiac arrest. Only 2 of 11 were removed for elevated liver enzymes. Twenty subjects were on daily THA as of February 15, 1988.

Put another way, the number of subjects exposed to tacrine in optimal dosage per increment of time was:

          Less than 3 months   = 14 subjects
          3 to 11 months   = 12 subjects
          12 to 23 months   = 9 subjects
          24 to 35 months   = 9 subjects
          Greater than 36 months = 1 subject.

In this population there were scattered minor chronic adverse effects such as "sinusitis," increased salivatin and occasional loose stools. Full analysis of these complaints has not been completed.

Major chronic anticipated side effects would be:

1. Hepatoxicity
2. Bone marrow suppression
3. Carcinogenicity.

To date, no cancers have been noted in these patients. Full analysis of serial complete blood counts has not been conducted; however, general review of the data has not shown any significant drop in red blood cell counts or white blood cell counts in any specific patient, resulting in holding administration of THA.

The first case of confirmed THA-induced elevations of liver function studies was reported to the U.S. Food and Drug Administration by our group in January, 1988. The second case was almost concomitantly. Liver biopsies were obtained in both cases. Autopsy material from the liver of the third case exposed to THA for six months was obtained.

CASE STUDIES

Case SD-21:

History: 75-year-old white married housewife with 5-year history of progressive dementia unresponsive to Hydergine. The patient has a concomitant history of clinical depression which responded to Elavil. The patient had substantial deficits in psychometric testing, including a verbal IQ of 84; performance IQ equal to 73, full-scale IQ equal to 79, Mini-Mental Status Examination 14/30; Orientation Test 3/12; Names Learning Test 1/60; Global Alzheimer's Scale Stage 4/6; laboratory evaluation demonstrated cortical atrophy by cranial magnetic resonance scan which was consistent with Alzheimer's. 2-D echocardiogram was within normal limits; 24-hour Holter monitor showed 56 premature contractions per 24 hours. Carotid duplex scan was unremarkable. Chest x-ray was within normal limits. Chemistry profile was within normal limits with a glucose of 105 mg/100 cc, SGOT of 35 U/L, SGPT at 36 U/L, alkaline phosphatase of 115 U/L, blood sugar taken two hours after eating was 135 mg/100 cc. Urinalysis was negative. FTA was negative. A CBC was within normal limits. Serum B-12, folic acid, thyroid profile, TSH and protime were all within normal limits.

Clinical Follow-up: The patient was placed on THA 100 mg per day as her maximum dose. Beyond this, she had unacceptable nausea. She had increased salivation on 100 mg per day and was somewhat more comfortable at 75 mg a day with minimal drop in her psychometric test results. In the fifth month of tacrine therapy, elevation of liver enzymes was noted. The SGOT was 102 U/L and the SGPT was 144 U/L. No change in alkaline phosphatase or lactic dehydrogenase was noted. Clinically, the patient appeared quite robust and healthy. THA was discontinued and the enzymes returned to normal within six weeks. She was re-initiated on 50 mg of THA a day, resulting in elevation of SGOT to 66 U/L and SGPT of 70 U/L. Liver spleen sonogram showed no abnormal findings. An abdominal magnetic resonance imaging showed no liver abnormalities. A liver biopsy was performed in January of 1987.

Histology of the biopsy was benign with regard to hepatocellular disease. Mild steatosis, manifest by diffuse distribution of empty round vacuoles in the cytoplasm of the liver cell, was noted. This is a change seen in diabetics and pre-diabetics. Focal areas of minimal mononuclear cells infiltrate were noted in scattered centrolobular areas. There were no appearent eosinophils.

Case SD-27:

History: 59-year-old white married right-handed male with a 2-year history of rapidly progressive memory loss. Psychometric tests showed substantial deficits. Mini-Mental Status was 18/30 points. Verbal IQ was 82 and performance IQ was 78. This represented an approximately 30-point drop from the examination two years previously. Names Learning Test was 36/60. His Orientation Test was 9/12. His Global Stage was Stage 2/6. His Alzheimer's Deficit Scale score was 72/100. Laboratory evaluation: Regional cerebral blood flow was within normal limits and an I-131 brain scan was consistent with Alzheimer's disease. Cerebral magnetic resonance scan was within normal limits without evidence of infarction. His electrocardiogram showed a sinus bradycardia of 53 beats/minute, but a 24-hour Holter monitor was unremarkable. His EEG was abnormal with frontal slowing in the Delta range. A BEAM was, likewise, abnormal with slowing in the Theta and Delta ranges in the frontal and mid vertex regions. Bilateral carotid duplex scans were within normal limits. CBC, protime, thyroid profile, serum B-12, folate and FTA were all within expected limits. Chemistry scan showed an SCOT of 15 U/L (0-40), SGPT was 17 U/L (0-45), LDH was 108 (60-200), alkaline phosphatase equaled 76 U/L (30-115), total bilirubin equaled 1.0 (0.2-1.2).

CLINICAL FOLLOW-UP:

On the 10th week of THA treatment, the patient was noted to have an elevation of SGOT to 63 U/L with an SGPT of 149 U/L. THA was discontinued. His liver enzymes fell to the normal range within 2 months. He was re-challenged with THA. On the 15th day his SGOT elevated to 55 U/L with an SGPT of 126 U/L. His alkaline phosphatase was 40 U/L and the total bilirubin was 0.9 mg %. Liver spleen sonogram showed no abnormal findings. An abdominal magnetic resonance imaging showed no liver abnormalities. A liver biopsy was done.

Hepatic histology was consistent with benign drug-induced changes with preserved lobular architectural acute focal inflammation in periportal areas. This inflammation was predominantly mononuclear cells with occasionally eosinophiles.

Case SD-20:

History: This is a 94-year-old white separated male with known insulin-dependent diabetes mellitus, diminished visual acuity and presbycusis. The patient had a precipitous deterioration of mental function within the past two years, unexplained by these conditions. Full evaluation was consistent with Alzheimer's disease. The patient did show substantial improvement on THA in a double-blind study. Abnormal liver function tests were **not** noted in this case. In the 6th month of THA treatment, the patient sustained a vertebrobasilar cerebrovascular accident. THA was discontinued. Two weeks later the patient, who had been living at home, expired in a nursing home. An autopsy was performed and liver tissue was obtained.

Histology was similar to Case SD-27, in that there were discrete areas of periportal mononuclear inflammation with rare eosinophiles. Additionally, there was a mild degree of diffuse vacuolization seen in diabetes. This was actually less pronounced than SD-21.

In review of **our total experience**, 8 of 45 subjects exposed to THA had elevated liver function tests. Six of 8 occurred by the 10th week of exposure. Six of 8 subjects were on other drugs known to be hepatotoxic at the time of the elevated liver function tests. In three cases THA was discontinued and the patient was re-challenged with THA. In 5 cases reduction of THA or removal of concomitant potentially offending drug resulted in an acceptable lowering of liver function abnormalities.

Figure 1 displays the mean SGOT values $\pm$ 1.96 standard error of the mean in the 31 subjects on this protocol for more than 90 days. The horizontal line at SGOT = 80 U/L represents twice the upper limit of normal. This is the "Panic Line." Superimposed on this graph are the Peak Values of the eight subjects with abnormal hepatic function tests. These peak values are located at the month of onset of serum enzyme abnormality. Both cases with onset after the first two months were related to use of an additional drug with known hepatic side effects. The meaning of the different symbols is as follows: (*) denotes the highest SGOT value in subjects on the protocol three or more months. (□) indicates highest SGOT value in subjects not on the study for three or more months. (*) denotes peak SGPT in two subjects whose SGOT was either not significantly elevated or did not reach the panic line.

Review of the literature (Ockner 1982) shows striking similarities between THA and an old friend to neuropsychiatry -- chlorpromazine (thorazine).

Classic chlorpromazine-induced jaundice is a serious medical complication occurring in 0.5 to 1.0% of exposed cases. Histologically there is a mononuclear periportal inflammation with local eosinophilia and centrolobular cholestasis. Lobular hepatocellular necrosis with formation of acidophilic bodies occurs. Treatment is discontinuation of chlorpromazine, monitoring and supportive care. Prognosis is excellent. Death is rare. Seventy-five percent recover within 3 months. The remainder recover within one year. Classically, this was felt to be an idiosyncratic and immune phenomenon.

Recently, this opinion has been altered. It is now recognized that **transient and subclinical chlorpromazine-induced hepatic dysfunction** occurs in 50% of chronic users. The changes histologically are occasional clusters of periportal mononuclear cells with variable local eosinophilia. This is the same reaction noted for THA. The thorazine reaction is considered a benign drug reaction, as it is transient and often asymptomatic. They may go unnoticed. Chlorpromazine-induced changes and liver function tests tend to occur within 3 to 5 weeks after treatment has begun. THA-related changes in liver functions in our small sample tend to occur between 3 and 8 weeks. It is now known that chlorpromazine forms 1:1 complexes with bile salts, reversibly inhibits bile secretory function, inhibits ATPase reactions. This is a dose-related phenomenon. Thus, reduction of chlorpromazine dosage improves the clinical picture. In our clinical work it was noted that reduction in THA dose resulted in reduction of abnormal liver function abnormalities.

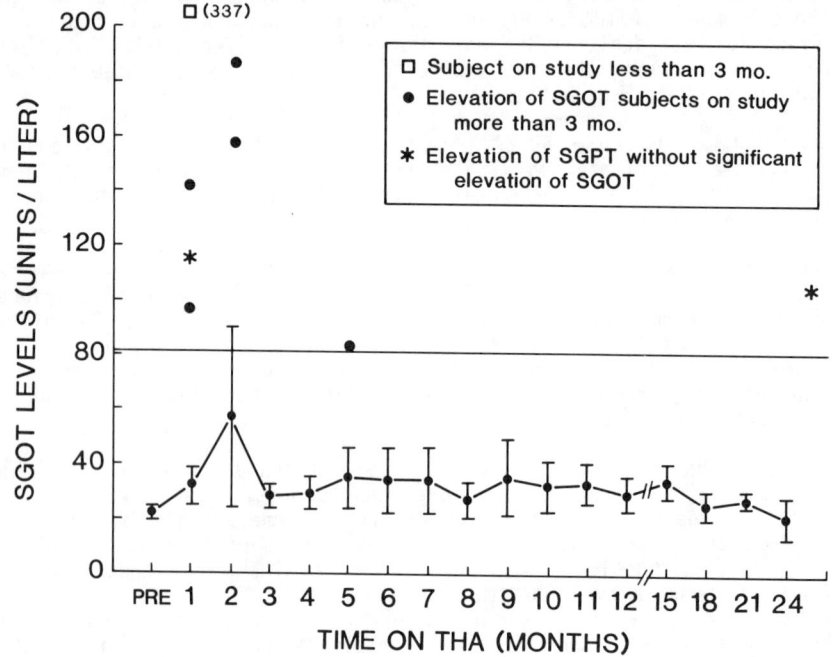

In conclusion, our current data would indicate that THA with lecithin results in evidence of hepatic dysfunction in 8 or 45 subjects (17%). The reaction histologically is similar to the classic benign chlorpromazine hepatocellular dysfunction. Clinically it occurs early in the course of treatment (2-12 weeks). The two exceptions to this were noted when other drugs with known hepatotoxic properties were added to the drug regimen. Once present, the elevations of serum hepatic enzymes subside by withdrawal of THA, withdrawal of a concomitantly administered hepatotoxic drug (e.g. Thoridiazide), or both. The THA reaction may be dose dependent; and we speculate, may be a transient benign phenomenon in some patients.

It is clear that **tacrine will "go back to the future."** THA may be discarded as the diamino-acradines were in the 1930's. They were found more toxic than other promising agents. Alternatively, tacrine may be like thorazine where the apparent benefit far outweighs the risk. In this case it is hoped it may become rapidly available to patients with Alzheimer's disease.

BIBLIOGRAPHY

Albert, A. and W. Glendhill. 1945. Improved synthesis of amino acridines. Part IV substituted 5-aminoacridines. J. Soc. Chemical Industry 44:169-172.

Itil, T. and M. Fink. 1966. Anticholinergic drug-induced delirium: experimental modification, quantitative EEG and behavioral correlation. J. Nerv. Ment. Dis. 143:492-507.

Ockner, R.K. 1982. In Hepatology, eds. D. Zakim and T.D. Boyer, 691-722. Philadelphia; W. B. Saunders Co.

Summers, W.K. and T.C. Reich. 1979. Delirium after cataract surgery: Review and two caes. Am. J. Psych. 136:386-391.

Summers, w.K., M.R. Kaufman, F. Altman, J.M. Fisher. 1980. THA - A review of the literature and its use in treatment of five overdose patients. Clin. Toxicol. 16:269-281.

Summers, W.K., L.V. Majovski, G.M. Marsh, K. Tachiki, A. Kling. 1986. Oral tetrahydroamino acridine in long-term treatment of senile dementia, Alzheimer's type. NEJM 315:1241-1245.

THE HISTORY OF THA

Joe E. Thornton and Samuel Gershon
*Western Psychiatric Institute and Clinics, Dept. of Psychiatry,
Univ. of Pittsburgh School of Medicine, Pittsburgh, PA*

THA (1,2,3,4, tetrahydro - 5 aminoacridine) has been a promising drug for almost 40 years (Figure 1). Studies on the basic pharmacology of THA were done in the 1950's at Australian universities. However, this work has not been extended appreciably and thus, despite the long history of THA, more work is needed to understand its full pharmacologic profile. In this historical review, we will suggest that many of the clinical actions of THA may not be mediated solely through its potent central anticholinesterase activity.

Figure 1. <u>THA [1,2,3,4, tetrahydro-5-aminoacridine]</u>

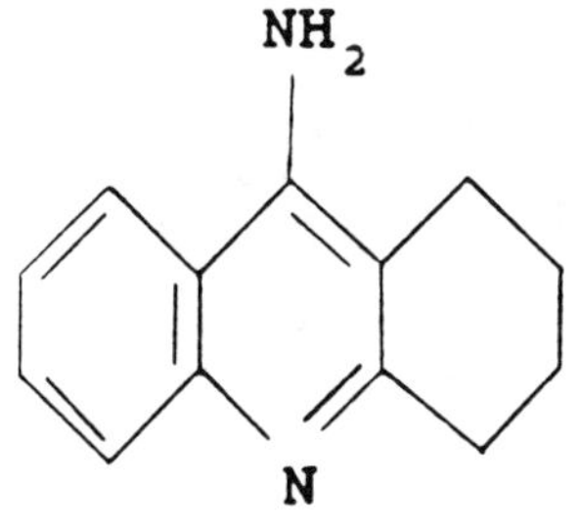

Rediscovery of THA in Screening for Antibiotics

Rubbo, Albert, and Maxwell (1942) set out to study the relationship between structure and action of the acridine antiseptics. They found that increasing basicity of the compounds increased their antiseptic activity and 5-amino-acridine was the most potent. Mono-amino-acridine caused necrosis of the skin and the animals died in about 24 hours after injection. Mice injected with 2- and 3-amino-acridine died about 18 hours without any marked symptoms. These authors related the efficacy of the test compounds to their basicity and stated that this facilitates the union between the drug and the acid group of the bacterial protein. This hypothesis seemed to be consistent with the test drugs' affinity for gram positive organisms. The alternative hypothesis was that the activity is dependent upon the molecular structure of the compound such that "resonant amino structures" are more potent than structures with hydrogen bonds, giving greater weight to the reactive condition to the nitrogen atoms. Also, another property as shown by investigators Albert, Goldacre, and Rubbo (1941) was that if the di-amino-acridines and mono-amino-acridines are ranked in order, it is dependent not only on basicity but on the ability to pass from oil to water. Albert, Rubbo, Goldacre, Davey, and Stone (1945) attempted to discriminate between the hypothesis of high basicity, "extra ionic resonance" or combination of both. These authors examined 30 examples of 130 acridines and synthesized 77 new compounds. They studied not only amino derivatives of the acridines but chloral derivatives, nitro-substituted derivatives of 5-amino-acridine, methyl derivatives, hydroxy derivatives and miscellaneous derivatives. What these authors found was that the question was difficult to answer, since the compound's antiseptic activity still was very dependent on high basicity. But those compounds with high basicity also had high ionic resonance. The authors also discussed the "dimensional factor." For example, 5-amino-acridine is flat due to its resonant structure. 5-amino-6,7,8,9 tetrahydroacridine (THA) retains the basicity of its parent compound but does not have resonant bonding, has a "bent ring" structure, and most interesting, has no antibacterial activity. The authors, thus, systematically showed the structure function relationship of the test antibiotics.

Use of THA as a Morphine Antagonist

Shaw and Bentley (1949) were searching for a drug that could reverse many of the untoward effects of morphine. The first series of reports were based on the acridine series. In 1952, they extended the series to drugs with known anticholinesterase activity (Shaw & Shulman, 1955). These authors were screening a wide variety of compounds for the ability to antagonize morphine. They went through the aminoacridine series among others. Through their screening, they found that all the analeptics they had tested that were active had either a muscarine-like action or were anticholinesterase drugs. Testing THA and physostigmine in morphinized dogs, they noted respiratory stimulation immediately after injection of these drugs before other signs of arousal. This finding was most striking with THA. THA and physostigmine then began to cause various degrees of excitement and even possibly a prolonged period of extreme agitation. They also suggested that these analeptic activities were specific towards morphine and hyoscine. In their experiments, THA and physostigmine did not arouse rats, cats, and guinea pigs treated with pentobarbitone. However, this may be due to the timing of administration of the drugs. They also noted that the drugs did not antagonize the analgesic effects of morphine. Based on their animal experiments, they suggested that clinical trials in patients were warranted. They hypothesized the most potent to be THA, then aminacrine, and then either physostigmine or a mixture of physostigmine and aminacrine.

In a different paper by Shaw and Bentley in 1953, they showed that there was no connection between cholinesterase (CHe) activity, anti-CHe activity, and morphine antagonism. Also, the morphine antagonism was not due solely to the basicity or lipophilicity of THA. They concluded that the action of the drugs in potentiating acetylcholine may not reside entirely upon their anti-CHe activity but may be due to a postulated receptor activity. (Remember that in the 1950's and early 1960's the cell receptor theory had not been fully developed or accepted).

Christie, Gershon, Gray, Shaw, McCance, and Bruce (1958) found that, with three drugs (nalorphine, cyclizine and amiphenazole), they could control the certain side effects of morphine, such as depressed level of consciousness, depressed respiration, nausea and vomiting. They note that in an earlier paper by Gershon and Shaw (1958a) that a large element of the pharmacology of morphine is related to its histamine like actions. Thus, they hypothesized that antihistamines should abolish morphine narcosis. They found that nalorphine is a drug of choice to relieve respiratory depression, cyclizine was useful in eliminating nausea and vomiting, and amiphenazole was useful in counteracting sedation. The patients could be treated with doses of morphine as high as 600 mg a day for the control of chronic pain and this could be done safely. Note in perspective that most volunteers who take 30 mg of morphine have severe side effects (the most frequent being nausea and vomiting; also, dizziness and ataxia were noted initially, followed by drowsiness; then occasionally, itch, tremor, spasm of urinary sphincter; then weakness or shakiness). Stone, Moon, and Shaw (1961) used the single drug THA along with morphine to treat patients with intractable pain. These authors note that the most commonly used drug was amiphenazole, which was given to antagonize some of the side effects of morphine. However, amiphenazole was unreliable as a respiratory stimulant and was unstable in solution. Here, they refer to the historical work by Shaw and Bentley (1952), who discovered a group of substances which restored consciousness to dogs narcoticized with morphine. They were particularly interested in drugs which would wake the animals up but not interfere with analgesic effects. These drugs were called partial antagonists, in distinction to complete antagonists such as levallorphan. They also found that THA was effective as a partial antagonist to morphine. The authors describe a case report of treatment using large doses of morphine along with THA for intractable pain. They conclude that there are two partial antagonists for morphine and that THA is preferable, and that it is a more reliable respiratory stimulant, does not interfere with sleep, and is chemically stable. In some instances, it actually increases morphine analgesia. THA also blocks euphoria and seems to block the addiction potential of morphine. Particularly interesting is that it was possible to reduce morphine without withdrawal symptoms. These authors reported that they had treated 54 patients sucessfully with combined morphine and THA for intractable pain.

Many years later, Albin, Orr, Bunegin, and Henderson (1975) wrote that, with all the advantages reported for THA, particularly on intractable pain with morphine, they could not understand why more work had not been done. They had a research unit in anesthesiology at Pittsburgh, and studied THA for its effects on ketamine. So they began pilot studies to repeat the experience in this country and in animals. They were successful in attenuating physical dependence to morphine and decreased abstinence syndromes in Rhesus monkeys. However, the use of THA to study morphine addiction never progressed beyond that work.

Table 1. Antagonist Activity of THA

Barbiturate Coma (mice & rats)
Uremic Coma (nephrectomized cats)
Magnesium Chloride (Induced Coma dogs)
Prevents ventricular fibrillation by Digitalis Electroshock
Reverses tubocurarine   equipotent with
Potentiates suxemethonium   neostigmine
Protects against morphine withdrawal ppted by Nalorphine in
    addicted monkeys

THA as an Anti-Psychotomimetic

In 1960, one of the authors (S.G.) of this paper reported on the blocking effect of tetrahydroaminacrin on JB-329 (Gershon 1960). JB-329 (Ditran) had significant anticholinergic activities and THA from the compounds that had been tested, both for the bactericidal activities and in the reversal of morphine side effects, had been shown to be the most effective anticholinesterase inhibitor. Ditran 0.05mg/kg I.V. induced psychotic symptoms such as confusion, disorientation, and visual, tactile, and auditory hallucinations. THA was given I.V. about 30 minutes after induction of these symptoms to return the patients to clarity. The author observed that the underlying diagnosis influenced the nature of the psychosis induced by Ditran, but no matter what the clinical picture was, THA reversed the effects. The author noted even then, that although both Ditran and THA work on cholinergic systems, this may not be their sole mechanism of action. This work was extended to the effects of several psychotomimetcs (Gershon & Olariu, 1960). They tested the psychological and physiological effects of LSD, mescaline, sernyl, and Ditran. They found that succinic acid antagonized mescaline and LSD and exhibited some antagonism to the central and peripheral effects of sernyl but not Ditran. Anticholinesterases were ineffective against LSD, mescaline, sernyl, and the anticholinesterase drugs physostigmine, neostigmine and DFP were ineffective against the central effects of Ditran. They noted that physostigmine dosage was limited due to side effects such as nausea, vomiting and weakness. THA was effective in reversing the central effects of Ditran. This was postulated at the time in the belief that THA was more effective since it is more lipophilic centrally and had fewer peripheral side effects. Sixty mg of THA in 20 ml of water given I.V. over 5 minutes showed a rapid improvement within 2 minutes in Ditranised subjects. The antidotal effects of THA were rapid in onset, but after about 4 hours the psychotic symptoms would reappear in some cases.

Animal studies were conducted to further examine the effects of THA reversal of this class of psychotomimetic (Bell & Gershon, 1964). Ditran was given to conscious and unrestrained dogs to produce a model of disturbances in which to test for antagonists. The authors studied yohimbine because it had cholimimetic and anti-adrenergic properties and they also compared its ability to antagonize atropine. The authors note that just because Ditran is anticholinergic and that THA blocks acetylcholinesterase, does not mean that these are the mechanisms by which the drugs produce their physiologic actions. They note that other centrally active cholinesterase inhibitors have not been as effective in blocking the effects of Ditran nor have other anticholinergic drugs been as psychotomimetic as Ditran, i.e. atropine. The authors proposed that THA also acts at receptors to block the Ditran binding site. They presented evidence that THA may disrupt the binding of Ditran to a nitrogenous receptor site. They noted that yohimbine could not bind at that same site and that clinically the antagonistic effects of yohimbine were different, and had a different time course than that of THA. They compared the relative binding of atropine, Ditran, and 3-quinuclidol benzilate. They compared its relative affinities for nitrogenous binding sites and showed that THA was most effective against Ditran and atropine and less effective against the 3-quinuclidol benzilate. Yohimbine was similarly effective for all three. Thus, the Ditran antagonist activities of THA and yohimbine could not be explained solely by anticholinesterase activities nor by binding to similar receptor sites.

The clinical work on the reversal of the effects of Ditran was extended by developing the Ditran Rating Scale (DRS) (Gershon & Olariu, 1960). In retrospect, the DRS could be useful in monitoring patients with dementia and behavioral disturbances, or in patients with delirium. The DRS has items from the neurologic exam, such as speech, tremors; autonomic findings, perceptual distortions, disturbances of consciousness, attention, concentration, orientation, anxiety, mood lability, picking at invisible objects, and disturbances in drawing (see Appendix).

Extending the antipsychotomimetic studies, Neubauer, Sundland, and Gershon (1966a) used the DRS to monitor the effects of Ditran in patients. Fifteen to 20 minutes after Ditran 0.05mg/kg I.V., the patients would become confused or stuporous, along with increased anxiety and panic, hyperreflexia, dry mouth, some increase in pulse and blood pressure, often perceptual disturbances, occasionally mood lability, hallucinations, and almost always impaired ability to accurately complete the Bender-Gestalt figures. Peak effects were usually 20 minutes after administration, and would last for 1 to 2 hours if not antagonized, and take up to 8 to 24 hours to completely recover without an antidote. Most patients reported partial to complete amnesia for the peak effects. THA was effective in counteracting most of these effects, being least effective on depression and being most effective on reversing neurological impairment and impairments on the Bender-Gestalt figures. Yohimbine showed some effect, but not nearly as much as THA and also it was effective on different symptoms. Yohimbine had minimal effect on Bender-Gestalt figures, but it was almost 100 percent effective in reversing any hallucinations that occurred. Neostigmine was effective in reversing most of the autonomic effects of Ditran but only weakly effective in reversing depression of consciousness and impairment of Bender-Gestalt, and even then was associated with serious side effects. Thus, Ditran exerted central anticholinergic effects and THA was an effective antidote for these effects.

THA as an Anti-Delirium Agent.

Itil and Fink (1966) studied the effects of experimental drug induced delirium and its EEG changes. Ditran induced the reduction of alpha activity and caused an increase in delta and theta waves, also occasional epileptic spikes, and these changes correlated with the observed clinical symptoms. These symptoms were impaired visuospatial tasks and of interest "phylogenetically older neurological reflexes appeared." The clinical and EEG effects of Ditran were reversed by THA. The patient's confusional state cleared and there was inhibition of psychomotor activity. EEG patterns were more organized and alpha waves recurred. Yohimbine showed a slight decrease in the delirious state, especially in improving levels of consciousness. The patients were more alert, and speech was more relevant, but the patients were more agitated. They continued to have perceptual distortions, EEG patterns remained disorganized and alpha activity did not recur. Thorazine and Promazine clinically quieted the patient, eliminating agitated, panicked behavior, but the EEG changes were consistent with a coma or stupor. Physostigmine improved consciousness, but did not significantly alter the psychotic state, however the dose may have been too low. There was some improvement in the EEG, but peripheral side effects, such as muscle fasciculations and cardiac irregularities limited systematic studies with physostigmine. The effects of atropine and ditran on the EEG did depend somewhat on the existing or pre-drug state (Itil 1966). Atropine-like drugs in low doses may induce acceleration in the EEG if it is predominantly slow to begin with, and this happened in patients who had head injury or who had prior ECT.

Ketchum, Sidell, Crowell, Aghajanian, and Hayes (1973) studied three centrally active anticholinergic compounds (atropine, scopolamine and Ditran) in 158 normal young males in a dose range study. Their findings indicated that there were not qualitative differences in the drugs, just differences in potency, relative central affinity, and time course of effects. The toxicity of these belladonna related drugs was reversed by certain anticholinesterase drugs such as physostigmine, THA, and to a lesser degree DFP and Sarin, but not neostigmine (which does not cross blood brain barrier), methylphenidate or the phenothiazines. What is most significant here is that both THA and DFP are centrally active, both preferentially inhibit pseudocholinesterase but THA was much more effective than DFP as an antagonist to belladonnoid intoxication. Also of note is that physostigmine, which is also centrally active, and an inhibitor of true cholinesterase, is effective. Of some interest is a phenomenon of relative refractoriness, i.e., when atropine is given, if physostigmine follows 15 to 30 minutes later, it is not as effective as when it follows 2 hours after atropine administration. The therapeutic effects of physostigmine were striking but short-lived, i.e., about 2 hours. THA was effective for over 4 hours. These authors deferred on further speculation on the significance of these findings. They also commented in their study that although phenothiazines would quieten patients, the phenothiazines also potentiated the impairment of cognitive functioning in the experimental anticholinergic delirium.

Albin, Bunegin, Massopust, and Jannetta (1974) used THA to reduce the delirium induced by ketamine, which is very similar to PCP. Ketamine was used as a parenteral anesthetic agent but was limited by the side effects, i.e., psychosis and delirium when people woke up. These investigators were reviewing the literature for some type of antagonist to ketamine and discovered that THA reverses the psychotic symptoms of the anticholinergic psychotomimetics, and that it was also useful as a mild neuromuscular blocking agent, stimulated respiration, and was a partial antagonist to morphine. They speculate that the THA antagonism of ketamine probably involves both adrenergic and cholinergic receptor sites in brain. They also speculate that THA may be useful in the treatment of acute intoxication with drugs of abuse, such as PCP.

Maayani, Weinstein, Ben-Zvi, Cohen, and Sokolovsky (1974) studied sernyl and ten of its derivatives for their activities on acetylcholinesterase and butyrylcholinesterase. They also studied drug interactions with antagonists such as Sarin and THA. They correlated the acetylcholinesterase activity and CNS activity as well as the structural relation of the drugs to agonist and antagonist effects on the cholinergic system. They conclude that the pronounced effects of sernyl are central rather than peripheral, and that these effects are efficiently antagonized by THA and the reversal of the hyperactivity in mice or the reversal of anesthesia in guinea pigs is mediated by acetylcholine. They were still unable to identify or distinguish the roles between acetylcholinesterase and butyrylcholinesterase. They also conclude that the psychotomimetic activities of sernyl and its derivatives were due to its anticholinergic effects caused by direct interaction with the cell receptor.

Summers, Kaufman, Altman, Fischer (1980) reported on the use of THA in the treatment of 5 tricyclic overdose patients. In the introduction, they comment that Slovis, Ott, Teitelbaum, and Lipscomb (1972) popularized the use of physostigmine for this in 1972. But there were some problems with the use of physostigmine, mainly bradycardia, seizures, short duration of action, and so, for those reasons, they chose THA because it was relatively effective, had longer duration of action, had a high therapeutic index, and had minimal cardiac or peripheral cholinergic effects. They report on the treatment of their five cases and then review the pharmacology of THA. In this paper Summers commented that THA may be useful for the amelioration of some of the symptoms of dementia.

Hannington-Kiff (1987) reported using THA for the central anticholinergic syndrome caused by anesthestic agents. He observed that when THA was given to prolong the muscle relaxing effects of suxamethonium, that the patients were more alert and less confused after anesthesia than patients who had not received THA. This was particularly true for elderly patients.

THA as an Adjunct in Anesthesia

Gershon (1958b) first demonstrated THA's effect as a decurarizing agent. Hunter (1975) listed three factors which should make THA useful as adjunctive medication in anesthesia: 1) prolongs muscle relaxing action of suxamethonium; 2) as a decurarizing agent; and 3) as an analeptic to stimulate respiration without antagonizing the analgesic activity of morphine. In Great Britain, THA had some popularity among anesthesiologists for extending the muscle paralysis (phase I depolarization) of suxemethonium. Physostigmine is also used for this purpose but many practitioners felt that THA was better tolerated with fewer problems with bradycardia and hypotension. However, even in Britain, THA is now difficult to obtain; its use has waned.

The Mechanism of Action of THA

Table 2. Actions of THA

Inhibits acetylcholinesterase
Partial antagonist of morphine
Inhibitor of cyclic AMP PDE
Blocks potassium channels and inhibits sodium inactivation
Inhibits monoamine oxidase
Prolongs action of suxamethonium
?Stimulates cholinergic firing

Based on this historical overview we conclude that THA is a broad spectrum analeptic and has effects clinically comparable to physostigmine but is better tolerated and longer acting. The mechanisms of action of THA are summarized in Table 2. The literature provides little information on the basic pharmacokinetics and pharmacodynamics of THA, safety in chronic usage, mechanisms of action especially in terms of receptor binding, and effects on neuronal activity. THA warrants further systematic investigation.

REFERENCES

Albert, A., Rubbo, S.D., Goldacre, R.J., Davey, M.E., & Stone, J.D. 1945. The influence of chemical constitution on antibacterial activity. Part II: A general survey of the acridine series. British J. Exp. Path. 26: 160-192.

Albin, M.S. 1978. Tetrahydroaminoacridine and phencyclidine intoxication. JAMA 240: 529.

Albin, M.S., Orr, M.D., Bunegin, L., & Henderson, P.A. 1975. Tetrahydroaminoacridine antagonism to narcotic addiction. Experimental Neurology 46: 644-648.

Albin, M.S., Bunegin, L., Massopust, L.C., Jr., & Jannetta, P.J. 1974. Ketamine-induced postanesthetic delirium attenuated by tetrahydroaminoacridine. Experimental Neurology 44: 126-129.

Bell, C. & Gershon, S. 1964. Experimental anticholinergic psychotomimetics: Antagonism of yohimbine and tacrine (THA). Med. Exp. 10: 15-21.

Brown, M.L., Gershon, S., Lang, W.J., & Korol, B. 1966. The effects of psychoactive drugs on the behavioral response to ditran in dogs. Arch. Int. Pharmacodyn. Ther. 160: 407-423.

Christie, G., Gershon, S., Gray, R., Shaw, F.H., McCance, I., & Bruce, D.W. 1958. Treatment of certain side-effects of morphine. British Medical Journal 1: 675-680.

Curley, W.H., Standaert, F.G., & Dretchen, K.L. 1984. Physostigmine inhibition of 3', 5'-cyclic AMP phosphodiesterase from cat sciatic nerve. J. of Pharmacology & Experimental Therapeutics 228: 656-661.

Gershon, S. 1960. Blocking effect of tetrahydroaminacrin on a new psychotomimetic agent. Nature 186: 1072-73.

Gershon, S. 1966. Behavioral effects of anticholinergic psychotomimetics and their antagonists in man and animals. Recent Advances in Biological Psychiatry 8: 141-149.

Gershon, S. & Olariu, J. 1960. JB 329 - A new psychotomimetic. Its antagonism by tetrahydroaminacrin and its comparison with LSD, mescaline and sernyl. J. of Neuropsychiatry 1: 283-292.

Gershon, S., Neubauer, H., & Sundland, D. 1966. Ditran and its antagonists in a mixed psychiatric population. J. Nerv. Ment. Dis. 142: 265-277.

Gershon, S. & Shaw, F.H. 1958a. Morphine and histamine release. J. Pharm. Lond. 10: 22-29.

Gershon, S. & Shaw, F.H. 1958b Tetrahydroaminacrin as a decurarising agent. J. Pharm. 10: 638.

Hannington-Kiff, J.G. 1987. Tetrahydroaminoacridine and the central anticholinergic syndrome. The Lancet i: 862.

Hunter, A.R. 1965. Tetrahydroaminacrine in anaesthesia. British J Anaesth 37: 505-513.

Itil, T., Fink, M., Neubauer, H., & Gershon, S. 1964. Drug induced dissolution of cortical electrical activity and its correlation to psychopathological phenomena. Paper presented at the Collegium Internationale Psychopharmakologicum Meeting, Birmingham, England, 1964.

Itil, T. & Fink, M. 1966. Anticholinergic drug-induced delirium: Experimental modification, quantitative EEG and behavioral correlations. J. Nerv. Ment. Dis. 143: 492-507.

Itil, T. 1966. Quantitative EEG changes induced by anticholinergic drugs and their behavioral correlates in man. Recent Advances in Biological Psychiatry 13: 151-173.

Ketchum, J.S., Sidell, F.R., Crowell, E.B., Jr., Aghajanian, G.K., & Hayes, A.H., Jr. 1973. Atropine, scopolamine, and ditran: Comparative pharmacology and antagonists in man. Psychopharmacologia (Berl.) 28: 121-145.

Maayani, S., Weinstein, H., Ben-Zvi, N., Cohen, S., & Sokolovsky, M. 1974. Psychotomimetics as anticholinergic agents -- I. Biochemical Pharmacology 23: 1263-81.

Neubauer, H., Gershon, S., & Sundland, D.M. 1966. Differential responses to an anticholinergic psychotomimetic (Ditran) in a mixed psychiatric population. Psychiatria et Neurologia (Basel) 151: 65-80.

Neubauer, H., Sundland, D., & Gershon, S. 1966a. Ditran and its antagonists in a mixed psychiatric population. J. of Nerv. Ment. Dis. 142: 265-277.

Neubauer, H., Sundland, D.M., & Gershon, S. 1966b. Sernyl, Ditran, and their antagonists: Succinate and THA. Intl. J. of Neuropsychiatry 2: 216-222.

Rowntree, D.W., Nevin, S., & Wilson, A. 1950 The effects of diisopropylfluorophosphonate in schizophrenia and manic depressive psychosis. J. of Neurology, Neurosurgery & Psychiatry 13: 47-59.

Rubbo, S.D., Albert, A., & Maxwell, M. 1942. The influence of chemical constitution on anti-septic activity. I: A study of the mono-amino-acridines. Brit. J. Exp. Path 23: 69-83.

Shaw, F.H. & Bentley, G. 1949. Some aspects of the pharmacology of morphine with special reference to its antagonism by 5-aminoacridine THA and other chemically related compounds. Med. J. Australia 2: 868.

Shaw, F.H. & Bentley, G. 1952. Morphine antagonism. Nature 169: 712-713.

Shaw, F.H., & Bentley, G.A. 1953. The pharmacology of some new anti-cholinesterases. Austral. J. Exp. Biol. 31: 573-576.

Shaw, F.H. & Shulman, A. 1955. Treatment of intractable pain with large doses of morphine and diamino-phenylthiazole. British Medical Journal June 4: 1367-69.

Slovis, T.L., Ott, J.E., Teitelbaum, D.T., & Lipscomb, W. 1972. Physostigmine therapy in acute tricyclic antidepressant poisoning. Clin. Toxicol. 4: 451.

Stone, V., Moon, W., & Shaw, F.H. 1961. Treatment of intractable pain with morphine and tetrahydroaminacrine. British Medical Journal February 18: 471-473.

Summers, W.K., Kaufman, K.R., Altman, F., Fischer, J.M. 1980. THA - A review of the literature and its use in treatment of five overdose patients. Clinical Toxicology 16: 269-281.

APPENDIX

Ditran Rating Scale for Acute Investigations

A priori cluster	Item
1. Neurological impairment (NEUR)	1. Knee-jerk reflex 2. Ankle clonus 3. Extensor plantar response 4. Finger-nose test 5. Tremors of hands 6. Impairment of fine movements 7. Speech impairment 8. Impaired visual accommodation
2. Autonomic changes (AUT)	9. Pulse rate 10. Blood pressure-systolic 11. Blood pressure-diastolic 12. Pupillary dilation 13. Impaired light reflex 14. Dryness of mouth 15. Nausea 16. Vomiting
3. Perceptual distortion (PERC)	17. Dizziness 18. Feels not normal 19. Feels as if in a dream 20. Body-image distortion
4. Consciousness (CONS)	21. Insensitivity to pricking 22. Drowsiness 23. Confusion-steady 24. Confusion in waves 25. Add 3 + 6 26. Three similarities 27. Attention span (6 figures) 28. Recite six months forward 29. Orientation-time (day, month, year) 30. Orientation-space (ward, hospital, city) 31. Orientation-person (own name, name of nurse and physician) 32. Orientation-time intervals

Figure 3 - cont'd

A priori cluster	Item
	33. General information (president, governor, capitol)
	34. Reaction time
	35. Understanding questions
	36. Impaired concentration
	37. Reduced cooperation
	38. Withdrawn
	39. Stupor
5. Anxiety and restlessness (ANX)	40. Anxiety
	41. Panic
	42. Restlessness
	43. Disturbed by hallucinations
6. Change of expression (EXP)	44. Staring into space
	45. Vacuous smile
	46. Flat affect
	47. Passive
7. Depression (DEPR)	48. Depression-sadness
	49. Cheerfulness
	50. Disinhibition
8. Hallucinations (HALU)	51. Visual images
	52. Picking at invisible objects (e.g., smoking imaginary cigarettes)
9. Bender-Gestalt (B-G)	53. Bender-Gestalt
Other Effects	54. Jerking, twitching, spontaneous clonus
	55. Irrelevant answers

Part V

New Approaches to Pharmacotherapy of Alzheimer's Disease

METRIFONATE: A Review

Ingrid Nordgren and Bo Holmstedt
*Dept. of Toxicology, Karolinska Institute, Box 60400,
S-104 01, Stockholm, Sweden*

The organophosphorous compound 2,2,2-trichloro-1-hydroxyethyl dimethyl phosphonate was introduced as an insecticide, trichlorfon, in 1952 (Lorenz et al., 1955) and as a drug, metrifonate, in the treatment of schistosomiasis in 1960 (Lebrun and Cerf, 1960). Metrifonate, by now, has been given to millions of patients in the tropics. This organophosphorous compound is unique in that it is not an active cholinesterase inhibitor per se, but is transformed non-enzymatically into the active component dichlorvos, 2,2-dichlorovinyl dimethyl phosphate (DDVP).

$$(CH_3O)_2\overset{O}{\underset{}{P}}-\underset{OH}{CH}-CCl_3 \longrightarrow (CH_3O)_2\overset{O}{\underset{}{P}}OCH=CCl_2$$

Metrifonate   Dichlorvos

The mechanism for this transformation has been extensively studied (Kharash et al., 1955; Bengelsdorf, 1956; Sohr and Lohs, 1965; and Dedek et al., 1969). The reaction may occur under both aqueous and anhydrous conditions. Positive chemical evidence exists for the conversion metrifonate - DDVP not only in solution but also in the living organism (Nordgren et al., 1978, 1980 and 1981).

Metrifonate has gained a reputation for being a safe drug in the treatment of infections with Schistosoma haematobium (Talaat, 1964a,b; Shoeb et al., 1965; Abdalla et al., 1965; Hanna et al., 1966; Forsyth and Rashid, 1967a,b; Katz et al., 1968; Davis and Bailey, 1969; Plestina et al., 1972; Reddy et al., 1975; and Jewsbury et al., 1977). It is given orally in the form of capsules or tablets. The optimum dose in clinical trials has been found to be 7.5 mg/kg, given once every 14 days or once monthly to a maximum of three doses if necessary. Other therapeutically used doses have been previously tabulated by Holmstedt et al. (1978) in a review article summarizing the toxicological and pharmacological information then available. The dose selected in the treatment of schistosomiasis is generally well tolerated. Millions of patients in the tropics have been administered metrifonate. Accurate observations may be difficult under field conditions but serious acute side-effects most likely would have been reported. Mild side-effects, however, are on record. Lebrun and Cerf (1960), with the dose subsequently adopted (7.5 mg/kg), reported nausea, colic and sweating in one out of 15 adults. Using higher doses Talaat et al. (1963) reported that the subjects complained of lassitude and colic lasting up to six hours. Vomiting occurred in a very small percentage of cases. Davis and Bailey (1969) and Plestina et al. (1972) reported side-effects at a dose of 10 mg/kg body weight but only five complaints of trivial nature in school children given 7.5 mg/kg body weight in a trial in Tanzania. In the treatment of onchocerciasis doses as high as 10 mg/kg consecutively during six days have been given. Fifteen cases subjected to this treatment experienced vertigo and fall in blood pressure during the last two days. 10 mg/kg body weight during three days, repeated several times at two weeks intervals, caused no complaints from the patients (Thylefors, personal communication).

In 1973, the World Health Organization summarized (WHO Report):

"Cholinergic symptoms such as nausea, vomiting, bronchospasm, abdominal discomfort, diarrhoea, and a feeling of weakness are rare at recommended dosage levels but were encountered in early trials, when doses now considered to be excessive, were given. Atropine can be used for symptomatic relief and oximes are available for enzyme reactivation. Symptoms occurring at recommended dose levels usually disappear spontaneously in 12 to 14 hours."

However, the doses used cause a fall in blood cholinesterase activities. There are indications that patients with liver damage may be more susceptible than normals (Cavagna et al., 1969).

The transformation of metrifonate into DDVP, as well as the effect on brain acetylcholine have been studied in experimental animals by Nordgren et al. (1978). When DDVP is injected i.p. to mice in a dose of 10 mg/kg body weight, a peak level in brain is reached within one minute. The level decreases very rapidly approaching three minutes. By contrast, Figure 1 demonstrates the slow formation of DDVP when metrifonate (125 mg/kg) is injected i.p. Depicted is also the level of metrifonate. DDVP and metrifonate were analyzed by gas chromatography – mass spectrometry using deuterated internal standards (Nordgren et al., 1978). The level of DDVP in brain roughly follows the concentration of metrifonate. There is, however, a delay in the peak concentration of DDVP, which occurs at about 12-13 min, in other words, much later than when DDVP is injected directly.

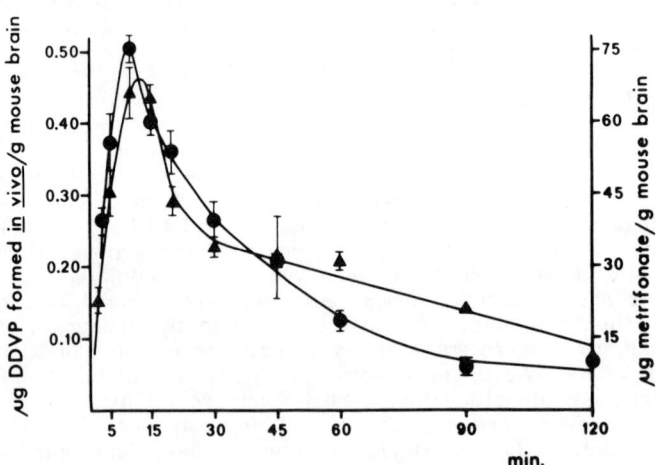

Figure 1. Concentration of metrifonate (dots) and DDVP (triangles) in mouse brain after i.p. injection of metrifonate 125 mg/kg. Vertical bars indicate mean and standard error of six to nine mice.

Figure 2, upper curve, shows the result of i.p. injection of 10 mg DDVP/kg body weight in mice on acetylcholinesterase activity and accumulation of acetylcholine (ACh) in whole brain versus time. The symptoms at this dose are salivation, diarrhoea and in some cases difficulties with respiration. They are most clearly recognized about 15 min after administration and disappear almost completely towards the end of 60 min. As seen in the figure, the enzyme activity and acetylcholine levels reach their minimum and maximum, respectively, at 15 min. Acetylcholine was analyzed according to Karlén et al. (1974) and the acetylcholinesterase activities were determined according to Ellman et al. (1961). Results with metrifonate are presented in Figure 2, lower curve. Metrifonate was injected i.p. in a dose of 125 mg/kg body weight. Symptoms such as salivation, diarrhoea and difficulties in breathing showed a maximum around 30 min with almost complete recovery towards the end of 120 min. A complete reversal to normal had not occurred towards the end of 180 min. As expected, the effect on the enzyme, and thereby on acetylcholine, is delayed and prolonged after metrifonate administration as compared to DDVP administration. The same is seen on the synthesis rate of brain acetylcholine versus time following an i.v. injection of deuterium labelled choline. Metrifonate needs a longer time after treatment than DDVP to cause a certain degree of decrease in acetylcholine turnover (Nordgren et al., 1978).

The formation of DDVP from metrifonate has also been demonstrated in man. Metrifonate and DDVP levels, as well as cholinesterase activity, have been studied in plasma and erythrocytes in patients treated with Bilarcil^R (metrifonate) against schistosomiasis (Nordgren et al., 1980 and 1981). The dosage of metrifonate varied between 7.5 - 10 mg/kg body weight in a single dose repeated after two weeks. Metrifonate and DDVP were analyzed according to Nordgren et al. (1978 and 1980). Cholinesterase activity was determined according to Augustinsson et al. (1978). Figure 3 and 4 show the results from one of the patients. Peak levels of both compounds occur within two hours, and the compounds are detectable during at least eight hours. The plasma cholinesterase is inhibited almost completely and recovers slowly after one day. The erythrocyte cholinesterase activity dropped to about 20 - 40 per cent of the pre-exposure value and recovered more slowly than the activity in plasma. The plasma and erythrocyte cholinesterase activity had not recovered completely at the second treatment. The activity varied between 47 - 78 per cent of the enzyme values before the first treatment for both types of esterases. Erythrocyte cholinesterase activity was found to reach a lower level after the second dose than after the first one and it also seemed to recover more slowly after the second dose. The activity in plasma was next to zero as was found already after the first dose.

Among the patients studied some milde side-effects (nausea and mild vertigo subsiding in a few hours) were noted. None of the patients felt any need to rest in bed from the side-effects.

The shape of the blood curves confirm the prediction that metrifonate would act as a slow release precursor for the active cholinesterase inhibitor DDVP. Theoretically, the comparably mild and prolonged effect of metrifonate compared to some other cholinesterase inhibitors may be of essential importance in the treatment of Alzheimers disease.

ACKNOWLEDGEMENTS

The work was supported by the Swedish Medical Research Council, Grant No. 14X-00199; by funds from the Karolinska Institute and by the U.S. Army Medical Research and Development Command.

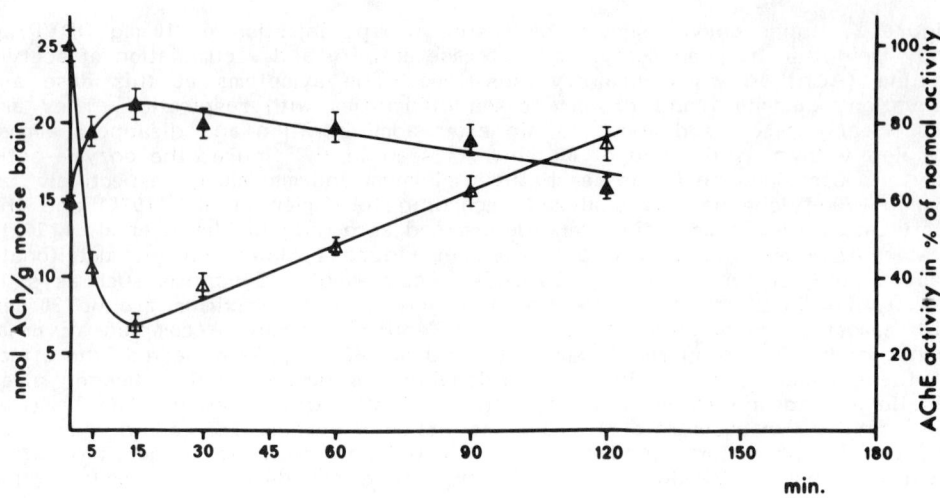

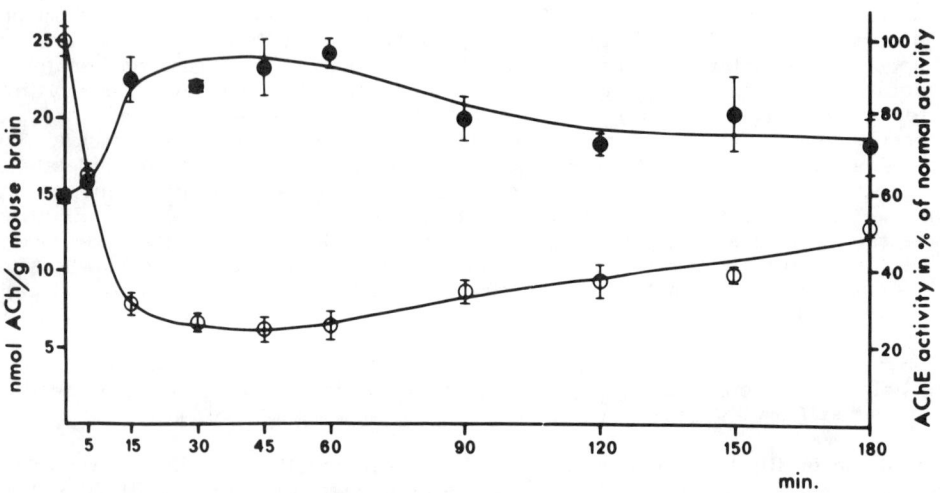

Figure 2. Upper curve: Accumulation of ACh (filled symbols) and inhibition of acetylcholinesterase (open symbols) in mouse brain after i.p. injection of DDVP 10 mg/kg. Lower curve: Accumulation of ACh (filled symbols) and inhibition of acetylcholinesterase (open symbols) after i.p. injection of metrifonate 125 mg/kg. Vertical bars indicate mean and standard error of six to eight mice.

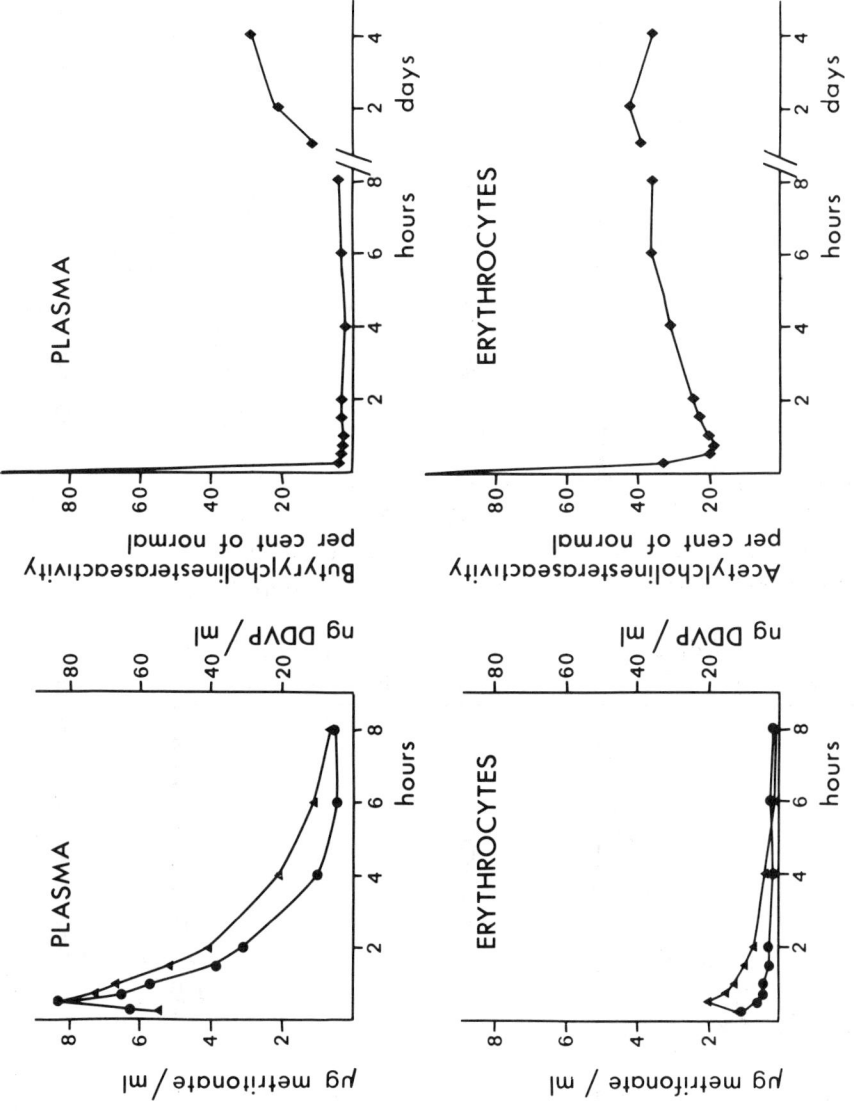

Figure 3. Cholinesterase activities and levels of metrifonate (triangles) and DDVP (dots) in plasma and erythrocytes after administration of metrifonate 10 mg/kg to a patient (I.A.). First treatment.

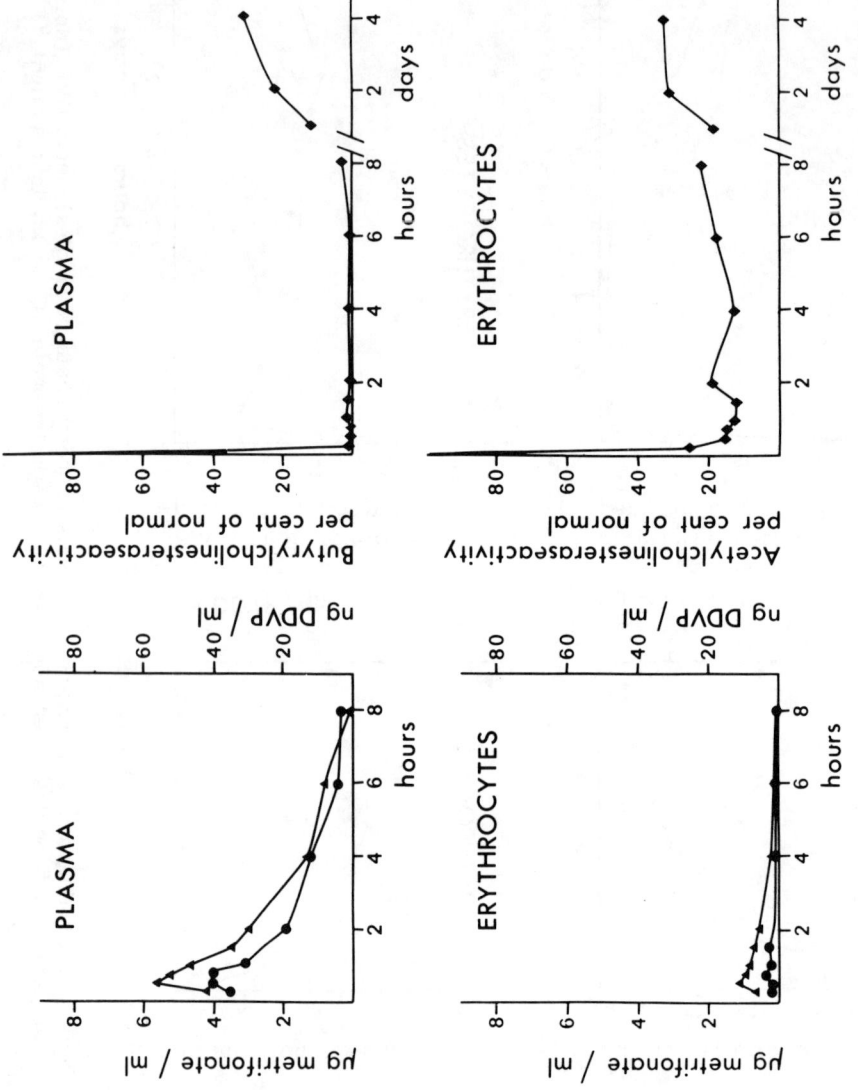

Figure 4. Cholinesterase activities and levels of metrifonate (triangles) and DDVP (dots) in plasma and erythrocytes after administration of metrifonate 7.5 mg/kg to a patient (I.A.). Second treatment.

REFERENCES

Abdalla, A., Saif, N., Taha, A., Ashmawy, H., Tawfik, J., Abdel-Fattah, F., Sabet, S. and Abdel-Meguid, M. 1965. Evaluation of an organo-phosphorus compound, Dipterex in the treatment of bilharziasis. J. Egypt. Med. Ass. 48:262-273.

Augustinsson, K.-B., Eriksson, H. and Faijersson, Y. 1978. A New approach to determining cholinesterase activities in samples of whole blood. Clin. Chim. Acta, 89:239-252.

Bengelsdorf, I.S. 1956. Dialkyl β-chlorovinyl phosphates. A proposed scheme to predict the product of the action of base with dialkyl α-hydroxyphosphonic esters. J. Org. Chem. 21:475-476.

Cavagna, G., Locati, G. and Vigliani, E.C. 1969. Clinical effects of exposure to Vapona insecticide in hospital wards. Arch. Environm. Hlth. 19:112-123.

Davis, A. and Bailey, D.R. 1969. Metrifonate in urinary schistosomiasis. Bull. Wld. Hlth. Org. 41:209-224.

Dedek, W., Koch, H., Uhlenhut, G. and Bröse, F. 1969. Aur Kenntnis der Umsetzung von ^{3}H-Trichlorphon zo DDVP. Z. Naturforsch. 24b:663-664.

Ellman, G.L., Courtney, K.D., Andres, V., Jr. and Featherstone, R.M. 1961. A new and rapid colorimetric determination of acetylcholinesterase activity. Biochem. Pharmacol. 7:88-95.

Forsyth, D.M. and Rashid, C. 1967a. Treatment of urinary schistosomiasis. Practice and theory. Lancet, 1:130-133.

Forsyth, D.M. and Rashid, C. 1967b. Treatment of urinary schistosomiasis with trichlorphone. Lancet, II:909-912.

Hanna, S., Basmy, K., Selim, O., Shoeb, S.M. and Awny, A.Y. 1966. Effects of administration of an organophosphorus compound as an antibilharzial agent, with special reference to plasma cholinesterase. Brit. Med. J. 1:1390-1392.

Holmstedt, B., Nordgren, I., Sandoz, M. and Sundvall, A. 1978. Metrifonate: Summary of toxicological and pharmacological information available. Arch. Toxicol. 41:3-29.

Jewsbury, J.M., Cooke, M.J. and Weber, M.C. 1977. Field trial of metrifonate in the treatment and prevention of schistosomiasis infection in man. Ann. Trop. Med. Parasit. 71(1):67-83.

Karlén, B., Lundgren, G., Nordgren, I. and Holmstedt, B. 1974. Ion-pair extraction and gas-phase analysis of acetylcholine and choline. In Choline and Acetylcholine: Handbook of chemical assay methods, ed. I. Hanin, pp. 163-175. New York: Raven Press.

Katz, N., Pellegrino, J. and Pereira, J.P. 1968. Experimental chemotherapy of schistosomiasis. III. Laboratory and clinical trials with trichlorphone, an organophosphorus compound. Rev. Soc. Bras. Med. Trop. II, 5:237-245.

Kharash, M.S. and Bengelsdorf, I.S. 1955. The reaction of triethyl phosphite with α-trichloromethyl carbonyl compounds. J. Org. Chem. 20:1356-1362.

Lebrun, A. and Cerf, C. 1960. Note préliminaire sur la toxicité pour l'homme d'un insecticide organophosphoré (Dipterex). Bull. Wld. Hlth. Org. 22:579-582.

Lorenz, W., Henglein, A. and Schrader, G. 1955. The new insecticide 0,0-dimethyl-2,2,2-trichloro-1-hydroxy ethylphosphonate. J. Amer. Chem. Soc. 77:2554-2556.

Nordgren, I., Bergström, M., Holmstedt, B. and Sandoz, M. 1978. Transformation and action of metrifonate. Arch. Toxicol. 41:31-41.

Nordgren, I., Holmstedt, B., Bengtsson, E. and Finkel, Y. 1980. Plasma levels of metrifonate and dichlorvos during treatment of schistosomiasis with Bilarcil^R. Am. J. Trop. Med. Hyg. 29(3):426-430.

Nordgren, I. Bengtsson, E., Holmstedt, B. and Pettersson, B.-M. 1981. Levels of metrifonate and dichlorvos in plasma and erythrocytes during treatment of schistosomiasis with Bilarcil^R. Acta pharmacol et toxicol. suppl. V, 79-86.

Plestina, R., Davis, A. and Bailey, D.R. 1972. Effect of metrifonate on blood cholinesterases in children during the treatment of schistosomiasis. Bull. Wld. Hlth. Org. 46:747-759.

Reddy, S., Oomen, J.M.V. and Bell, D.R. 1975. Metrifonate in urinary schistosomiasis. A field trial in Northern Nigeria. Ann. Trop. Med. Parasit. 69(1):73-76.

Shoeb, S.M., Basmy, K. and Haseeb, N.M. 1965. Studies on the use of an organophosphorus compound "Dipterex" in the treatment of urinary bilharziasis. J. Egypt. Med. Ass. 48:389-395.

Sohr, H. and Lohs, Kh. 1965. Zum Verhalten von Dipterex in vässriger Lösung. Mber. dtsch. Akad. Wiss. Berlin, 7:708-711.

Talaat, S.M., Amin, N. and El Massry, B. 1963. The treatment of bilharziasis and other intestinal parasites with Dipterex. A preliminary report on one hundred cases. J.Egypt. Med. Ass. 46:827-832.

Talaat, S.M. 1964a. Dipterex. An oral therapeutic agent in the treatment of schistosomiasis and other intestinal parasites. J. Egypt. Med. Ass. 47:589-593.

Talaat, S.M. 1964b. A further report on the treatment of schistosomiasis with Dipterex using 10 mg/kg body weight for six doses. J. Egypt. Med. Ass. 47:312-315.

Thylefors, B. Personal communication.

World Health Organization Report: Schistosomiasis control. 1973. Wld. Hlth. Org. Techn. Rep. Ser. No. 515.

STUDIES ON THE NOOTROPIC EFFECTS OF HUPERZINE A AND B:
Two Selective AChE Inhibitors

Xi Can Tang, Xiao Dong Zhu, and Wei Hua Lu
*Shanghai Institute of Materia Medica, Chinese Academy of Sciences,
Shanghai 200031*

The apparent losses in cholinergic function in the cortex and hippocampus in senile dementia of the Alzheimer type (SDAT) are of particular interest with regard to the defects of memory in this condition, since there is considerable evidence linking ACh and these areas to memory. The severe destruction of cholinergic cerebral system in patients with SDAT has led to the consideration of cholinomimetic treaments (Coyle et al. 1983). Administration of cholinomimetics such as physostigmine (Phys) can temporarily influence memory tasks in normal subjects (McGeer and McGeer 1984) and in Alzheimer dementia patients (Giacobini 1987). Cognitive improvement following Phys therapy is believed to be related to inhibition of the activity of AChE in the central nervous system (Whitehouse et al. 1981). However, Phys is not an ideal drug for clinical use because of its short half-life and very narrow therapeutic window (Drukarch et al. 1987).

Both huperzine A (Hup-A) and huperzine B (Hup-B) are new alkaloids isolated from the Chinese herb <u>Huperzia serrata</u> (Liu et al. 1986). It has been demonstrated in this laboratory that these two alkaloids exhibited inhibitory activity on ChE. The anti-ChE effects of these two alkaloids were evaluated <u>in vitro</u> using both acetyl- and pseudo-cholinesterases. Enzyme sources were erythrocyte membrane of rat, caudate nuclei of pig, rat serum and human serum. Acetylthiocholine iodide and butyrylcholine iodide were used as substrates for respective enzymes. Enzyme inhibition was measured according to the method of Ellman et al. (1964). As Table 1 shows, the pI_{50} (negative logarithm of molar concentration causing 50% inhibition of ChE) values indicate that Hup-A and Hup-B exerted more potent inhibitory effects on AchE than BuChE. The anti-AChE effect of Hup-A was about 3 times more potent than Phys, but less than Phys with BuChE.

Table 1. Comparison of the Anti-ChE Activities of 5 ChE Inhibitors

	pI_{50}			
	Erythrocyte Membrane	Caudate Nuclei	Serum	
			Rat	Human
Huperzine A	7.2	6.9	4.1	2.9
Huperzine B	6.1	6.2	3.9	2.8
Physostigmine	6.9	6.7	5.6	6.1
Galanthamine	5.7	5.6	4.9	4.7
Neostigmine	6.8	6.3	5.2	4.9

Hup-A and Hup-B are mixed and reversible inhibitors of AChE (Wang et al. 1986; Xu et al. 1987).

In experiments performed on unanesthetized rabbits with intact brains, Hup-A 0.05 mg/kg produced an alert EEG pattern which started within 6.6 ± 1.4 min after i.v. and lasted 11 ± 8 min. Atropine 2 mg/kg antagonized this effect, but methylbromide atropine did not. The same effect was also observed with Hup-B at the dose of 0.5 mg/kg, but its action duration was longer than Hup-A (Yan et al. 1987).

The blood level of Hup-A following i.v. or i.g. of ^{3}H-Hup-A in rats was found to decline in two phases, the distribution phase and the elimination phase, with half-lives of 6.7; 121.6 min (i.v.) and 9.8; 247.5 min (i.g.), respectively. In mice the radioactivities were detected in various tissues 15 min after i.v. ^{3}H-Hup-A. The majority (73.2%) of the radioactivity was excreted in the urine 24 hr, while only 2.4% was recovered from the feces. Paper chromatograms of urine revealed that ^{3}H-huperzine was excreted partially in prototype and its metabolite, respectively (Wang et al. 1988).

Dose-response curves for salivation indicated that both Hup-A and Hup-B were less potent than Phys. Tested in rats with a single intravenous injection, the therapeutic indexes of Hup-A and Hup-B are superior to that of Phys (Yan et al. 1987).

These observations prompted us to further investigate the actions of Hup-A and Hup-B on cognitive functions by using Y-maze and step-down passive avoidance methods. Phys was used as comparison.

FACILITATORY EFFECT ON LEARNING AND MEMORY
Brightness Discrimination Performance in Y-Maze

Rats were placed on an electrified grid in a Y-maze and learned to run into the light arm (safe). The criterion of learning or retrieval was met after they had chosen the light arm 10 trials in succession. Hup-A injected intraperitoneally with 100, 167 µg/kg 20 min before training caused a significant decrease in the number of trials to criterion. Hup-A (0.1 mg/kg i.p. or 0.3 mg/kg i.g.) given 20 min before hypercapnia prevented the CO_2-induced impairment of learning of brightness discrimination tasks in rats. As is shown in Table 2, rats were tested 48 hr later after training, Hup-A was administered 20 min before test, facilitation of retrieval was produced dose dependently at doses of 36-167 µg/kg i.p. under the same conditions, Phys at doses of 80-180 µg/kg i.p. improved learning and retrieval processes, but neostigmine did not. The improving effects of Hup-A on learning, memory retention and retrieval processes are superior to those of Phys. Scopolamine 0.2 mg/kg s.c., atropine 5 mg/kg s.c. or hemicholinium 20 µg/10 µl i.c.v. antagonized the positive effects of Hup-A 0.1 mg/kg on retrieval process, but methyl-atropine did not (Tang et al. 1986). The results suggested that the facilitation actions of Hup-A were due to an effect on the central cholinergic system, especially the muscarinic system.

Spatial Discrimination in Performance in Y-Maze

Mice were placed on an electrified grid in a Y-maze and learned to run to a safe area which was the counter arm defined according to the mice running directly into the arm during the first shock. The criterion of learning, memory retention and memory retrieval was met when the mice had chosen the safe area in 10 consecutive trials. Hup-A (75 µg/kg) and Hup-B (0.5 mg/kg) injected intraperitoneally before training caused significant decrease in the number of trials. Administration of Hup-A (75-125 µg/kg) or Hup-B (0.6-0.8 mg/kg) 10 min before hypercapnia significantly prevented the CO_2-induced impairment of acquisition. The facilitatory effects on acquisition were also seen after oral Hup-A dose of 100 µg/kg and Hup-B dose of 0.8 mg/kg. Hup-A (75-150 µg/kg i.p.) or Hup-B (0.4-1.2 mg/kg i.p.) improve memory retention and memory retrieval when administered immediately after training or 24 hr after training. Phys was also active in all the above-mentioned experiments and was less potent than Hup-A, but about 3-5 times more potent than Hup-B (Zhu et al. 1987).

Table 2. Effects of Anticholinesterase Agents on the Memory Retrieval of A Visual Discrimination in Rats ($\bar{x} \pm SD$)

Drug	Dose (µg/kg)	Rats	Trials to Criterion	
			Training	Test
Saline	-- i.p.	40	13 ± 4	14 ± 9
Huperzine A	36 i.p.	8	11 ± 4	7 ± 7**
	60 i.p.	8	12 ± 5	4 ± 4**
	100 i.p.	8	10 ± 4	2.5 ± 1.4**
	167 i.p.	8	10 ± 4	5 ± 5**
Physostigmine	120 i.p.	8	13 ± 4	13 ± 2
	150 i.p.	8	9 ± 4	6 ± 4**
	180 i.p.	8	11 ± 6	3 ± 3**
Neostigmine	30 i.p.	8	14 ± 4	17 ± 7
Saline	-- i.g.	7	13 ± 6	15 ± 9
Huperzine A	300 i.g.	9	12 ± 4	6 ± 4*

*$p < 0.05$, ** $p < 0.01$ as compared with the saline control

Step-Down Passive Avoidance

Each mouse was placed gently on the platform set in the center of the grid floor. As soon as the mouse moved off the platform with its paws on the grid floor, electric shocks were delivered continuously until the mouse returned to the platform for the purpose of escape from the electric shocks. The step-down latency (SDL) and escape latency (El) were measured. Mice were treated with drugs immediately after the training test; the test trials were performed 24 hr after the passive avoidance training. Figure 1 shows Hup-A, Hup-B and Phys improve retention in aged mice, the SDL were prolonged in a dose-related manner.

Hup-A at doses of 0.1-0.125 mg/kg i.p. or 0.14-0.4 mg/kg i.g. significantly reversed $NaNO_2$ (120 mg/kg s.c.), cycloheximide (150 mg/kg s.c.) or electroconvulsive shock-induced disruption of the memory retention of a passive avoidance task in mice (Zhu et al. 1988). Hup-B reversed the disruptive effects of $NaNO_2$, cycloheximide and electroconvulsive shock at the doses of 1.0-1.5 mg/kg (i.p.) and 2.4 mg/kg (i.g.).

Mice were given Hup-A 0.1 mg/kg i.p. and scopolamine 2 mg/kg s.c. immediately after training and tested 24 hr after medication. As is shown in Figure 2, the scopolamine-induced amnesia was reversed by Hup-A. Under the same condition, the effective doses of Hup-B and Phys were 1.25 mg/kg and 0.125 mg/kg i.p., respectively.

These results indicate that, in addition to their ability to counteract memory impaired by cholinergic blockade, Hup-A and Hup-B beneficially influence memory processes thought to be dependent on protein synthesis. Hup-A and Hup-B may prove clinically useful in treating the memory deficits of Alzheimer's disease and improving cognitive functions impaired by various noxious influence. Clinical studies showed that Hup-A did improve memory in patients of cerebral arteriosclerosis with memory impairment. Hup-A produced memory improvement 1-4 hr after injection and sustained for 8 hr or so. The therapeutic value of Hup-A 30 mg is even superior to hydergine 600 mg (Zhang 1986).

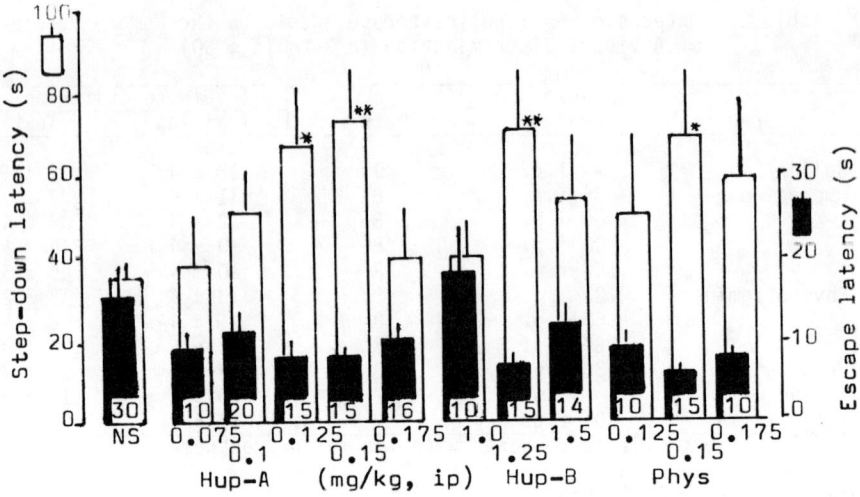

Figure 1. Facilitation of memory retention of a passive avoidance task in aged mice ($\bar{x} \pm$ SE).

*$p < 0.05$, ** $p < 0.01$ compared with NS (saline) group. Figures in bars refer to number of mice.

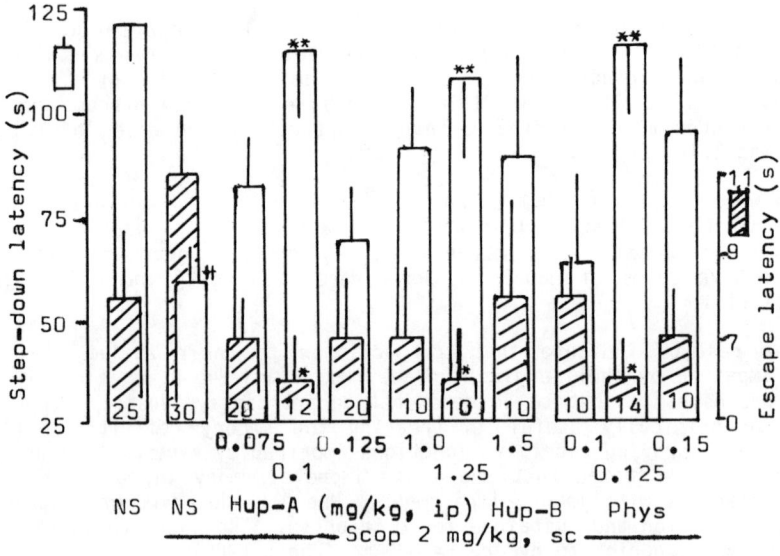

Figure 2. Reversal of scopolamine (Scop)-induced disruption of the memory retention of a passive avoidance task in mice ($\bar{x} \pm$ SE).

+$p < 0.05$, ++$p < 0.01$ compared with Scop + NS group.

REFERENCES

Coyle, J.T. D.L. Price and M.R. DeLong. 1983. Alzheimer's disease: a disorder of cortical cholinergic innervation. Science 219:1184-1190.
Drukarch, B., K.S. Kits, E.G. VanderMeer, J.C. Lodder and J.C. Stoof. 1987. 9-amino-1,2,3,4-tetrahydroacridine (THA), an alleged drug for the treatment of Alzheimer's disease, inhibits acetylcholinesterase activity and slow outward K^+ current. Eur. Journal Pharmacol 141:153-157.
Ellman, G.L., K.D. Courtney, V. Jr. Andre and R.M. Featherstone. 1961. A new and rapid colorimetric determination of acetylcholinesterase activity. Biochem. Pharmacol. 7:88.
Giacobini, E., S. Somani, M. McIlhany, M. Downen and M. Hallak. 1987. Pharmacokinetics and pharmacodynamics of physostigmine after intravenous administration in beagle dogs. Neuropharmacology 26:831-836.
Liu, J.S., C.M. Yu, Y.Z. Zhou. 1986. The structures of huperzine A and B, two new alkaloids exhibiting marked anticholinesterase activity. Can. Journal Chem. 64:837.
Lu, W.H., J. Shou and X.C. Tang. 1988. Improving effect of huperzine A on discrimination performance in aged rats and adult rats with experimental cognitive impairment. Acta Pharmacologica Sinica 9:11-15.
McGeer, P.L. and E.G. McGeer. 1984. Cholinergic system and cholinergic pathology. In Handbook of Neurochemistry, Vol. 6, 2nd ed., ed. L. Abel, 379-410. New York:Plenum Press.
Tang, X.C., Y.F. Han, X.P. Chen and X.D. Zhu. 1986. Effects of huperzine A on learning and retrieval process of discrimination performance in rats. Acta Pharmacologica Sinica 7:507-511.
Wang, Y.E., D.X. Yue and X.C. Tang. 1986. Anti-cholinesterase activity of huperzine A. Acta Pharmacologica Sinica 7:110-113.
Wang, Y.E., J. Feng, W.H. Lu and X.C. Tang. 1988. The physiological disposition and pharmacokinetics of huperzine A in rats and mice. Acta Pharmacological Sinica, In Press.
Whitehouse, P.J., D.L. Price, A.W. Clark, J.T. Coyle and M.R. DeLong. 1981. Alzheimer disease: evidence for selective loss of cholinergic neurons in the nucleus basalis. Ann Neurol. 10:122-126.
Xu, H. and X.C. Tang. 1987. Cholinesterase inhibition by huperzine B. Acta Pharmacologica Sinica 8:18-22.
Yan, X.F., W.H. Lu, W.J. Lou and X.C. Tang. 1987. Effects of huperzine A and B on skeletal muscle and electroencephalogram. Acta Pharmacologica Sinica 8:117-123.
Zhang, S.L. 1986. Therapeutic effects of huperzine A. on the aged with memory impairment. New Drug and Clinical Remedies 5:260-262.
Zhu, X.D. and X.C. Tang. 1987. Facilitatory effects of huperzine A and B and learning and memory of spatial discrimination in mice. Acta Pharmaceutica Sinica 22:812-817.
Zhu, X.D. and X.C. Tang. 1988. Effect of huperzine A and B on impaired memory retention in mice. Acta Pharmacological Sinica, Accepted.

GALANTHAMINE:
Another Look at an Old Cholinesterase Inhibitor

Edward F. Domino
Dept. of Pharmacology, Univ. of Michigan, Ann Arbor, MI

I. INTRODUCTION

At the present time the cholinergic deficiency hypothesis involving nucleus basalis and related areas in the basal forebrain of Alzheimer's patients provides the most practical means of developing new therapies. In view of the interest in experimental treatments using physostigmine and tetrahydroaminoacridine, it seems appropriate to search for other cholinesterase inhibitors (ChEI) that readily penetrate the blood-brain-barrier. Galanthamine is such an agent. The purpose of this brief review is to stimulate further examination of this agent.

II. CHEMICAL ISOLATION AND CHARACTERIZATION

The early chemical and pharmacognosy literature is a little confusing on the precise details of the chemical isolation and characterization of the alkaloid galanthamine. This reviewer has not had the opportunity to check all of the original literature in the native language of the author(s). What follows is an attempt to briefly summarize some of the papers that have been quoted in this field.

Galanthamine apparently was first isolated in the USSR from the caucasian snowdrop *Galanthus woronowii* by Proskurnina and Yakovlena (1947) and Proskurnina and Areshkina (1952, 1953). Proskurnina and Areshkina (1947, 1948) reported on the presence of galanthamine in *Galanthus nivalis* which also contained other alkaloids including lycorine, tazettine, etc. Proskurnina and Yakovlena (1952, 1953) reported on the further isolation of galanthamine from *Galanthus woronowii*. In these early years there was some confusion over the precise chemical structures of the alkaloids isolated. For example, Uyeo and Kobayashi (1953) identified galanthamine as identical with lycoramine extracted from *Lycoris radita*. The identity of these compounds was also emphasized by Boit and Ehmke (1955, 1956) and Briggs *et al*. (1956). Kobayashi *et al*. (1956) suggested that the original assigned structures of galanthamine and lycoramine differed in that the structure first assigned to galanthamine was inconsistent with their experimental results. Galanthamine was again extracted from *Galanthus nivalis var. gracilis* in Bulgaria by Bubeva-Ivanova (1957). Bossier *et al*. (1960) reviewed the botanical and chemical data then available. They discussed at length the different botanical sources of galanthamine and related alkaloids as well as their chemical structures. In 1962 Barton and Kirby reported the chemical synthesis of galanthamine. In 1964 Williams and Rogers described the structure of galanthamine methiodide. Kametani *et al*. (1971) reported on a new alternative synthesis of galanthamine. The chemical structure of galanthamine as described in the Merck Index (1983) is shown on the following page.

Galanthamine

Because of its heterocyclic structure as a phenantridine derivative and its crude relationship in chemical structure to codeine, the structure of galanthamine has been compared to that of codeine. However, their pharmacological properties are very different. Galanthamine is also known by several other generic names including galantamine(sic) and lycoremine(sic). According to the Merck Index (1983), the chemical composition of galanthamine is 4a,5,9,10,11,12- hexahydro- 3-methoxy- 11-methyl- 6H-benzofuro [3a,3,2-ef][2] benzazepin-6-ol. As can be seen from its chemical structure, galanthamine is a tertiary amine that would be expected to penetrate readily the blood-brain-barrier.

III. SOURCES

Currently galanthamine hydrobromide is available under the trade name Nivalin, from Pharmachim State Economic Association, 16 Iliensko, Chaussée, Sofia, Bulgaria. The clinical indications for the use of Nivalin as specified by the company are obviously overly inclusive, including neuritis, especially facial, radiculitis, radiculoneuritis, polyneuritis, poliomyelitis, myopathies including myasthenia gravis pseudoparalytica, progressive muscular dystrophy, spinal and neuritic muscular atrophy. Additional indications are spastic pareses and paralysis, and other sequelae of lesions of the central nervous system of vascular, inflammatory, toxic, and traumatic origin including cerebral apoplexy, meningitis, meningoencephalitis and myelitis. As would be expected from a ChEI, Nivalin very early was shown to be effective in neuromuscular diseases, especially myasthenia gravis (Pestel, 1961). Gopel and Bertram (1971) also have reported on their favorable experiences using Nivalin in some neurologic diseases. In addition, Nivalin has been indicated for glaucoma and as an anticurare agent in anesthesia as well as in diseases with decreased tone of smooth musculature of the urinary bladder and gastrointestinal tract. In view of the cholinergic deficit hypothesis in Alzheimer's disease, one would expect that galanthamine was tried in this condition although it is not well described in the literature available to this reviewer.

The contraindications for Nivalin are said to include epilepsy, bronchial asthma, cardiac diseases with bradycardia, and hyperkinesia. Interestingly, side effects are said to occur very rarely and include increased salivation, nausea, dizziness and bradycardia (Cozanitis et al., 1973). Although a number of pharmaceutical companies in western Europe have been interested in Nivalin, to date, to our knowledge, none of these companies are currently supplying galanthamine or its salts.

In the USSR galanthamine is recommended to be administered s.c. as the hydrobromide in doses of 10 and 20 mg maximally. The agent is used primarily as an antimyasthenic and antagonist of curare-like effects as well as for a variety of central nervous system disorders.

IV. PHARMACOLOGY

A. Cholinesterase inhibition

Irwin and Smith (1960a,b) and Boissier and Lesbros (1962) compared the degree of acetylcholinesterase inhibition (AChEI) in skeletal muscle and brain, and butyrylcholinesterase inhibition (BChEI) in plasma. They were able to show that muscle acetylcholinesterase (AChE) activity was inhibited by galanthamine to a greater extent than that produced by pyridostigmine, although clearly less than that of neostigmine. These investigators concluded that galanthamine and related alkaloids are of interest because of their potential therapeutic usefulness as ChEI. Galanthamine was more effective in inhibiting plasma BChE than it was muscle AChE. Pyridostigmine was much less potent in inhibiting AChE.

B. Antagonism of Central Anticholinergic Effects

Plaitakis and Duvoisin (1983) collected evidence from the literature that Homer's Moly is *Galanthus nivalis* of which the active ingredient is galanthamine. They suggested that Circe's poisonous plants included stramonium which was thought to be used by Circe to induce amnesia in Odysseus' crew. Plaitakis and Duvoisin suggest that the description of Moly as an antidote in Homer's Odyssey may represent the oldest recorded use of a ChEI to reverse the central anticholinergic syndrome.

Baraka and Harik (1977) determined the effectiveness of 0.5 mg/kg of galanthamine hydrobromide given i.v. in reversing the central anticholinergic syndrome induced by 2 mg of scopolamine i.v. in 10 adult volunteers. Following the administration of i.v. scopolamine, the volunteers became drowsy within 10 min, reaching a peak effect within 30-40 min. The drowsiness was sometimes associated with disorientation, visual hallucinations, and delirium. The administration of 0.5 mg/kg of galanthamine hydrobromide i.v. reversed the central effects of scopolamine. The subjects became alert within 5-10 min and were completely awake after 30 min. As would be expected following the administration of scopolamine, the pulse rate increased from a control of 60-80 to 110-130/min. After galanthamine was given, the heart rate returned to control levels of 60-70/min. Two hr later, the subjects felt more alert than usual and did not feel sleepy. These investigators also showed that the EEG changes associated with scopolamine (which included a decrease in *alpha* rhythm) reached their maximum within 30 min and, in addition, the *alpha* rhythm was replaced by disorganized 4-6 Hz *theta* activity of moderate to low amplitude. Again, galanthamine promptly reversed the EEG effects of scopolamine. Although the authors concluded that galanthamine produced a long lasting reversal of the central anticholinergic syndrome, from their study one could only determine that the effect of the drug lasted 2 hr with no evidence that it had a much longer duration of action.

Cozanitis (1977) also reported on the effectiveness of galanthamine hydrobromide in treating the central anticholinergic syndrome induced by scopolamine. A 31 year old man who ingested an unknown quantity of scopolamine time release tablets exhibited a marked anticholinergic syndrome. He was flushed, semicomatose and only slightly reactive to pain. In addition, his respiration was slow and shallow. His breath smelled of ethyl alcohol. After gastric lavage a large mass of tablets was recovered, but the patient became delirious requiring restraints. Galanthamine hydrobromide, in a dose of 20 mg i.v., within 10 min produced a dramatic improvement. The patient was observed over the next 10 hr and was free of psychiatric and physiological manifestations of poisoning. His breathing rate rose from 14 to 20/min. Cozanitis emphasized that galanthamine hydrobromide appeared to be a long acting ChEI in antagonizing the central effects of scopolamine.

C. Antagonism of Skeletal Neuromuscular Blockade

The anticurare action of galanthamine was described by Mashkovsky (1955). Boissier et al. (1960) showed the potency of antagonism by galanthamine of nondepolarizing muscle relaxants is 1/10 that of neostigmine. Galanthamine seemed to act faster but was shorter acting. It prolonged the action of succinylcholine neuromuscular blockade. Bradycardia was seen in the frog, rabbit, and dog, and hypotension in the dog. As expected, galanthamine enhanced acetylcholine (ACh) provoked contraction of smooth muscle and increased the contraction of striated muscle after maximal direct and indirect stimulation.

When given s.c., the effect of galanthamine is somewhat slower and less potent than neostigmine but is said to be longer lasting and less toxic (Pestel, 1961). The quaternary derivative of galanthamine is more potent than its non-quaternary hydrobromide, being about four times as potent as a ChEI. Paskov et al. (1964) showed in experimental animals that galanthamine was effective in antagonizing morphine induced depression. The LD_{100} of galanthamine in rats is 45 mg/kg, in rabbits 12 mg/kg, and in cats 60 mg/kg (Stoyanov, 1964a,b). Stoyanov and Vulchanova (1963) reported on the efficacy of galanthamine as a curare antagonist in a large number of patients in Bulgaria. In addition, Mayrhofer (1966) reported his favorable clinical experiences with galanthamine as a curare antagonist.

Wislicki (1967) provided additional evidence that galanthamine hydrobromide, which has advantages as an antagonist of nondepolarizing muscle relaxants should be studied in Western countries. He described the use of galanthamine in 24 patients. Galanthamine was about 1/10 as potent as neostigmine. Changes in pulse rate and blood pressure were slight and it was rarely necessary to inject atropine before galanthamine. Salivation had to be suppressed by atropine in only 1/7th of the cases studied.

Cozanitis and his colleague (Baraka and Cozanitis, 1973; Cozanitis, 1971, 1974) have provided extensive data that galanthamine is a very useful agent in reversing nondeplolarization block in man. Cozanitis et al. (1977) compared the potency of galanthamine and neostigmine in reversing d-tubocurarine blockade in the isolated rat diaphragm. Neostigmine appeared to be about 18 times as potent on a weight basis in the rat whereas the potency ratio in vivo in man is about 20 times greater than galanthamine. These investigators indicated that the potency differences appear to be determined on the basis of drug effects on neuromuscular transmission and not on pharmacokinetic differences.

D. General Pharmacology and Toxicity

Paskov summarized much of the general pharmacology of galanthamine in 1959. Boissier and his colleagues (1960) also described the pharmacological actions of galanthamine which indicate its effectiveness as a cholinergic agonist. Compared to neostigmine, galanthamine was much less potent but also appeared to be less toxic. In the mouse the toxicity of galanthamine hydrobromide compared to neostigmine was 16.5 times less after i.v. and 21.3 times less when given i.p. Subsequently, Boissier and Lesbros (1962) compared the AChEI actions of galanthamine with physostigmine, neostigmine, lycoramine methyl iodide, and galanthamine methyl iodide. They were able to show that galanthamine itself was a ChEI of both serum BuChE as well as AChE from red blood cells and skeletal muscle. Galanthamine itself was relatively weak compared to neostigmine and physostigmine. The quaternary methyl iodide preparation was more potent relatively as a true AChEI compared to tertiary galanthamine hydrobromide. Kostowski and Gumulka (1968) suggested that galanthamine had ganglionic actions more like neostigmine than physostigmine and that more than muscarinc agonist actions were involved. However, in unanesthetized cats galanthamine caused marked EEG desynchronization in the neocortex and increased *theta* activity in the hippocampus which was completely abolished by atropine or benactyzine, suggesting a predominant muscarinic action in the brain. After mesencephalic lesions, the hippocampal *theta* rhythm previously induced by galanthamine was blocked (Mashkovsky and Illuchenok, 1961).

Cozanitis et al. (1983) studied the effects of galanthamine in a variety of animals and isolated organ preparations. They observed that galanthamine given s.c. had antinociceptive activity. This was compared to that induced by physostigmine and morphine. Naloxone blocked the antinociceptive effects of galanthamine but not that of physostigmine, suggesting that galanthamine had an opioid narcotic component not possessed by physostigmine. Both ChEIs produced analgesia in the mouse writhing test and potentiated the effect of morphine in the rat hotplate analgesic test. Although galanthamine produced analgesia in the intact animal, it failed to produce opioid-like activity in isolated organ preparations such as the longitudinal muscle strip of the guinea pig ileum, the mouse vas deferens, and the cat nictitating membrane. The overall data suggest that galanthamine may release endogenous opioids in the intact animal and that galanthamine-induced analgesic effects are partially antagonized by naloxone. These investigators also noted that galanthamine has a molecular configuration similar to codeine (Williams and Rogers, 1964) but galanthamine was not acting through a classic *mu* opioid mechanism but rather apparently as an unusual ChEI different from physostigmine. Cozanitis and Rosenberg (1974) compared the effects of galanthamine hydrobromide on dextromoramide-depressed respiration in rabbits. They studied this phenomenon in view of their previous experience (see below) of apnea reversal by galanthamine in a patient who ingested a large dose of dextromethoramide, methaqualone and diphenhydramine. Cozanitis and Rosenberg showed that a dose of 1 mg/kg or more of galanthamine was effective in antagonizing the respiratory depression. However, the action was slower in onset than that produced by 1 mg/kg of nalorphine.

Rupreht et al. (1983) studied the involvement of the central cholinergic and opioid systems in nitrous oxide withdrawal in mice. Mice were exposed to a mixture of 1.4 atm of nitrous oxide and 0.6 atm oxygen for a period of 60 min. After 60 min of exposure to this mixture, the pressure was reduced to the ambient level over a period of 1 min and each animal was tested for withdrawal convulsions. Physostigmine, galanthamine, and naloxone were compared in their ability to modify the withdrawal syndrome. Physostigmine, in a dose of 0.4 μg/g decreased to 17.5 min the period of susceptibility during which withdrawal convulsions were observed. A similar significant decrease was observed when the same dose of physostigmine was given 1 min after the nitrous oxide was discontinued. Over a 10 fold dose of galanthamine (5 μg/g) had similar properties and decreased to 33.3 min the period of time in which seizures could be elicited. The same dose of galanthamine, in contrast to physostigmine, failed to significantly affect the period of withdrawal convulsions when the galanthamine was given in the first min after withdrawal. This suggests that physostigmine has a more rapid onset of action than galanthamine. In addition, naloxone, 0.8 μg/g before nitrous oxide exposure significantly decreased the predisposition to seizures. However, the same dose of naloxone administered after the first min after nitrous oxide was discontinued failed to influence the duration in which the convulsant phenomenon could be elicited.

E. Antagonism of Drug-Induced Respiratory Depression

Paskov et al. (1964) found that galanthamine hydrobromide antagonized respiratory depression induced by morphine, meperidine, and dextromoramide irrespective of whether the animals' carotid bodies had been removed, suggesting that this was a direct central nervous system effect. These investigators also noted that intracisternal injections of galanthamine hydrobromide antagonized the respiratory depression induced by these narcotics. Stoyanov et al. (1965) and Stoyanov and Statkov (1968) reported galanthamine hydrobromide to antagonize steroid and fentanyl depression. Cozanitis and Toivakka (1974) used galanthamine hydrobromide to antagonize the respiratory depression of a 44 year old man who ingested a large quantity of methaqualone, diphenhydramine and dextromoramide. Respiration ceased almost immediately after admission and an endotracheal tube was passed to facilitate artificial ventilation. The i.v. administration of galanthamine hydrobromide, in a dose of 20 mg, within 20 min caused respiratory stimulation. An additional dose of 10 mg was even more effective. Cozanitis and Toivakka (1971) did a comparative clinical pharmacologic study of the effects of galanthamine hydrobromide and atropine-neostigmine in conscious volunteers. They reported that the electroencephalographic pattern after i.v. galanthamine bromide fulfills the requirement of an analeptic.

Tassonyi et al. (1976) showed that galanthamine hydrobromide in doses of 20-70 mg rapidly reversed postoperative apnea after neurolept analgesia. The patients regained consciousness, followed instructions, and moved well. However, 15-40 min after galanthamine hydrobromide administration suddenly the patients lost consciousness and breathing stopped. After about 20-60 min of artificial ventilation the patients gradually regained consciousness and their breathing became normal. Tassonyi et al. (1976) discussed the various possibilities for this phenomenon. Perhaps the most likely explanation is that galanthamine hydrobromide has a much shorter duration of action in man than has previously been recognized.

F. Endocrine effects

Cozanitis et al. (1980) compared the effects of galanthamine and neostigmine on plasma ACTH in patients undergoing surgical anesthesia. The patients required skeletal muscle relaxation in which galanthamine and neostigmine were indicated to reverse neuromuscular blockade. Following the administration of these agents to reverse nondepolarizing neuromuscular blockade, galanthamine, in contrast to neostigmine, statistically significantly elevated plasma ACTH, suggesting that the rise in plasma cortisol they also observed is ACTH dependent. They concluded that a peripheral cholinergic mechanism was not involved because galanthamine is a tertiary amine whereas neostigmine is a quaternary amine and therefore the latter would not be expected to penetrate the blood-brain-barrier.

G. Pharmacokinetics

Mihailova and Yamboliev (1986) described the pharmacokinetics of galanthamine hydrobromide (Nivalin) following single i.v. and oral doses to rats. The plasma samples were collected and the concentrations of the drug determined spectrophotofluorometrically. A two compartment open model was found to best describe the experimental data. Galanthamine had an elimination half life of about 40-50 min, a volume of distribution of over 2 l/kg, a plasma clearance of about 2 l/hr/kg and an oral availability of about 65%. Westra et al. (1986) studied the pharmacokinetics of galanthamine in 8 anesthetized patients. They showed that after an i.v. injection of 0.3 mg/kg the decrease in the serum concentration of galanthamine followed a biexponential curve. Serum concentrations decreased very rapidly between 2 and 30 min with a $t_{\frac{1}{2}}$ of the *alpha* phase of 6.42 and then declined more slowly with a $t_{\frac{1}{2}}$ of the *beta* phase of 264 min. Total serum clearance of galanthamine amounted to 5.37 ml/min/kg and the renal clearance was 1.36 ml/min/kg. Cumulative urinary excretion of galanthamine between 0 and 48 hr after injection amounted to 28.0% of the administered dose. Biliary excretion of galanthamine during 24 hr amounted to 0.2% of the dose. There was no evidence of glucuronide or sulfate conjugation of galanthamine. These authors pointed out that the elimination half life of galanthamine was 264 min whereas the ChEIs neostigmine, pyridostigmine and endrophonium, currently used as antagonists of nondepolarizing neuromuscular blockade, have elimination half lives in man of respectively 80, 46, and 114 min. Paskov et al. reported that serum BuChE is inhibited for about 3 hr after the administration of galanthamine. In contrast, the inhibition of BuChE activity by neostigmine and pyridostigmine have been reported to be shorter.

Westra et al. (1986) concluded that from a pharmacokinetic point of view, galanthamine is suitable to reverse the unwanted and prolonged side effects of skeletal neuromuscular blocking agents but obviously further pharmacological studies are needed. Especially clear, however, from their pharmacokinetic analysis is the fact that galanthamine is not a very long acting compound. This finding is consistent with our own unpublished studies on the comparative effects of physostigmine, tetrahydroaminoacridine, and galanthamine in suppressing self-stimulation behavior in the rat where galanthamine is only as long acting as physostigmine but much less potent.

V. SUMMARY

Galanthamine is a tertiary cholinesterase inhibitor that has central nervous system actions. It is well known and widely used in eastern European countries. Although less potent than physostigmine and with a similar duration of action, it is claimed to be less toxic. Galanthamine deserves further study as a possible indirect cholinergic agonist treatment of the cognitive deficits in Alzheimer's disease.

VI. REFERENCES

Baraka, A. and Cozanitis, D. 1973. Galanthamine versus neostigmine for reversal of nondepolarizing neuromuscular block in man. Anesth. Analg. Cur. Res. 52: 832-836.
Baraka, A. and S. Harik. 1977. Reversal of central anticholinergic syndrome by galanthamine. J. Am. Med. Assoc. 238: 2293-2294.
Barton, D.H.R. and G.W. Kirby. 1962. Phenol oxidation and biosnthesis. Part V. The synthesis of galanthamine. Proc. Chem. Soc.: 806-817.
Boissier, J.-R., G. Combes, and J. Pagny. 1960. La galanthamine, puissant cholinergique naturel. I. Sources. Structure chimique. Caracterisation. Extraction. Toxicite. Action sur les fibres lisses. Extrait des Anneles pharmaceutiques francaises. 18: 888-900.
Boissier, J.-R. and J. Lesbros. 1962. La galanthamine, puissant cholinergique naturel. II. Activite anticholinesterasique de la galanthamine et de quelques derives. Extrait des Annales pharmaceutiques francaises 20: 150-155.
Boit, H.G. and H. Ehmke. 1955. Chem. Ber. 88: 1590.
Boit, H.G. and H. Ehmke. 1956. Chem. Ber. 89: 163 and Chem. Ber. 90: 725, 2197.
Briggs, C.K., P.F. Highet, R.S. Highet, and W.C. Wildman. 1956. J. Amer. Chem. Soc. 78: 2899.

Bubeva-Ivanova, L. 1957. Fidochemical Investigation of Galanthus nivalis var. gracilis. Farmatsiya (Sofia) 2: 23.
Cozanitis, D.A. 1971. Experiences with galanthamine hydrobromide as curare antagonist. Der Anaesthetist 20: 226-229.
Cozanitis, D.A. 1974. Galanthamine hydrobromide versus neostigmine. A plasma cortisol study in man. Anaesthesia 29: 163-168.
Cozanitis, D.A. 1977. Galanthamine hydrobromide, a longer acting anticholinesterase drug, in the treatment of the central effects of scopolamine (hyoscine). Anaesthesist 26: 649-650.
Cozanitis, D., A. Dessypris, and K. Nuuttila. 1980. The effect of galanthamine hydrobromide on plasma ACTH in patients undergoing anaesthesia and surgery. Acta anaesth. scand. 24: 166-168.
Cozanitis, D.A., T. Friedmann, and S. Fürst. 1983. Study of the analgesic effects of galanthamine, a cholinesterse inhibitor. Arch. Int. Pharmacodynamie Therap. 266: 229-238.
Cozanitis, D.A., K. Nuuttila, P. Karhunen, and A. Baraka. 1973. Changes in cardiac rhythm with galanthamine hydrobromide. Anaesthesist 22: 457-459.
Cozanitis, D.A. and P. Rosenberg. 1974. Preliminary experiments with galanthamine hydrobromide on depressed respiration. Anaesthesist 23 320-305.
Cozanitis, D.A., A.H.J. Schaf, and J. Van Den Akker. 1977. The potency ratio of galanthamine and neostigmine on the reversal of the d-tubocurarine block in the isolated rat diaphragm. Acta Anaesthesiologica Belgica 28: 53-60.
Cozanitis, D.A. and E. Toivakka. 1971. A comparative study of galanthamine hydrobromide and atropine/neostigmine in conscious volunteers. Der Anaesthesist 20: 416-421.
Cozanitis, D.A. and E. Toivakka. 1974. Treatment of resiratory depression with the anticholinesterase drug galanthamine hydrobromide. Anaesthesia 29: 581-584.
Gopel, W. and W. Bertram. 1971. Erfahrungen mit Nivalin in der neurologischen Therapie. Psychiat. Neurol. med Psychol. Leipzig 23: 712-718.
Irwin, R.L. and H.J. Smith III. 1960a. Cholinesterase inhibition by galanthamine and lycoramine. Biochem. Pharmacol. 3: 147-148.
Irwin, R.L. and H.J. Smith III. 1960b. The activity of galanthamine and related compounds. Arch. int. Pharmacodyn. 127: 314.
Kametani, T., C. Seino, K. Yamaki, S. Shibuya, K. Fukumoto, K. Kigasawa, F. Satoh, M. Hiiragi, and T. Hayasaka. 1971. Studies on the syntheses of heterocyclic compounds. Part CCCLXXVI. Alternative total syntheses of galanthamine and N-benzylgalanthamine iodide. J. Chem. Soc. 1043-1047.
Kobayashi, S., T. Shingu, and S. Uyeo. 1956. Structure of galanthamine and lycoramine. Chem. and Ind. (London) 177-178.
Kostowski, W. and W. Gumulka. 1968. Note on the ganglionic and central actions of galanthamine. Int. J. Neuropharmacol. 7: 7-14.
Mashkovsky, M.D. 1955. Effects of galanthamine on the acetylcholine sensitivity of skeletal musculature. Pharmacol. and Toxic. 18: 21.
Mashkovsky, M.D. and R.J. Illuchenok. 1961. On the problem of action of galanthamine on the central nervous system. Z. Nevropatol. i. Psichiatrii 61: 166-175.
Mayrhofer, O. 1966. Clinical experiences with diallyl-nor-toxiferine and the curare antidote galanthamine. Southern Med. J. 59: 1364-1368.
Merck Index, 10th Edition. 1983. Windholz, M., S. Budavari, R.F. Blumetti, and E.S. Otterbein, eds., No. 4210, galanthamine, pg. 620. Merck and Company, Rahway, NJ.
Mihailova, D. and Yamboliev, I. 1986. Pharmacokinetics of galanthamine hydrobromide (Nivalin[R]) following single intravenous and oral administration in rats. Pharmacology 32: 301-306.
Paskov, D.S. 1959. Nivalin-Pharmakologicheska Karakteristika Sofia.
Paskov, D., H. Dobrev, and N. Nikoforonov. 1964. Antagonistic effect of Nivalin (galanthamine) versus morphine upon the respiratory center. Acta Inst.

Superioris Med. (Sofia) 43: 1.
Pestel, M. 1961. Une nouvelle médication de la myasthénie et des dystrophies neuromusculaires: la Nivaline. Presse méd. 69: 182.
Plaitakis, A. and R.C. Duvoisin. 1983. Homer's moly identified as Galanthus nivalis L: Physiologic antidote to stramonium poisoning. Clin. Neuropharmacol. 6: 1-5.
Proskurnina, N.F. and L.Y. Areshknina. 1947. J. Chim. Gen. USSR 17: 1216 and 1948. Chem. Abst. 42: 1595h.
Proskurnina, N.V. and L.Y. Areshkina. 1952. Alkaloids of Galanthus woronovi II. Isolation of a new alkaloid. Zhur. Obshchei Khim. 22: 1899; and 1953. Chem. Abst. 47: 6959c.
Proskurnina, N.F. and A.P. Yakovlena. 1947. On the alkaloid Galanthus woronowi. zh. Obshch. Kalim 17: 1216.
Proskurnina, N.F. and A.P. Yakovlena. 1952. J. Gen. Chem. USSR 22: 1899; and 1953. Chem. Abst. 47: 6959c.
Rupreht, J., B. Dworacek, R. Ducardus, P.I.M. Schmitz, and M.R. Dzoljic. 1983. The involvement of the central cholinergic and endophinergic systems in the nitrous oxide withdrawal syndrome in mice. Anesthesiol. 58: 524-526.
Stoyanov, E.A. 1964a. Galanthamine hydrobromicum "Nivalin", ein neues Antidot der nichtpolarisierenden Muskelrelaxantien. Der Anaesthetist 13: 217.
Stoyanov, E.A. 1964b. Spirographic evidence on the anticurare effect of Nivalin. Acta Inst. Sup. Med. (Sofia) 43: 39.
Stoyanov, EA. and S. Vulchanova. 1963. The clinical application of Nivalin as an antidote of curare. Acta Inst. Sup. Med (Sofia) 42: 45.
Stoyanov, E.A., H. Dobrev, and D. Paskov. 1965. The effect of Nivalin on respiration in steroid narcosis. Acta Inst. Sup. Med. (Sofia) 44: 1.
Stoyanov, E.A. and P. Statkov. 1968. Experimental study of the antagonism between fentanyl and Nivalin. Med. Arch. (Sophia) 6: 105.
Tassonyi, E., K. Labencz, L. Vimlati, and I. Kiss. 1976. Postoperative apnoea following Nivalin administration. Anaesthetist 25: 529-531.
Uyeo, S. and S. Kobayashi. 1953. Pharm. Bull. Tokyo 1: 139.
Westra, P., M.J.S. van Thiel, G.A. Vermeer, A.M. Soeterbroek, A.H.J. Scaf, and H.A. Claessens. 1986. Pharmacokinetics of galanthamine (a long-acting anticholinesterase drug) in anaesthetized patients. Br. J. Anaesth. 58: 1303-1307.
Williams, D.J. and D. Rogers. 1964. The structure of galanthamine methiodide. Proc. Chem. Soc.: 357.
Wislicki, L. 1967. Nivalin (galanthamine hydrobromide), an additional decurarizing agent. Some introductory observations. Br. J. Anaesth. 39: 963-968.

METHANESULFONYL FLUORIDE:
A CNS Selective Cholinesterase Inhibitor

D.E. Moss[1], H. Kobayashi[2], G. Pacheco[1], R. Palacios[1], and R.G. Perez[1]
[1] Lab. of Psychobiology, Dept. of Psychology, Univ. of Texas, El Paso, TX
[2] Dept. of Veterinary Medicine, Iwate Univ., Ueda, Morioka 020, Japan

INTRODUCTION

Senile dementia of the Alzheimer type (SDAT) is a complex disease process that produces gross neuropathology and generalized deterioration of brain function (Wisniewski, Merz, Wen, Iqbal, and Grundke-Iqbal, 1985). The severity of the pathology is correlated with widespread reduction in regional cerebral glucose metabolism (Metter, Riege, Benson, Kuhl and Phelps, 1985). In addition to reduction in cholinergic markers, SDAT has also been associated with specific changes in many neurotransmitters including somatostatin (Davies, Katzman, and Terry, 1980; Rossor, Emson, Mountjoy, Roth and Iverson, 1980; Beal, Mazurek, Tran, Chattha, Bird, and Martin, 1985), monoamines (Adolfsson, Gottfries, Roos, and Winblad, 1979; Gibson and Ball, 1983), and various amino acids (Arai, Kobayashi, Ichimiya, Kosaka, and Iizuka, 1984). It is, therefore, an oversimplification to characterize the dementia associated with SDAT as anticholinergic dementia.

Even though SDAT involves much more than deterioration of cholinergic function in the central nervous system (CNS), the cholinergic hypothesis of dementia has stimulated clinical and basic research. The cholinergic hypothesis is simply that dementia is the result of insufficient cholinergic function within the CNS. The rationale for this hypothesis is based on the classic discovery that indicators of cholinergic neurotransmission are markedly reduced in brains from Alzheimer's patients as compared to age-matched controls (Davies and Maloney, 1976; Perry, Gibson, Blessed, Tomlinson, and Perry, 1977). Because the cells in the nucleus basalis of Meynert and related nuclei are thought to provide important cholinergic projections to the hippocampus and cortex, loss of these neurons could reduce cholinergic activity in the cortex of Alzheimer's patients (Whitehouse, Price, Struble, Clarke, Coyle, and DeLong, 1982). The neuropathological evidence demonstrating an extensive loss of cholinergic function in the basal forebrain and cortex in SDAT has been widely confirmed by many investigators and is a relatively common feature of the disease.

Treatment strategies intended to facilitate cholinergic function in the CNS to treat dementia assume that a reduction in cholinergic function is the cause of dementia. There is, in fact, some support for this hypothesis. The degree of dementia and memory impairment that occurs in SDAT is correlated with the decrement in cortical cholinergic transmission (Perry, Tomlinson, Blessed Bergmann, Gibson, and Perry, 1978). In addition, however, loss of the cholinergic system related to the nucleus basalis of Meynert and other basal forebrain nuclei is associated with dementia in other disorders including Parkinson's and Korsakoff's diseases (Nakano and Hirano, 1984; Arendt, Bigl, Arendt and Tennstedt, 1983), parkinsonism-dementia complex of Guam (Nakano and Hirano, 1983) and dementia pugilistica (Uhl, McKinney, Hedreen, White, Coyle, Whitehouse, and Price, 1982). These studies show that the loss of basal forebrain cholinergic neurons is not a neuropathological feature specific to SDAT. Therefore, cholinergic therapies developed for SDAT may also be useful in other clinical applications.

It is the apparent association between deterioration of the cholinergic system and dementia that is the basis for the cholinergic strategies for treating SDAT. In fact, some marginal facilitation of memory performance has been obtained in clinical tests with SDAT patients with physostigmine (Brinkman and Gershon, 1983; Davis and Mohs, 1982), a cholinesterase inhibitor that crosses the blood-brain barrier and has, therefore, central as well as peripheral effects. In another study, twelve patients with SDAT were treated with various oral doses of physostigmine, up to 2 mg, every 2 hours for 3-5 days. Of the ten patients who completed the study, three showed significant improvement at the highest dose that could be tolerated. Cortisol measures showed that clinical improvement was correlated with enhanced central cholinergic activity (Mohs, Davis, Johns, Mathe, Greenwald, Horvath, and Davis, 1985). In addition, there has been a report that tacrine (tetrahydroaminoacridine, THA), a reversible cholinesterase inhibitor with peripheral as well as central effects, produced cognitive improvement in several patients with SDAT (Summers, Majovski, Marsh, Tachiki, and Kling, 1986). At the present time, cholinesterase inhibitor therapy appears to be a promising approach to treating SDAT.

One of the major problems related to the development of anticholinesterase therapy in SDAT is the great potential for severely toxic effects. The somatic motor and autonomic nervous systems are greatly influenced by cholinesterase inhibitors which produce very disturbing and toxic peripheral side effects. One way to limit peripheral toxicity has been to propose direct intraventricular delivery of cholinergic drugs (Harbaugh, Roberts, Coombs, Saunders, and Reeder, 1985; Mattio, McIlhany, Giacobini, and Hallak, 1986). However, in the dog, more rapid and higher levels of cholinesterase inhibition in the cortex were produced by intravenous rather than intraventricular injection of physostigmine. Not surprisingly, intraventricular injections produced higher levels of physostigmine in regions located closely to the ventricular surface (Mattio et al., 1986; Andjelkovic, Beleslin, and Vasic, 1971). Another, more practical suggestion may be, however, the development of a cholinesterase inhibitor that has an inherent selectivity for CNS enzyme (Davies, 1981; Moss, Rodriguez, Selim, Ellett, Devine, and Steger, 1985b).

SULFONYL FLUORIDES
The purpose of the research reported here was to develop very long-lasting CNS selective cholinesterase inhibitors that might be suitable for treatment of SDAT and other CNS diseases that involve a decline in cholinergic function. Some sulfonyl fluorides have been discovered to be cholinesterase inhibitors that meet these criteria and, furthermore, appear to have remarkably low general toxicity (Moss et al., 1985b; Moss, Rodriguez, Herndon, Vincenti

and Camarena, 1986). Because of these special characteristics, these compounds may have some potential as therapeutic agents in SDAT and related CNS diseases.

The sulfonyl fluorides that react with cholinesterase are irreversible inhibitors that form a covalent sulfonyl-enzyme complex similar to the phosphonyl-enzyme complex formed by the organic phosphates (Fahrney and Gold, 1963; Kitz and Wilson, 1962; Myers and Kemp, 1954). These substances are, however, generally less reactive than organic phosphates (Fahrney and Gold, 1962; Myers and Kemp, 1954) and this may account for their selectivity. The sulfonyl fluorides, being less reactive, also to produce fewer noncholinesterase side effects such as "Ginger Jake" paralysis, a delayed neurotoxic effect produced by certain organophosphorus anticholinesterase agents in humans. This organophosphate-induced condition is characterized by polyneuritis, flaccid paralysis of the arms and legs, degeneration of myelin sheaths and axons of the spinal cord, somatomotor neurons and medulla and it does not appear to be the result of cholinesterase inhibition (Brimblecombe, 1974). The sulfonyl fluorides do not, however, produce peripheral neuropathy (Caroldi, Lotti, Masutti, 1984) and can, in fact, be used to protect against neurotoxic organophosphates (Johnson, 1980).

EXPERIMENTS IN RODENTS
Of the thirty-six compounds tested in vivo in rats, only five show significant inhibition of cholinesterase and all of these inhibit brain enzyme to a greater degree than cholinesterases from heart, ileum and pectoral muscle (Moss et al., 1985b; 1986). The reason for the apparent selectivity toward the CNS is unknown and, furthermore, there is no obvious relationship between the molecular structure and selectivity or activity against cholinesterase (Moss et al., 1986).

The two inhibitors that have been studied in greatest detail are phenylmethylsulfonyl fluoride (PMSF, phenylmethanesulfonyl fluoride, alpha-toluenesulfonyl fluoride) and methanesulfonyl fluoride (MSF). Both of these compounds produced an average of 90% inhibition of brain cholinesterase with less than 30 to 35% inhibition of enzyme in peripheral tissues when the drug was administered daily at a low dose (Figure 1)(Moss et al., 1985b). It is important to note that MSF produced this effect at 0.5 mg/kg as compared to 85 mg/kg for PMSF. MSF may, therefore, be the most clinically useful compound.

The remarkable difference between the central and peripheral effects observed in these experiments is the result of two factors: 1) the inherent selectivity of MSF and PMSF to inhibit CNS enzyme; and 2) the relatively slow rate at which CNS enzyme is synthesized. The direct effects produced by a single dose of PMSF, including the selectivity toward brain and the relatively slow resynthesis of CNS enzyme are shown in Figure 2. As with recovery from organophosphates, the rates of recovery are limited by the rates of synthesis of new enzyme in each tissue. From Figure 2, the synthesis of new enzyme in brain appears to occur with a mean half-time of approximately 11 days compared to 1, 3 and 6 days for ileum, heart and pectoralis, respectively.

In vital organs, there is a substantial excess of cholinesterase above the amount required for normal function, and less than 50%

inhibition is generally regarded as not pharmacologically significant (Brimblecombe, 1974). Using this figure as a rough guide, direct measurements of enzyme activity show that MSF and PMSF both produce pharmacologically significant CNS cholinesterase inhibition with limited peripheral effects. These data are supported by observations in rats which show there are no losses of strength or coordination and remarkably low general toxicity even after high doses over a long period of time (Moss et al., 1985b).

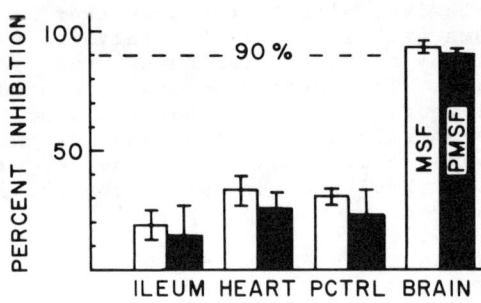

Figure 1. Cholinesterase inhibition in rat ileum, heart, pectoral muscle and brain produced by five daily administrations of PMSF (85 mg/kg) and MSF (0.5 mg/kg) by gavage The error bars represent one SEM. [Redrawn from Moss et al., 1985b]

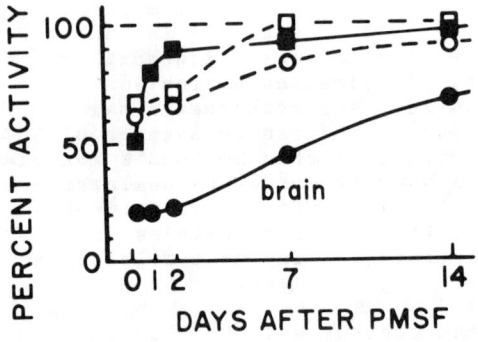

Figure 2. Time course of cholinesterase inhibition produced in rats by one injection of PMSF (100 mg/kg) in various tissues A time point of 4 hours is shown between 0 and 1 day. Ileum is represented by filled squares, heart by open squares, skeletal muscle (pectoralis) by open circles, and brain by filled circles. [Redrawn from Moss et al., 1985b].

In addition, the effects of MSF on mouse brain cholinesterase and acetylcholine content have been assessed. Treatment with 1.5 mg/kg MSF per day for three days caused up to 87% inhibition of forebrain cholinesterase and a significant increase in free acetylcholine levels for more than five days after the end of treatment[1]

---

[1]Personal communication, H. Kobayashi, Iwate University, Morioka Japan. To be presented to the Japanese Pharmacological Society (March 1988), by T. Nakano, H. Kobayashi, D.E. Moss and A. Yuyama.

EXPERIMENTS IN MONKEYS

In order to prepare for possible human tests with MSF as a therapeutic agent, preliminary experiments were undertaken to assess the efficacy and toxicity of this compound on young male monkeys (M. fasicularis). In this experiment on the effects of MSF on cerebrospinal fluid (CSF) cholinesterase, the monkeys were anesthetized (ketamine) and CSF was taken by lumbar puncture from each subject. Control samples, separated by several days, were taken to establish the level of cholinesterase activity present in each monkey prior to MSF treatment. Two days after the second control CSF sample was taken, all four monkeys were injected with one dose of 1.5 mg/kg MSF. CSF was sampled and analyzed for cholinesterase activity for several days after MSF treatment in order to establish the degree of inhibition produced by a single injection of 1.5 mg/kg MSF and, in addition, the rate at which enzyme is replaced in primates.

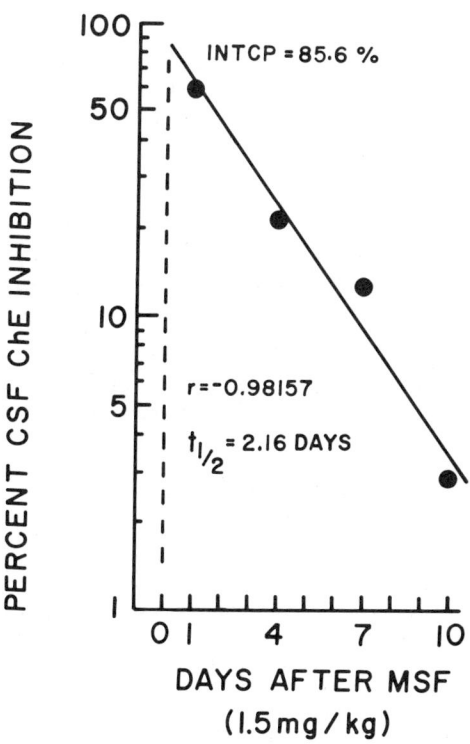

Figure 3. A pseudo-first order plot of recovery of CSF cholinesterase activity in CSF after 1.5 mg/kg MSF. 86% inhibition was produced by the single injection and the half-life for the recovery of enzyme activity (rate of new synthesis) was approximately 2.2 days.

After several months of no drug treatments, a long-term toxicity experiment was conducted. Intramuscular injections of MSF in sterile peanut oil were begun in two of the monkeys. Two others were injected with vehicle only and served as controls. The dose was increased over two months to 1.5 mg/kg which was continued for the following twelve injections (approximately five weeks). During treatment with 1.5 mg/kg by intramuscular injections two to three times per week, blood samples were drawn several times. Erythrocyte cholinesterase levels in the MSF treated monkeys were so low that they could not be determined reliably on some days. However, there was no loss of appetite, no loss of body weight, and no general behavioral indications of toxicity (e.g., no lethargy or unusual behaviors).

Two to three days after the last injection, all of the monkeys were subjected to cortical biopsy to determine the CNS effect of MSF treatment. The subjects were anesthetized and a sample of about 0.05 grams of cortex was removed. Cortex samples taken from MSF treated monkeys showed approximately 80% inhibition of cholinesterase activity relative to the cortex samples taken from the untreated controls. In view of the data presented in Figure 3 showing 2.2 days as the half-time for the recovery of CSF cholinesterase, 80% inhibition of cortex enzyme up to three days after the last dose of MSF can only be explained by cortex enzyme being replaced more slowly than cholinesterase in the CSF.

At the time the cortical biopsies were taken, a sample of blood was again drawn from the femoral vein for the determination of a full clinical chemical blood profile identical to pretreatment profiles. There were no significant changes in the clinical blood profiles of the treated monkeys as compared to their pretreatment profiles or the untreated monkeys.

After a drug-free period of several weeks, a pilot dose-response experiment was conducted. In this experiment, monkeys were injected with various doses of MSF from 0.5 mg/kg to 3.0 mg/kg. Twenty four hours after the injection, CSF samples were taken and analyzed for cholinesterase activity. Using the pseudo first-order model (Figure 3), it was estimated that 54, 73, 86, and virtually 100 percent inhibition of CSF enzyme was produced by 0.5, 1.0, 1.5, and 3.0 mg/kg MSF. The level of CSF cholinesterase inhibition was linear with log-dose with a correlation of +0.9593 over the dose range.

The lowest dose tested, 0.5 mg/kg MSF, produced an effect similar to that estimated to be necessary for an optimum therapeutic effect in humans insofar as about 50% inhibition of CSF cholinesterase has been correlated with maximum memory improvement (Thal, Fuld, Masur, Sharpless, 1983.). At much higher doses, however, no toxic effects were observed. Even at the 3.0 mg/kg dose that resulted in an estimate of virtually 100 percent inhibition at the time of the injection and an experimental determination of 78 percent inhibition 24 hours later, there were no clear signs of cholinesterase inhibition toxicity. The highest dose, 3.0 mg/kg, did appear to produce lethargy and some illness but these symptoms were gone at 24 hours and did not require veterinary attention.

DISCUSSION
Methanesulfonyl fluoride was effective in inhibiting CSF cholinesterase well over 80% in primate cortex and CSF without toxic side effects as measured either by general behavior or clinical blood analysis. Except at 3.0 mg/kg MSF, the monkeys remained vigorous and active throughout the experiment. It is, therefore, possible to produce a pharmacologically significant level of CNS cholinesterase inhibition from MSF treatment without apparent toxicity

from peripheral cholinesterase inhibition. The monkeys used in the MSF toxicity tests have shown no motor signs of delayed neurotoxicity or other problems in the eleven months during which they received MSF.

The sulfonyl fluorides appear to be efficacious cholinesterase inhibitors with an inherent selectivity for the CNS. However, they have not been as thoroughly studied as other compounds. Although it has been shown that sulfonyl fluorides do not affect norepinephrine, dopamine, or serotonin content in the cortex of rats even after several weeks of administration (Moss et al., 1985b), other parameters have not been examined. The most important of these include assessment of the effects of MSF and PMSF on brain choline acetyltransferase, choline uptake, brain proteases, and direct effects on CNS acetylcholine receptors. Evaluation of these noncholinesterase effects will be informative insofar as PMSF and MSF do not, for example, produce the same extrapyramidal motor effects in rats as those observed with physostigmine (Moss, Rodriguez, and McMaster, 1985a; Rodriguez, Moss, Reyez, and Camarena, 1986). The reason for this behavioral difference between sulfonyl fluorides and physostigmine is not clear.

The sulfonyl fluorides as therapeutic agents in SDAT, of course, share some of the problems inherent in all cholinesterase inhibitor therapies. The first is that anticholinesterase therapy can be expected to be effective only in the presence of sufficient amounts of endogenous acetylcholine. Anticholinesterases can, therefore, be expected to be effective only relatively early in the disease, before deterioration of the CNS cholinergic system has progressed too far. Also, anticholinesterase agents may only have therapeutic effects when used in combination with other therapies. For example, the best results with anticholinesterase agents may be obtained when they are used with M2 muscarinic receptor antagonists to prevent the M2 receptors on presynaptic membranes from defeating the effect of the cholinesterase inhibitor by feedback inhibition of acetylcholine release (Mash, Flynn and Potter, 1985). Similarly, combining a cholinesterase inhibitor with precursors like choline or lecithin with drugs that enhance oxidative metabolism in the CNS may prove to produce significant enhancement of cognitive function (Bartus, Dean, Sherman, Sherman, Friedman, and Beer, 1981). Lastly, of course, the most successful use of anticholinesterase agents would be as an adjunct to other treatments that effectively stop the progression of the underlying pathophysiology of the disease, preferably early enough that the CNS cholinergic system is still largely intact.

Despite the strong rational basis and limited experimental support for the cholinergic hypothesis, research in cholinergic pharmacology has not produced an effective long-lasting treatment for Alzheimer's dementia. Several factors could account for this lack of success. Alzheimer's disease is a complex process that involves many neurotransmitter systems. Because of this, it may be that no therapeutic strategy based on one system can be successful. Similarly, the well documented loss of cholinergic function within the CNS is probably an epiphenomenon that occurs as a consequence of neuronal death. It may be unreasonable to expect any important therapeutic effects to be observed unless the underlying disease process can be arrested in an early stage, while enough of the essential neurotransmitter systems are still responsive to therapeutic interventions. In spite of the marginal results obtained to date, however, further research is warranted and the results may have significant future clinical applications in SDAT or other diseases characterized by dementia and reductions in CNS cholinergic function.

CONCLUSIONS

Certain sulfonyl fluorides, particularly methanesulfonyl fluoride, are highly effective long-lasting CNS selective cholinesterase inhibitors that are remarkably low in general toxicity. These special qualities may make these unique compounds useful in treating SDAT.

ACKNOWLEDGMENTS

Supported in part by the Meadows Foundation of Dallas, Texas; NIMH through the MBRS Program, DRR, NIH; and a gift from the Moss family.

REFERENCES

Adolfsson, R., C.G. Gottfries, B.E. Roos, and B. Winblad. 1979. Changes in brain catecholamines in patients with dementia of the Alzheimer type. Brit. J. Psychia. 135: 216-223.

Andjelkovic, D., D.B. Beleslin, and B.V. Vasic. 1971. Effect of eserine injected intraventricularly on behaviour and activity of cholinesterase in some structures of the cerebral ventricles of the conscious cat. J. Pharm. Pharmac. 23:984-985.

Arai, H., K. Kobayashi, Y. Ichimiya, K. Kosaka, and R. Iizuka. 1984. A preliminary study of free amino acids in the postmortum temporal cortex from Alzheimer-type dementia patients. Neurobiol. Aging 5: 319-321.

Arendt, T., V. Bigl, A. Arendt, and A. Tennstedt. 1983. Loss of neurons in the nucleus basalis of Meynert in Alzheimer's disease, paralysis agitans and Korsakoff's disease. Acta Neuropathol. 61: 101-108.

Bartus, R.T., R.L. Dean, K.A. Sherman, E. Friedman, and B. Beer. 1981. Profound effects of combining choline and piracetam on memory enhancement and cholinergic function in aged rats. Neurobiol. Aging 2: 105-112.

Beal, M.F., M.F. Mazurek, V.T. Tran, G. Chattha, E.D. Bird, and J.B. Martin. 1985. Reduced numbers of somatostatin receptors in cerebral cortex in Alzheimer's disease. Science 229: 289-291.

Brimblecombe R.W. 1974. Drug Actions on Cholinergic Systems. Baltimore: University Park Press.

Brinkman, S.D. and S. Gershon. 1983. Measurement of cholinergic drug effects on memory in Alzheimer's disease. Neurobiol. Aging 4: 139-145.

Caroldi, S., M. Lotti, and A. Masutti. 1984. Intra-arterial injection of DFP or PMSF produces unilateral neuropathy or protection, respectively, in hens. Biochem. Pharmacol. 33: 3213-3217.

Davies, P. 1981. Theoretical treatment possibilities for dementia of the Alzheimer type: the cholinergic hypothesis. In Strategies for the Development of an Effective Treatment for Senile Dementia, eds. T. Crook and S. Gershon, 19-34. New Canaan (Conn), Mark Powley.

Davies, P., R. Katzman, and R.D. Terry. 1980. Reduced somatostatin-like immunoreactivity in cerebral cortex from cases of Alzheimer's disease and Alzheimer senile dementia. Nature 288: 279-280.

Davies, P. and A.J.R. Maloney. 1976. Selective loss of central cholinergic neurones in Alzheimer's disease. Lancet 2: 1403.

Davis, K.L., R.C. Mohs. 1982. Enhancement of memory processes in Alzheimer's disease with multiple-dose intravenous physostigmine. Amer. J. Psychia. 139: 1421-1424.

Fahrney, D.E. and A.M. Gold. 1963. Sulfonyl fluorides as inhibitors of esterases. I. Rates of reaction with acetylcholinesterase, chymotrypsin and trypsin. J. Amer. Chem. Soc. 85: 997-1000.

Gibson, C.J. and M.J. Ball. 1983. Hippocampal monoamine deficits in Alzheimer's disease. J. Neurochem. Suppl. 41: S20B.

Harbaugh, R.E., D.W. Roberts, D.W. Coombs, R.L. Saunders, and T.M. Reeder. 1984. Preliminary report: intracranial cholinergic drug infusion in patients with Alzheimer's disease. Neurosurgery 15: 514-518.

Johnson, M.K. 1980. Delayed neurotoxicity induced by organophosphorus compounds. In Mechanisms of Toxicity and Hazard Evaluation, eds. Holmstedt, Lauwerys, Mercier, Roberfroid, 27-38. Amsterdam, Elsevier/North Holland.

Kitz, R. and I.B. Wilson. 1962. Esters of methanesulfonic acid as irreversible inhibitors of acetylcholinesterase. J. Biol. Chem. 237: 3245-3249.

Mash, D.C., D.D. Flynn, and L.T. Potter. 1985. Loss of M2 muscarinic receptors in the cerebral cortex in Alzheimer's disease and experimental cholinergic denervation. Science 228: 1115-1117.

Mattio, T., M. McIlhany, E. Giacobini, and M. Hallak. 1986. The effects of physostigmine on acetylcholinesterase activity of CSF, plasma and brain. A comparison of intravenous and intraventricular administration in beagle dogs. Neuropharmacol. 25:1167-1177.

Metter, E.J., W.H. Riege, D.F. Benson, D.E. Kuhl, and M.E. Phelps. 1985. Patterns of regional cerebral glucose metabolism in Alzheimer's disease patients. In Senile Dementia of the Alzheimer's Type, eds. J.T. Hutton and A.D. Kenny, 35-47. New York, Alan R. Liss.

Mohs, R.C., B.M. Davis, C.A. Johns, A.A. Mathe, B.S. Greenwald, T.B. Horvath, and K.L. Davis. 1985. Oral physostigmine treatment of patients with Alzheimer's disease. Amer. J. Psychia. 142: 28-33.

Moss, D.E., L.A. Rodriguez, and S.B. McMaster. 1985a. Comparative behavioral effects of CNS cholinesterase inhibitors. Pharmac. Biochem. Behav. 22: 479-482.

Moss, D.E., L.A. Rodriguez, S. Selim, S.O. Ellett, J.V. Devine, and R. Steger. 1985b. The Sulfonyl Fluorides: CNS Selective Cholinesterase Inhibitors with Potential Value in Alzheimer's Disease? In Senile Dementia of the Alzheimer Type, eds. J.T. Hutton and A.D. Kenny, 337-350. New York, Alan R. Liss.

Moss, D.E., L.A. Rodriguez, W.C. Herndon, S.P. Vincenti, and M.L. Camarena. 1986. Sulfonyl Fluorides as possible therapeutic agents in Alzheimer's Disease: Structure/activity relationships as CNS selective cholinesterase inhibitors. In Alzheimer's and Parkinson's Disease: Strategies in Research and Development, eds. A. Fisher, C. Lachman, and I. Hanin, 551-556. New York, Plenum Press.

Myers, D.K. and A. Kemp. 1954. Inhibition of esterases by the fluorides of organic acids. Nature 173: 33-34.

Nakano, I. and A. Hirano. 1983. Neuron loss in the nucleus basalis of Meynert in parkinsonism-dementia complex of Guam. Ann. Neurol. 13: 87-91.

Nakano, I. and A. Hirano. 1984. Parkinson's disease: neuron loss in the nucleus basalis without concurrent Alzheimer's disease. Ann. Neurol. 15: 415-418.

Perry, E.K., P.H. Gibson, G. Blessed, B.E. Tomlinson, and R.H. Perry. 1977. Neurotransmitter abnormalities in senile dementia: choline acetyltransferase and glutamic acid decarboxylase activities in necropsy brain tissue. J. Neurol. Sci. 34:247-265.

Perry, E.K., B.E. Tomlinson, G. Blessed, K. Bergmann, P.H. Gibson, and R.H. Perry. 1978. Correlation of cholinergic abnormalities with senile plaques and mental test scores in senile dementia. Brit. Med. J. 2: 1457-1459.

Rodriguez, L.A., D.E. Moss, E. Reyez, and M.L. Camarena. 1986. Perioral behaviors induced by cholinesterase inhibitors: A controversial animal model. Pharm. Biochem. Behav. 25, 1217-1221.

Rossor, M.N., P.C. Emson, C.Q. Mountjoy, M. Roth, and L.L. Iverson. 1980. Reduced amounts of immunoreactive somatostatin in the temporal cortex in senile dementia of the Alzheimer type. Neurosci. Lett. 20: 373-377.

Summers, W.K., L.V. Majovski, G.M. Marsh, K. Tachiki, and A. Kling. 1986. Oral tetrahydroaminoacridine in long-term treatment of senile dementia, Alzheimer type. N. Eng. J.Med. 315: 1241-124.

Thal, L.J., P.A. Fuld, D.M. Masur, and N.S. Sharpless. 1983. Oral physostigmine and lecithin improve memory in Alzheimer's disease. Ann. Neurol. 13: 491-496.

Uhl, G.R., M. McKinney, J.C. Hedreen, C.L. White, III, J.T. Coyle, P.J. Whitehouse, and D.L. Price. 1982. Dementia pugilistica: loss of basal forebrain cholinergic neurons and cortical cholinergic markers. Ann. Neurol. 12: 99.

Whitehouse, P.J., D.L. Price, R.G. Struble, A.W. Clarke, J.T. Coyle, and M.R. DeLong. 1982. Alzheimer's disease and senile dementia: loss of neurons in the basal forebrain. Science 15: 1237-1239.

Wisniewski, H.M., G.S. Merz, G.Y. Wen, K. Iqbal, I. Grundke-Iqbal. 1985. Morphology and biochemistry of Alzheimer's disease. In Senile Dementia of the Alzheimer's Type, eds. J.T. Hutton and A.D. Kenny, 263-274. New York, Alan R. Liss.

INTRA-CEREBRO-VENTRICULAR BETHANECHOL: Dose and Response

Stephen Read

Dept. of Psychiatry and Neurology, Harbor-UCLA Medical Center, and
John Douglas French Center, 3951 Katella Ave., Los Alamitos, CA

INTRODUCTION

Although the cholinergic deficit remains the best established neurochemical lesion in Alzheimer's Disease (AD), the reasons for the disappointing results of cholinergic enhancement therapy in AD treatment remain obscure (Bartus, et al, 1987). The literature contains several hopeful reports, but systematic studies have not been encouraging. Sustained general benefit is claimed in only one study (Summers, 1986), and these data are now controversial. Many technical and procedural questions (Table 1), complicate the assessment of any potential therapy for AD. These present methodological difficulties which few, if any studies have addressed sufficiently to warrant firm conclusions concerning the potential for therapeutic benefit.

TABLE 1: SOURCES OF DIFFICULTY IN ASSESSING THE RESPONSE IN ALZHEIMER'S DISEASE TO CHOLINERGIC ENHANCEMENT

I. Disease factors	II. Pharmacologic
Accuracy of diagnosis	Compliance
Clinical definition of problem	Side effects from peripheral cholinergic systems
Lack of objective (biological) measures of response	Side effects from non-memory related CNS cholinergic systems
Variability in symptomatology ?presence of sub-types	Variable penetration of the blood brain barrier
Variability in rate of progression	Variable requirement for intact cholinergic synapse
Stage of illness	Dose-response relationship

Harbaugh and colleagues (1984) reported the feasibility of intra-cerbro-ventricular drug delivery in patients with AD. Surgery was well tolerated in these initial four patients. Although the family members reported improvement over the narrow range of doses examined, the authors carefully eschewed claims of efficacy. Intracerebro-ventricular infusion of bethanechol chloride (ICVBC), the only muscarinic agonist with a safety record in humans, can potentially control many of the pharmacokinetic variables listed in Table 1. In addition, biopsy at the time of cannula placement allows pathological confirmation of clinical diagnosis as well as other studies, such as neurochemical analysis, which could address the question of variability in treatment response.

Following this initial report, we were intrigued with ICVBC as a technique to address the basic question of the role of cholinomimetics in the treatment of AD. In view of the inconsistencies in the literature concerning the response to cholinergic drugs, we undertook an open study to investigate parameters of response. We chose a dose-response paradigm for three reasons:
    1. Graded change with dose may be the most convincing evidence of drug effect.
    2. In animal studies, response has an "inverted-U" shape with respect to dose (Cherkin and Flood, 1983) and the "width" of this U is particularly narrow for bethanechol (Cherkin, personal communication).
    3. There might be significant variability among patients either in the quality or kinetics of response.
Although small, the results from this study may be informative for interpreting and modulating the unenthusiastic results from subsequently reported double-blind crossover studies (Penn et al, 1988; Harbaugh, personal communication).

METHODOLOGICAL CONSIDERATIONS AND PROCEDURES

Recruitment: Candidates for the study had impaired memory, language, visual-spatial skills, cognition, and personality and a history of insidious onset with gradual progression typical for AD (Cummings and Benson 1983). Extrapyramidal signs, focal neurologic or cognitive findings, and depression, although not incompatible with a diagnosis of AD clinically, are atypical (McKhann et al, 1984) and constituted exclusions in our study. We also excluded patients with medical diagnoses (e.g., peptic ulcer disease, low blood pressure, bradycardia) which might confer especial risk from cholinergic drugs.

Candidates had to score 16-27 on the Mini-mental State (MMS) (Folstein,1975). We sought patients with mild dementia in order to avoid floor effects in the test battery, to have a relatively high level of function (to preserve) at the outset, and to allow patients meaningfully to participate in the informed consent procedure. To minimize the risk of aphasia following surgery, all were right-handed. Each candidate, together with his spouse and any other requested family members, was informed of the risks and benefits foreseeable in the proposed study as reviewed and approved by our Institutional Review Board. Participation in the study required the consent of both patient and spouse.

Evaluation: Many familiar neuropsychological tests are poorly suited for a dose-response study. Repeated administration would not only be tedious because of their length, but would raise questions about their validity. The core of the battery employed for the weekly visits required for the dose-response phase was the memory test of Randt et al (1980). Its relatively brief subtests facilitate use in an impaired population and the availability of five equivalent forms make it suitable for weekly measures. Supplementing the Randt were the MMS, Trails A (Reitan, 1979) and word list generation (Benton and Hamscher, 1967), chosen to allow some compatibility with the other concurrent clinical trials. Unfortunately absent in this series is a detailed measure of visual-spatial skills.

Each subject and spouse made at least five weekly visits to clinic after diagnostic evaluation and consent. At each visit, this short battery was administered followed by other assessments including the spouse's rating of status, inquiry about physical, cognitive, and psychiatric symptoms, vital signs (orthostatic pulse and blood pressure, temperature, and weight), and overall clinician's assessment of progress. During the dose response phase, the pump was also emptied and refilled at each visit.

Surgery: Each participating patient was admitted for surgery after serial clinical and neuropsychological evaluations. Right frontal craniotomy and biopsy of prefrontal cortex were performed under anesthetic consisting of intravenous diazepam and local lidocaine (without epinephrine). The patient was then placed under general endotracheal anesthesia and redraped for placement of intra-cerebro-ventricular cannula, abdominal placement of model 4000 Infusaid pump, and their connection by subcutaneous silastic tube. After recovery, patients were discharged home and followed in clinic.

Dose-Response Phase: After discharge, each patient resumed weekly clinic visits, following the pre-operative assessment protocol. After stable performance was regained, we began bethanechol chloride at an initial dose of 0.025 mg/ml. The drug concentration was increased each week until side-effects emerged or performance declined for three weeks in a row, at which time dose was decreased incrementally until an optimal dose was defined.

Continuation Phase: Following the dose-response phase, patient and spouse could elect to continue ICVBC at the optimal dose or to discontinue drug infusion. Patients were then followed in clinic as indicated. Continuing patients were seen approximately monthly, the iterative battery was scored and the pump refilled, with dose adjusted per clinical situation.

RESULTS

Recruitment: The five subjects, designated A, B, C, D, and E in order of their enrollment, were 53-64 (mean 60) years old. All had so-called "pre-senile" onset and had retired because of their dementia. They scored 18-27 on the MMS and 0.5 or 1.0 on the Clinical Dementia Rating Scale (Berg, et al, 19). Three had established family history of probable or definite Alzheimer Disease. None was depressed at entry.

Baseline Evaluation Phase: Clinical status remained stable for the several weeks of baseline assessments; neither clinician nor spouse noted a step change in overall function. In testing, the MMS, Randt, and Word List generation were acceptably stable, i.e., the standard deviation of baseline readings was substantially less than the mean score. The time to complete Trails A vacillated to a greater degree. It also proved to be an unstable measure in the dose response phase.

Table 2. Baseline Measures--Mean(S.D.)

Subj	MMS	Word List	Randt	Trails A
A	17.8 (0.5)	11 (0)	54.5 (3.4)	251 (76)
B	21.6 (1.8)	19.6 (4.2)	45.6 (10.3)	222 (80)
C	26.1 (2.4)	42.6 (7.2)	85.6 (7.2)	114 (25)
D	24.8 (1.7)	47.5 (7.8)	94.8 (8.5)	94 (14)
E	14.0 (1.5)	18.8 (3.5)	28.5 (1.9)	105 (39)

Surgical Phase: Biopsy confirmed the presence of Alzheimer's Disease in all five patients. X-ray CT scan confirmed that the cannula penetrated the ventricular system in all five. The cannula tip appeared to abut left hemisphere tissue in patients A, B, and D. Patients B, C, and D had no complications and returned to baseline clinical and neuropsychological function within three weeks.

Patient A had irritability and distress after surgery. He was found to have purulent cerebrospinal fluid four weeks after surgery, despite the absence of fever, meningismus, or localizing signs. The Staph. epidermidis infection was successfully treated with a three week course of oral rifampin and oxacillin and vancomycin administered into the Rickham reservoir or through the pump side-port.

Immediately post-operatively, Patient E had left upper extremity monoplegia, focal seizures with generalization and language impairment. X-ray CT scan revealed a large intracerebral hematoma of the right hemisphere. His course was further complicated by fever and severe allergic reactions to antibiotics and phenytoin.

Dose-Response Phase: Bethanechol chloride 0.025 mg/ml was started when each patient had stabilized after surgery. Patients B, C, and D were nearly at baseline two weeks post-op and began drug on week three. Patients A and E began drug only after their extended convalescences as described above, and at the initiation of drug neither had returned to pre-surgical performance on clinical or test performance. In Figure 1 MMS, Randt scores, and dose are plotted over time for each patient. Both pre- and post surgical baselines are shown for Patients A and E.

Patients B, C, and E showed a positive clinical response with optimal doses of 0.35, 0.95, and 0.20 mg per day, respectively. For B and especially C, neuropsychological tests correlated well with clinical impression. Performance improved somewhat before family and clinician perceptions and held up after psychomotor slowing began to intervene at higher doses. Interpretation of E's course was most difficult because of the superimposed stroke recovery, but interestingly, observations from concomitant speech therapy corroborated our clinical impressions. Responsitivity was ablated past some dosage level in all three (e.g. Patient C, Weeks 10-14), but better performance returned upon reduction of dose from the maximum. There was little overlap in the beneficial dose range for these three.

For the other two patients, the emergence of psychiatric symptomatology dominated the clinical picture. Patient A had some reported improvement in memory, and this observation had some support from neuropsychological test scores. However, he was distressed at the start of drug treatment and became increasingly agitated, dysphoric, and depressed as dose was increased. In the sixth week of ICVBC he had a generalized seizure and was placed on phenytoin, but dysphoria continued and abated only when ICVBC was withdrawn.

Patient D initially regained what his family recognized as his native critical nature, but soon developed irritability and began to hoard things and obsess about money. He eventually became frankly depressed. In addition, his neuropsychological test performance declined with increasing dose. He also developed generalized seizures which were controlled with medication.

Continuation Phase: Patients B, C, and E elected to continue ICVBC at the optimal dose defined above. In addition to his depression and agitation, patient A developed a sterile effusion at around the pump pocket in the superficial abdominal wall. Pump, tubing, and cannula were removed at a second operation. After further dose manipulations in Patient D, and disappointing results from courses of haloperidol and nortriptyline, we decided to discontinue drug treatment and he also had pump, tubing, and cannula removed without incident. Details of the further course are beyond the scope of this chapter, but Patients A and E have never returned home for any length of time, while the other three remain at home on ICVBC.

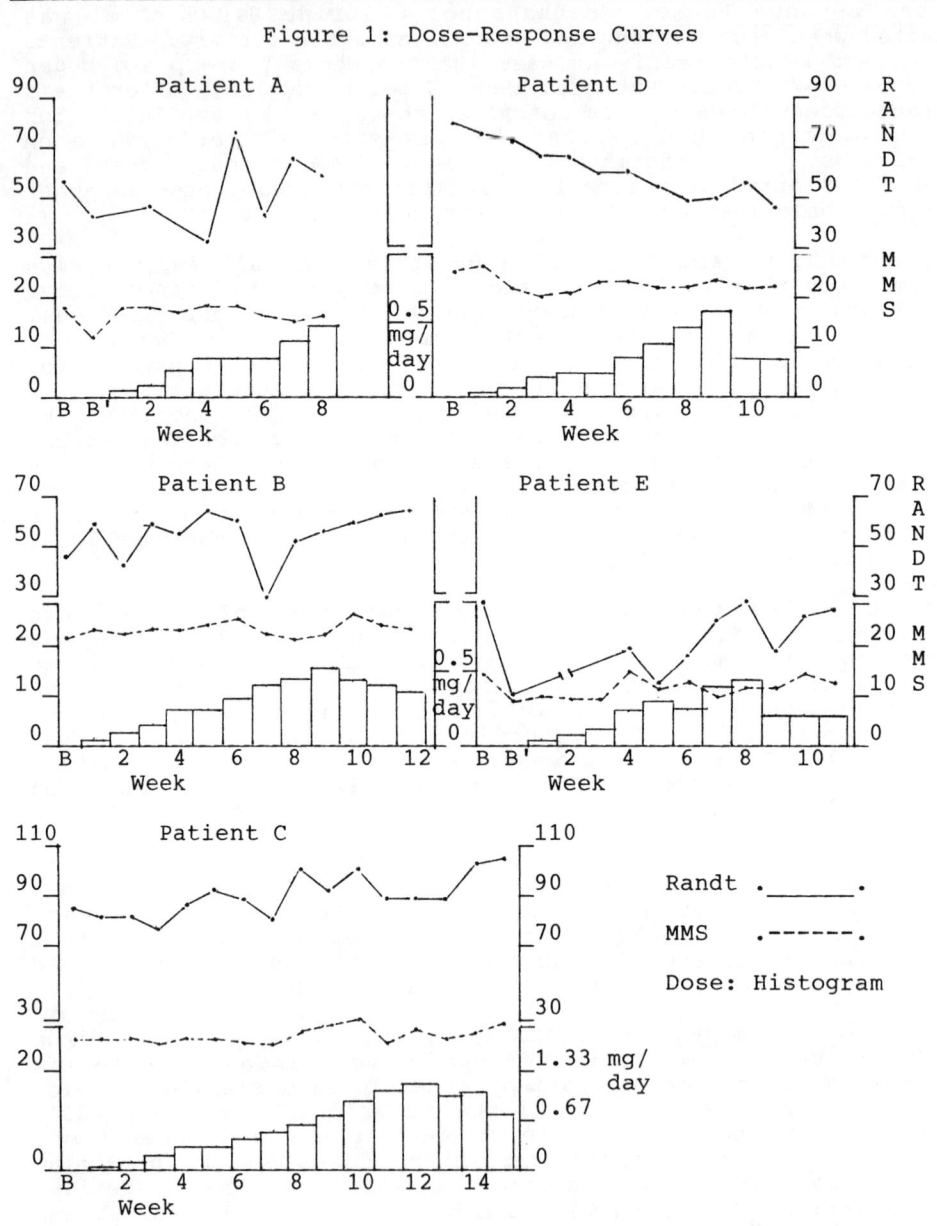

Figure 1: Dose-Response Curves

DISCUSSION

Since Harbaugh et al's 1984 report, there have been two studies of ICVBC utilizing a double-blind fixed dose crossover design. Harbaugh (personal communication) has also made available data from a multi-center double-blind crossover study of a single dose between 0.2 and 0.4 mg/d. Group differences in both protocols were very small; and were not felt to justify the surgical morbidity incurred by the procedure.

The major theme of this pilot study is the diversity of responses observed, even within a small group of subjects. Overall, three patients had positive responses to drug, and there was little overlap in the dose ranges showing response. Two patients had negative responses, dominated by the emergence of agitation and depression, although they showed opposite responses in their cognitive testing. Whatever the source of the variability in response, it must be considered in interpretating fixed dose studies of ICVBC, as well as other cholinergic and non-cholinergic drug trials.

The divergence in outcome does not appear to be explained by variability in patient sample, instability of the measures chosen, or surgical outcome. The sample was homogeneous with respect to conventional disease variables--all were male, caucasian, and had early onset of disease. Duration of disease did vary in the sample, but the two with relatively short duration (C and D at three years) had the most divergent outcomes. The three who showed an interpretable positive response had family history's of Alzheimer Disease, but the significance of this is not obvious.

Clinical and neuropsychological performance was stable on several measures at repeated visits prior to surgery. In particular, depression was observed in no subject. Surgical morbidity was serious and dominated the experience of patients A, and especially E, but did not appear to be determinate in the outcome of the drug trial. The two patients with adverse outcomes did both develop seizures during the drug trial. Seizure control was achieved in both, however, and the severity of the other, behavioral manifestations had been manifest before seizure and varied in proportion to dose.

Depression has been reported with the anticholinesterase physostigmine (Davis et al, 1976) and with the muscarinic agonist oxotremorine (Davis et al, 1987), but not previously with bethanechol. The dose-response paradigm allowed the accentuation of symptoms due to increasing dose, but even the mild symptoms at low dose (e.g. 0.2 mg/day, in the range of doses examined in the multi-center crossover study) intruded upon clinical and psychometric response. Inclusion of negative responders in an aggregate response assessment could possibly obscure a positive response occurring in another group of AD patients.

In the three patients in whom we observed beneficial effects of ICVBC, the effects were most marked over a rather narrow range of dose. Although the drug did not produce a dramatic amelioration of dementia, the responses were not inconsequential: Patients B and C had improvements on the Randt memory test two and three times respectively the standard deviation of pre-operative baseline, and patient C scored in the low normal range at his best dose. These effects appear to be sustained and correlate persuasively with family reports of increased alertness and attentiveness at home.

The only other systematic dose adjustment study is in the report by Penn et al (1988), who did not find a significant effect. Their trial, however, involved much larger dose increments (of 0.35 mg/day), used different instruments, and included no patient with MMS>23 prior to surgery. It is of interest that among our three responders, milder dementia, shorter duration of disease, magnitude of response, and size of optimal dose had the same rank order. Most surprising to us is that no adverse behaviors developed on the doses employed in the Chicago study. In our experience, this dosage interval could well have missed a dosage window.

Patients B, C, and E remain at home on drug more than two years post-operatively. Clinical observation and dose adjustment suggest there is continued benefit from ICVBC, although the evaluation of long term effects is, if anything, more complicated than determining response in the short term.

CONCLUSIONS

1. Careful application of current diagnostic criteria can lead to the identification of patients with Alzheimer's disease.
2. There is significant surgical risk in the current procedures for intra-cranial drug delivery.
3. Some patients with AD appear to benefit from bethanechol chloride. This effect is significantly dose-related in that doses above or below the optimal dose range lead to deteriorated performance.
4. In other patients with AD, bethanechol chloride causes dysphoria, agitation, and frank depression. Such a depressive reaction outweighs any effect on memory or cognitive function.
5. While there may be long term benefit from ICVBC, no current study is designed to assess this.
6. In view of Nos. 3 and 4, a fixed dose, doubleblind crossover study is likely to underestimate the benefit available from ICVBC to at least a subset of patients with AD.
7. A physiological marker of drug effect would greatly facilitate the conduct of a dose-response study.

ACKNOWLEDGEMENTS:

This work was supported by a grant from the John Douglas French Foundation for Alzheimer's Disease and by a Research Advisory Group award from the Veterans Administration. This paper does not necessarily represent the views of either organization. Intermedix-Infusaid provided pumps and technical assistance.

Collaborators include John Frazee, M.D., Jeffrey Cummings, M.D., Jill Shapira, R.N., M.N.-C., Cheryll Smith, Ph.D., and Uwamie Tomiyasu, M.D. Conservations with Bruce Miller, M.D., Donald Jenden, M.D., Roger Russell, M.D., Andrew Leuchter, M.D., and Arthur Cherkin, M.D. were enormously helpful, as well as with Robert Harbaugh, M.D. and the other members of the ICVBC group.

REFERENCES:

Bartus, R.T., R.L. Dean, and C. Flicker. 1987. Cholinergic Psychopharmacology: An integration of human and animal reseach on memory. In Psychopharmacology: The Third Generation of Progress, ed. H.Y. Meltzer, 219-232. New York: Raven Press.

Benton A.L., and K. Hamsher. 1967. Multilingual Aphasia Examination. Iowa City: University of Iowa Press.

Cherkin, A. and J.F. Flood. 1983. Remarkable potentiation among memory-enhancing cholinergic drugs in mice. Intervention in the Aging Process, Part A: Quantitation, Epidemiology, and Clinical Research. New York: Alan R. Liss, Inc., pp 225-245.

Cummings, J.L. and D.F. Benson. 1983. Dementia: A Clinical Approach. Boston: Butterworths.

Davis, K.L., L.E. Hollister, J. Overall, A. Johnson, and K. Train. 1976. Physostigmine:Effects on cognition and affect in normal subjects. Psychopharmacology 51:23-27.

Davis, K.L., E. Hollander, M. Davidson, B.M. Davis, R.C. Mohs, T.B. Horvath. 1987. Induction of depression with oxotremorine in patients with Alzheimer's disease. Am J Psychiatry 144:468-471.

Folstein, M.F., S. Folstein, and P.R. McHugh. 1975. The Mini-mental state. J Psychiatr Res 12:189-198.

Harbaugh, R.E., D.W. Roberts, D.W. Coombs, R.L. Saunders, and T.M. Reeder. 1984. Preliminary report: Intractanial cholinergic drug infusion in patients with Alzheimer's disease. Neurosurgery 15:514-518.

McKhann, G, D. Drachman, M.F. Folstein, R. Katzman, D. Price, and E.M. Stadlan. 1984. Clinical diagnosis of Alzheimer's disease. Neurology 39:939-944.

Penn R.D., E.M. Martin, R.S.Wilson, J.H. Fox, and S.M. Savoy. 1988. Intraventricular bethanechol infosion for Alzheimer's disease. Neurology 38:219-222.

Randt, C.T., E.R. Brown, and D.P. Osborne. 1980. A memory test for longitudinal measurement of mild to moderate deficiencies. Clinical Neuropsychology 2:184-194.

Reitan R.M., 1966. A research progam of the psychological effects of brain lesions in human beings. In Ellis N.R. (Ed.) International Review of research in mental retardation. New York: Academic Press.

Summers, W.K., L.V. Majovski, G.M. Marsh, K.Tachiki, and A. Kling. 1986. Oral tetrahydroaminoacridine in long-term treatment of senile dementia, Alzheimer type. N. Engl. J. Med. 315:1241-1245.

INTRAVENTRICULAR BETHANECHOL INFUSION FOR ALZHEIMER'S DISEASE: Results of Double-Blind and Escalating Dose Trials

R.D. Penn, E.M. Martin, R.S. Wilson, J.H. Fox, S.M. Savory, and D.A. Grosse
Rush-Presbyterian-St. Luke's Medical Center,
1653 West Congress Parkway, Chicago, IL

Controlled trials in Alzheimer's disease (AD) with a variety of cholinergic agents have produced mixed results (Bartus, Dean, Beer, and Lippa, 1982; Hollander, Mohs, and Davis, 1986; Little, Levy, Chuaqui-Kidd, and Hand, 1985). Peripheral cholinergic side effects are common in such trials and may limit delivery of sufficient doses of medication through the blood-brain barrier. The preliminary observations reported by Harbaugh and colleagues (Harbaugh, Roberts, Coombs, Saunders, and Reeder, 1984) on the effects of intraventricular infusion of bethanechol chloride, a muscarinic agonist, provided the impetus for the present study. In this paper, we report the results of double-blind crossover, open-escalating dose, and chronic maintenance trials with bethanechol chloride in AD.

METHODS. The initial study consisted of a 24-week, double-blind crossover trial of low-dose bethanechol. The second study was conducted in two phases: an 8-week, double-blind crossover trial of low-dose bethanechol using revised neuropsychological and behavioral assessment procedures, followed by an open escalating-dose trial of bethanechol at 0.35 to 1.75 mg/d. Following completion of the second study, patients have been maintained chronically on bethanechol at doses ranging from 0.35 to 1.75 mg/d. These patients have now been followed for an average of two years, and the course of their dementia over this period will be presented.

Subjects. The initial study sample consisted of seven women and four men, with a mean age of 64.9 years (SD = 7.1), a mean of 12.9 years of education, and a mean of 3.9 years of dementia history by family report (SD = 2.6). Subject selection procedures are described elsewhere (McKhann, Drachman, Folstein, Katzman, Price, and Stadian, 1984). All patients met NINCDS criteria for probable AD (Martin, Wilson, Penn, Fox, Clasen, and Savoy, 1987). Written consent to participate was obtained from each patient and family.

The sample for the second study consisted of nine of the ten patients who had participated in the initial double-blind study. A tenth patient entered a nursing home and was placed on haloperidol in sufficiently high doses to preclude further study participation.

Pump implantation. All infusions were delivered via an implanted drug pump (Intermedics Infusaid Model 400) placed in a subcutaneous pocket made in the abdominal wall under general anesthesia. The drug pump was joined to a Silastic outflow catheter which was tunneled subcutaneously to connect with the pump following the procedure employed by Harbaugh et al (Harbaugh, Roberts, Coombs, Saunders, and Reeder, 1984). Each patient had a 3- to 4-cm diameter left frontal craniotomy, approximately 1 cm posterior to the coronal

4 cm off the midline, for removal of a cortical biopsy specimen and passage of the catheter into the left lateral ventricle. Approximately 48 hours after implantation, the intraventricular flow pattern via the catheter was evaluated by In^{111} scan.

All biopsy specimens showed senile plaques in sufficient quantitites to meet research criteria for definite AD (Martin, Wilson, Penn, Fox, Clasen, and Savoy, 1987). One patient was withdrawn from study because of a transient postoperative right hemiparesis and dysphasia which apparently resulted from difficulty positioning the catheter in the ventricle. He recieved bethanechol in an open trial. A second patient developed a chronic subdural hematoma, which was drained; she entered the double-blind protocol after full recovery.

Procedures. Approximately 4 weeks following pump implantation, each patient entered a 24-week double-blind protocol. Each patient received three 4-week infusions of approximately 0.35 mg/d of bethanechol and three infusions of saline in a quasi-random order (ie, no more than two consecutive infusions of the same substance). Testing was performed once every 4 weeks, midway through each infusion period.

The neuropsychological test battery for the initial study included measures of global mental status (Folstein's Mini-Mental State Test Folstein, Folstein, and McHugh, 1975), language production (Controlled Oral Word Association Test, Benton and Hamsher, 1978), and memory (Benton's Serial Digit Learning, Benton, Hansher, Varney, and Spreen, 1983 and the Buschke-Fuld selective reminding procedure (Buschke and Fuld, 1974). The patients' families completed a 15-item activities of daily living (ADL) rating scale once per week. Behavioral abnormalities were monitored with the Daily Patient Report (DPR), a 26-item checklist; three times during the third week of each infusion period, a family member was interviewed over the telephone regarding the occurrence of each abnormal behavior during the past 24 hours. ADL data were collected more frequently than cognitive or DPR scores in order to insure regular participation in data collection by the families.

The second study was conducted in two phases. During the first phase, each patient received two 4-week infusions, one of bethanechol and one of saline placebo. Order of infusion was counterbalanced across patients. Testing was performed four times, during the second and fourth week of each infusion period. The purpose of this initial phase was to obtain preliminary data on the patients' responses to low-dose bethanechol (0.35 mg/d) using a revised battery of outcome measures.

At the completion of this initial phase, each patient entered an open escalating-dose trial during which bethanechol was infused continuously. Medical concerns regarding cholinergic toxicity precluded the use of a double-blind design for this phase of the study. Drug dosage was increased gradually, in increments of 0.35 mg/d every 2 weeks. Testing was performed at dosages of 0, 0.35, 1.05, and 1.75 mg/d, approximately 4 weeks apart for most patients.

The neuropsychological test battery for the second study included three measures retained from the initial set of outcome measures: The Mini-Mental State, Serial Digit Learning, and Controlled Oral Word Association. Three new measures were added to the cognitive battery; a 20-trial measure of simple auditory reaction time, designed to assess alertness; the Token Test (Benton and Hamsher, 1978) a measure of language comprehension; and experimental measures of pictorial naming and recognition memory. The behavioral outcome measures were revised as well. The Lawton-Brody Instrumental Activities of Daily Living Scale (IADL) (Lawton and Brody, 1969) replaced the ADL rating scale previously employed. In addition, the DPR was lengthened from 26 to 37 items to increase reliability and sample behavior problems more comprehensively. Finally, mood was assessed with a set of 19 items from the Katz Adjustment Scale (KAS)-Informant Version (Katz and Lyerly, 1963). The families filled out the IADL once weekly, the KAS items biweekly, and completed the DPR by telephone interview three times every 2 weeks (double-blind phase) or every 4 weeks (dose increase).

RESULTS. Initial study. The bethanechol was well-tolerated at 0.35 mg/d; no patients had side effects resulting in withdrawal from the study. Approximately one-half of the patients developed transient nausea during the initial drug infusion, but nausea did not occur reliably during subsequent drug infusions. One to 3 days were necessary for the drug to become effective as indexed by the appearance of nausea. Withdrawal periods appeared to be approximately the same length. Thus, the 2-week interval between the beginning of each new infusion and neuropsychological testing seems appropriate with sufficient time to allow for wash-out periods. Neurologic examinations following the double-blind trial were unchanged and no extrapyramidal symptoms or myoclonous occurred. Outcome scores for this initial study were evaluated by repeated-measures analyses of variance using a two-factor model (drug X testing period) for the cognitive scores and a three-factor model (drug X test X week) for the ADL and DPR scores. A significance probability of 0.05 was employed for all tests. The family of one patient did not cooperate with data collection procedures adequately, and these data were excluded from analysis.

No significant group difference in test scores for drug versus placebo infusions was found for the ADL or cognitive measures. There was a trend toward decreased abnormal behavior scores during drug infusion relative to placebo periods, but the difference fell short of statistical significance ($F[1,8] = 4.07$; $p < 0.08$).

The ADL and DPR measures provided sufficient data points for single-subject analyses; these scores were analyzed by matched-pairs t tests. Three of nine patients' ADLs and/or abnormal behavior scores were significantly ($p < 0.05$) improved during drug relative to placebo infusions; no patient significantly improved during placebo infusions. These a posteriori analyses, although suggestive of improvement in a subset of patients, must be regarded as exploratory.

Second study. All nine patients completed the 8-week double-blind phase. One patient died suddenly of unrelated causes during the course of the study. Her data from the escalating-dose trial, in progress at the time of her death, were not used. Seven of eight patients were able to tolerate (ie, did not show serious side effects) doses up to 1.75 mg/d, although two patients could not tolerate dose increments of 0.35 mg and required more gradual dose increases (one patient developed severe nausea, one developed hyperventilation). Consequently, the escalating-dose trial required approximately 12 to 16 weeks to completion for these two patients and 8 weeks for all others. An eighth patient developed myoclonus at 0.70 mg/d and was maintained on bethanechol at lower doses. One experimental patient refused to participate in cognitive testing. Thus, complete cognitive data sets were available for eight of nine patients for the double-blind phase and six of eight patients for the escalating-dose phase. Behavioral data (IADL, DPR, and KAS scores) were available for all patients completing each phase.

For the double-blind phase, a series of 2 X 2 repeated-measures analyses of variance was computed with treatment (drug/placebo) and testing as the within-subjects factors. With the exception of a significant treatment X test interaction for the Controlled Oral Word Association Test ($F[1,7] = 7.86$;

TABLE 1

Mean Scores on Outcome Measures: Dose Increase

Measure	Bethanechol Dosage (mg/day)			
	0	.35	1.05	1.75
Mini-Mental State[a]	15.83	16.25	15.17	13.83**
SD	4.88	5.72	5.75	5.49
Token Test	28.75	29.42	28.83	26.50
SD	8.05	8.81	7.84	7.71
Serial Digit Learning	40.64	42.36	30.50*	41.28
SD	22.45	30.69	24.85	24.30
Verbal Fluency	19.00	20.17	22.17	17.00
SD	9.75	8.65	13.22	12.74
Naming	35.17	34.92	32.83	32.33
SD	13.22	14.47	15.52	17.66
Recognition Memory	13.25	13.33	13.50	13.17
SD	3.36	3.09	3.78	5.88
Reaction Time[b]	2.64	2.53	2.78	2.96*
SD	.40	.22	.42	.54
Katz Adjustment Scale[c]	41.21	39.07	36.14*	41.57
SD	10.51	10.14	9.64	10.32
Lawton-Brody IADL[c]	46.64	46.54	47.88	49.04**
SD	5.11	4.64	6.73	6.19
Daily Patient Report[c]	12.62	12.05	9.55*	9.86*
SD	4.58	6.05	6.45	3.07

[a] $n=6$ unless indicated otherwise.

[b] Log-transformed.

[c] $n=7$.

* $p < .05$ (vs. placebo)

** $p < .01$ (vs. placebo)

$p < 0.05$), there were no significant main effects or interactions. An analysis of simple main effects of the treatment X test interaction indicated that scores on this measure were lower during placebo relative to drug at the end of the infusion period ($p < 0.05$), but scores did not differ at the middle of the infusion period. These negative results, obtained using both previously employed and new measures, were consistent with the results of the previous double-blind trial in that no significant improvements in functioning were observed when patients received 0.35 mg/d of bethanechol.

Table 1 shows the mean cognitive and behavioral scores obtained on placebo and 0.35, 1.05, and 1.75 mg/d bethanechol by the patients who completed the escalating-dose phase. All data were analyzed using repeated measures analyses of variance, with dose as the within-subjects factor. Reaction time, Mini-Mental State, and Serial Digit Learning scores showed significant dose effects (log-transformed reaction time, $F[3,15] = 4.85$, $p < 0.05$; Mini-Mental State, $F[3,18] = 4.86$, $p < 0.02$; Serial Digit Learning, $F[3,18] = 4.10$, $p < 0.05$). Post hoc analyses using the Duncan test indicated that mean reaction times were slower at 1.75 mg than at 0 or 0.35 mg ($p < 0.05$ for both tests). Similarly, Mini-Mental State scores were lower at 1.75 mg relative to 0 and 0.35 mg ($p < 0.01$ for both tests). Serial Digit Learning scores were lower at 1.05 mg relative to 0.35 and 1.75 mg ($p < 0.05$ for all tests). There were no significant dose effects for the remainder of the cognitive measures.

Analyses of the behavioral data showed significant dose effects for all measures (KAS, $F[3,18] = 4.24$, $p < 0.05$; IADL, $F[3,18] = 5.67$, $p < 0.01$; DPR, $F[3,18] = 3.11$, $p = 0.05$). Post hoc analyses of the mean KAS scores indicated that scores at 1.05 mg were significantly lower (improved) than scores at 0 or 1.75 mg ($p < 0.05$ for both tests). Finally, mean abnormal behavior frequency (DPR scores) was significantly reduced at 1.05 and 1.75 mg relative to placebo ($p < 0.05$ for both tests).

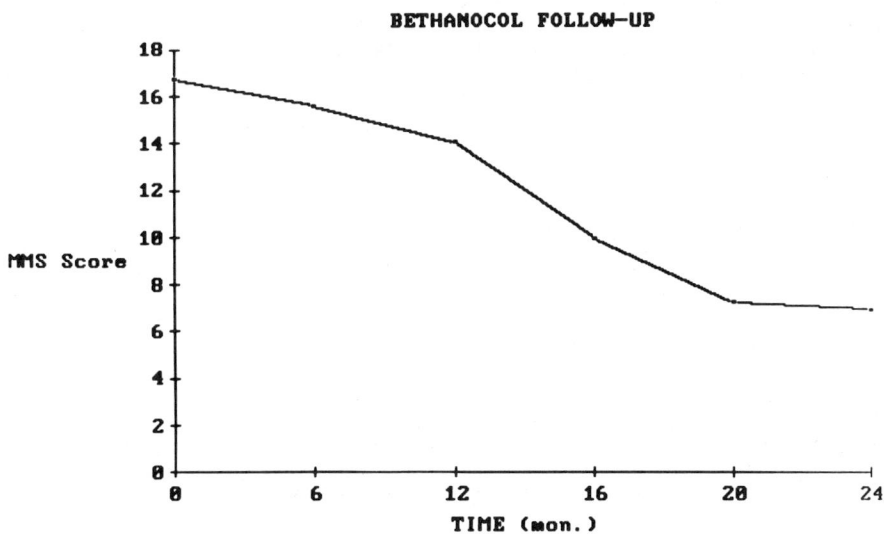

Two years of follow-up data are available on seven patients who participated in the dose escalation study. MMS scores from the baseline to most recent examination are presented in the figure. Despite chronic treatment with bethanechol in doses ranging from 0.35 to 1.05 mg/d, the dementia has continued to progress ($F (1,6) = 33.8$, $p < .05$).

DISCUSSION. We found no evidence that bethanechol reliably enhances cognitive functioning in patients with mild to moderately severe AD. Patients' scores on cognitive measures did not improve significantly during infusions of bethanechol at three different doses relative to placebo. In addition, low doses (0.35 mg/d) of bethanechol do not appear to affect patient behavior reliably.

Our data do suggest that a moderately increased bethanechol dosage (1.05 mg/d) may have a palliative effect on patient behavior, since rated mood disturbance and abnormal behavior frequency were decreased significantly at this dosage compared with placebo. However, doses as high as 1.75 mg/d appear detrimental to patient functioning since rated impairment in activities of daily living was significantly increased at this dosage compared with placebo and 0.35 mg/d. This trend toward increasing impairment at higher dosages is supported by the cognitive data: both global mental status and reaction time scores were significantly worse at 1.75 mg relative to placebo and 0.35 mg. The isolated decrease in Serial Digit Learning scores at 1.05 mg/d, however, is not readily explained since scores on this measure were not significantly decreased at the highest dosage.

This pattern of results suggests that bethanechol, like physostigmine (Davis, Hollister, Overall, Johnson, and Train, 1976) has a narrow therapeutic window. Our data indicate that patients' optimal doses should be determined individually. Although a dosage of 1.05 mg/d appeared optimal for most patients, one patient could not tolerate increases above 0.50 mg/d and two patients required extremely gradual increases. These experiences argue in favor of individualized dose-finding trials prior to data collection.

The primary clinical finding of this study is that intraventricular bethanechol, when carefully adjusted in dose, can improve behavior and mood in AD. There was, however, a suggestion of decreased memory at the dose producing the optimal behavioral effects. These findings await replication in an independent double-blind trial. Whether the degree of behavioral improvement is sufficient to justify an invasive neurosurgical procedure is unclear from our small study. Moreover, the long-term safety of this treatment is not known, and chronic maintenance on bethanechol has not apparently altered the progression of the dementia. Thus, the utility of intraventricularly administered bethanechol in the treatment of AD may be limited.

SUMMARY. Ten patients with biopsy-proven Alzheimer disease received low-dose (.35 mg/day) intraventricular bethanechol, a muscarinic agonist, and saline placebo in a 24-week double-blind crossover design. Eight of these ten patients later participated in an open escalating-dose (to 1.75 mg/day) trial of bethanechol. Patients' drug response was assessed by neuropsychologic examination and informant measures of activities of daily living, mood disturbance, and abnormal behavior. Bethanechol appears to have a narrow therapeutic window for positive effects; low doses did not reliably alter patient functioning, moderately increased doses appeared to have a palliative effect on patient mood and behavior and the highest dose was detrimental to patient functioning. Bethanechol does not appear to ameliorate the dementia of AD, but may exert a mildly positive effect on patient behavior and mood.

REFERENCES

Bartus RT, Dean RL, Beer B, Lippa AS. The cholinergic hypothesis of geriatric memory dysfunction. Science 1982;217:408-414.

Benton AL, Hamsher K deS. Multilingual aphasia examination. Iowa City, IA; University of Iowa, 1978.

Benton AL, Hamsher K deS, Varney N, Spreen O. Contributions to neuropsychological assessment. New York: Oxford, 1983.

Buschke H, Fuld PA. Evaluating storage, retention, and retrieval in disordered memory and learning. Neurology 1974;24:1019-1025.

Davis KL, Hollister LE, Overall J, Johnson A, Train K. Physostigmine: effects on cognition and affect in normal subjects. Psychopharmacology (Berlin) 1976;51:23-27.

Folstein MF, Folstein SE, McHugh PR. 'Mini-Mental State': a practical method for grading the cognitive state of patients for the clinician. J Psychiatr Res 1975;12:189-198.

Harbaugh RE, Roberts DW, Coombs DW, Saunders RL, Reeder TM. Preliminary report: intracranial cholinergic drug infusion in patients with Alzheimer's disease. Neurosurgery 1984;15:514-518.

Hollander E, Mohs RC, Davis KL. Cholinergic approaches to the treatment of Alzheimer's disease. Br Med Bull 1986;42:97-100.

Katz MM, Lyerly SB. Methods for measuring adjustment and social behavior in the community. Psychol Rep 1963;13:503-535.

Lawton MP, Brody E. Assessment of older people: self-maintaining and instrumental activities of daily living. Gerontologist 1969;9:179-186.

Little A, Levy R, Chuaqui-Kidd P, Hand D. A double-blind, placebo controlled trial of high-dose lecithin in Alzheimer's disease. J Neurol Neurosurg Psychiatry 1985;48:736-742.

Martin EM, Wilson RS, Penn RD, Fox JH, Clasen RA, SAvoy SM. Cortical biopsy results in Alzheimer's disease: correlation with cognitive deficits. Neruology 1987;37:1201-1204.

McKhann G, Drachman D, Folstein M, Katzman R, Price D, Stadian EM. Clinical diagnosis of Alzheimer's disease: report of the NINCDS-ADRDA Work Group under the auspices of the Department of Health and Human Services Task Force on Alzheimer's Disease. Neurology 1984;34:939-944.

CHOLINERGIC AGONISTS IN ALZHEIMER'S DISEASE PATIENTS

M. Davidson, E. Hollander, Z. Zemishlany, Lori J. Cohen, R.C. Mohs, and K.L. Davis
Mount Sinai School of Medicine, One Gustave Levy Place, New York, NY and
Bronx VA Medical Center, 130 W. Kingsbridge Road, Bronx, NY

Alzheimer's disease (AD), has been linked to abnormalities in several neurotransmitter and neuropeptide systems. Central noradrenergic, serotonergic, somatostatinergic and CRF deficits have been reported and generally replicated. (For reveiw see Rossor and Iversen, 1986.) The most consistent and pervasive degeneration, however, involves the cholinergic system. Neurochemical studies of brain specimens from patients with a clinical diagnosis of AD demonstrate large reductions of choline acetyltransferase and acetylcholinesterase in the cortex and hippocampus which are in part linked to degeneration of cholinergic neurons originating in the septum and nucleus basalis of Meynert (Davis and Maloney, 1976; Perry et al., 1977). The reduction of cortical and hippocampal choline acetyltransferase is correlated with both number of neuritic plaques and severity of dementia. These studies suggest that at least some symptoms of AD are related to the cholinergic deficit and might be ameliorated by manipulations of cholinergic neurotransmission. Pharmacological strategies aimed at increasing cholinergic neurotransmission may attempt to increase release or synthesis of acetylcholine, delay its synaptic degradation, or stimulate muscarinic receptors.

Both choline and lecithin (phosphatidylcholine) are ACh precursors and several lines of evidence support a central effect for these drugs. However numerous studies have demonstrated that administration of choline or lecithin to AD patients produces minimal or no symptomatic improvement (Etienne 1983; Davidson et al., 1987). Studies in which delay in Ach synaptic degradation was sought focused on physostigmine. Although most studies linked physostigmine with mild to moderate amelioration in AD symptoms in subgroup of patients, the limited degree of clinical improvements does not justify the large use of physostigmine in daily clinical practice. More recently, administration of the AchE inhibitor tetrahydroaminoacridine to AD patients produced moderate improvements in most patients included in the study (Summers et al., 1986). A multicenter study presently in course will try to establish if tetrahydroaminoacridine can indeed ameliorate AD symptoms.

However, studies of both presynaptic and synaptic cholinergic drugs are limited in that they require functional cholinergic neurons to provide a substrate for drug activity. In a disease characterized by degeneration of cholinergic neurons a strategy based on the use of cholinergic agonist appears at least theoretically superior to a synaptic or presynaptic strategy. The potential utility of agonists may however be limited by their tonic action on receptors, which is substantially dissimilar to physiological phasic events. Still, there are reports that cholinergic agonists are effective in enhancing performance in normal rodents (Haroutunian et al. 1985) and hypocholinergic animals (Haroutunian et al. 1985).

Given the widespread of neurotransmitter and neuropeptide deficits in Alzheimer's disease, problems of absorption, metabolism, and excretion of cholinergic agents in general, it could be anticipated that some patients would not respond to cholinergic drugs, some unrelated to the true efficacy of the agent. Consequently, to test a cholinergic strategy it is essential to attempt to obtain an indication that the compound is having a central cholinergic effect. Cortisol secretion is a potential indicator of central cholinergic activity. Systemic or intrahypothalamic administration of cholinomimetic drugs increase cortisol secretion (Krieger et al., 1967; Hillhouse et al., 1975), while anticholinergic drugs decrease cortisol secretion. Studies with oral physostigmine in AD patients have demonstrated correlations between symptom improvement and increases in mean nocturnal cortisol secretion (Mohs et al., 1985). Therefore plasma cortisol concentrations were measured in AD patients treated with RS 86.

This investigation presents two separate studies in which the cholinergic agonists RS 86 and oxotremorine were administered to two groups of AD patients.

RS 86 (2-ethyl-8-methyl-2,8-diazospiro-4,5-decan-1,3-dion-hydrobromide), is a long-acting and specific muscarinic agonist. In animals, RS 86 has cholinomimetic, analgesic, and sedating properties. In clinical trials with 486 patients of various diagnoses, RS 86 produced typical peripheral cholinergic side effects, including salivation, sweating, flushing, slight hypotension, nausea, vomiting, prolongation of P-R interval on EKG, and precipitation of asthma attack (Spiegal et al., 1984). Preliminary trials of RS 86 in AD patients have yielded mixed results. Some found no significant improvement in AD with doses up to 5 mg daily (Bruno et al., 1985), but others found mild improvements in cognitive function, mood, and social behavior in some AD patients (Wettstein and Spiegal, 1984).

Fifteen patients ages 54-78 (mean 66.1) meeting NINCDS criteria for "probable AD" (McKhann et al. 1986) received RS 86. All patients had a Memory and Information Test score of 10 or less, a Dementia Rating Scale score of 4 or more, and a Hachinski Ischemic Scale (Hachinski et al., 1975) score of 5 or less. All patients were drug free for 2 weeks prior to, and throughout, the study. The study design included an open dose-finding phase, followed by a double-blind, randomized, placebo crossover, replication phase. In the dose-finding phase, patients received placebo, and then progressively 0.5 mg, 1.0 mg, 1.5 mg of RS 86 orally, 3 times per day, for 7 days of each dose. The dose finding phase attempted to establish the best dose for each individual patient. Patients were assessed at the end of each week using the Alzheimer's Disease Assessment Scale (ADAS) (Rosen et al., 1984). At the end of the dose finding phase, an investigator determined the dose of RS 86 associated with the lowest severity, as measured by the ADAS. This dose of RS 86 and placebo were administered for 2 weeks each in a double-blind crossover study. Order of drug and placebo administration was random, and symptoms were again evaluated with the ADAS at the end of each week.

On the last night of each treatment condition in the replication phase, nine patients participated in a sleep study. Polyethylene catheters were inserted in the forearm vein at 10:00 p.m., and blood was sampled every half-hour from 11:00 p.m. to 11:00 a.m. the next day. To ensure that the study could detect RS 86's effects on the diurnal rise in cortisol level, treatment with either RS 86 or placebo was continued to 8:30 a.m. on the morning following the sleep study night.

Fifteen AD patients entered the dose-finding phase. Three patients could not tolerate RS 86 and dropped out of the study. One experienced syncope, another abdominal distress, and the third, a seizure. Although it is impossible to say whether or not these events were drug related, they are compatible with a cholinergic effect. Results of the replication phase for the 12 patients who completed the study were analyzed with an Analysis of Variance, treating order of drug administration as a between-subjects factor and with drug condition (RS 86 versus placebo) and time (week 1 versus week 2) as repeated measures factors. The analysis indicated no significant main effects or interactions (p 0.05 in all cases). Examination of the performance on the ADAS subscales (word recall, word recognition, cognitive behavior, and noncognitive behavior) revealed no substantial RS 86 effect on any particular symptom. As previous studies (Fuld et al. 1982) have suggested that decreased cholinergic activity may be associated with an increase in intrusion errors on memory test of the ADAS was analyzed separately. However, no evidence for a drug effect on this response was found. Despite the lack of any effect on mean scores, 7 of 12 patients show improvement on the overall ADAS scores in the replication phase, with improvements ranging from 3.4% to 36.2%, with a mean improvement of 17.4% for those 7 patients who improved, and on overall mean improvement of 7.03% for all 12 patients. Baseline severity of illness correlated with percent improvement (r - -.062, p = 0.03), such that the less severely impaired patients were more likely to improve on cognitive testing.

Side effects of RS 86 included chills, excessive sweat, flushing, depressed mood, prolongation of P-R interval on EKG, tremor, hypersalivation, and confusion in some not all patients, and was dose related, improving when the dose was lowered. There was no significant difference in mean noctrunal cortisol secretion for the period from 11:00 p.m. to 11:00 a.m. between the drug conditions (t = 1.43, p = 0.19). There was a marginally significant correlation of percent change in peak cortisol levels (measured from 3:30 to 5:30 a.m.) and improvement of symptom severity on ADAS (r = .61, p = .04 one tailed) suggesting that the limited symptomatic improvements can be related to the central drug effect.

Oxotremorine is another muscarinic receptor agonist shown to affect memory paradigms in rodents. Seven AD patients selected and evaluated as described in the previous study received single dose of oxotremorine and placebo. Each patient received placebo 0.25 mg, 0.5 mg, 1.0 mg, 1.5 mg and 2.0 mg of oxotremorine in random order and under conditions to which the patient and staff, were blind. The doses were given at 8:30 a.m., at 3-10-day intervals during a "dose-finding phase". This was followed by a replication phase in which the individualized, previously determined "best dose" of oxotremorine was compared to placebo in a double-blind manner. Two hours after oral administration of the oxotermorine or placebo dose, each patient was given a word recognition test.

Oxotremorine was found to have significant adverse effects making the assessment of this drug's cognitive effects impossible. Depressive symptoms; often severe were present in five of the seven patients and appeared to be dose dependent; the more severe depressive symptoms occurred 1-5 hours after the higher doses. No assessments of patients on depression rating scales were formally included in this study. On the basis of previous experience with physostigmine in AD, affective symptoms were not anticipated. However, the study design included a placebo-controlled, double-blind replication phase and a placebo dose during the dose-finding phase, and no patients developed an affective reaction while receiving placebo. Even the two patients who were not noted to be depressed became anxious and confused following administration of oxotremorine.

The cholinergic hypothesis in affective illness is well known and cholinomimetic drugs were reported to produce depressive symptoms (Davis et al., 1978; Sitaram et al., 1980) in normals and depressive patients, but not in AD patients. The severity of the affective response after administration of oxotremorine may be related to the tonic rather than phasic nature of receptor agonists compared to presynaptic and synaptic agents and their differential action on M_1 and M_2 muscarinic receptors. While presynaptic agents such as acetylcholinesterase inhibitors or releasers (e.g., 4 aminopyridine) may increase synaptic acetylcholinesterase in a phasic manner more physiologically equivalent to that of the intact neuron, agonists such as oxotremorine may stimulate receptors in a prolonged and tonic manner, which may be related to the drug's severe depressive side effects. No significant increase in blood pressure or pulse was noted in our patients with AD following administration of this cholinergic agonist, despite the increased central cholinergic activity that was manifested by the depressive side effects. Because of these side effects, only four patients completed the study, and thus the role of oxotremorine in enhancing memory and cognition in Alzheimer's disease remains untested.

Further studies with selective muscarinic agonists are necessary in order to establish the role of this class of drugs in the treatment of AD patients. Particular consideration should be given to the central and peripheral adverse effects associated with cholinergic agonists.

Acknowledgements: Thanks to D. Weston, R.N. and the Nursing Staff on 3B2.

Bruno, G., Mohr, E., Gillespie, M. et al. 1985. RS 86 therapy of Alzheimer's disease. Arch Neurol 43:659-661.

Davidson, M., Mohs, R.C., Hollander, E., Zemishlany, Z., Powchik, P., Ryan, T., Davis, K.L. 1987. Lecithin and piracetam in patients with Alzheimer's disease. Biological Psychiatry 77:112-113.

Davies, P., Maloney, A.J.F. 1976. Selective loss of central cholinergic neurons in Alzheimer's disease. Lancet ii: 1403.

Davis, K.L., Berger, P.A., Hollister, L.E. et al 1978. Physostigmine in mania. Arch. Gen. Psychiatry 35:119-122.

Etienne, P. 1983. Treatment of Alzheimer's disease with lecithin. In Reisberg B (ed), Alzheimer's Disease: The Standard Reference. New York: Free Press, pp. 353-354.

Fuld, P., Katzman, R., Davies, P. et al 1982. Intrusions as a sign of Alzheimer dementia: Chemical and pathological verification. Ann. Neurol. 11:155-159.

Hachinski, V.C., Iliff, L.D., Zilhka, E., et al 1975. Cerebral blood flow in dementia. Arch. Neurol. 32:632-637.

Haroutunian, V.H., Barnes, E., Davis, K.L. 1985. Cholinergic modulation of memory in rats. Psychopharmacology 87:266-271.

Haroutunian, V., Kanof, P.D., Davis, K.L. 1985. Pharmacological alleviation of cholinergic lesion induced memory deficits in rats. Life Sci. 37: 945-952.

Hillhouse, E.W., Burden, J., Jones, M.T. 1975. The effect of various putative neurotansmitters on the release of corticotrophin releasing hormones from the hypothalamus of the rat in vitro. I: The effect of acetylcholine and noradrenaline. Neuroendocrinology 17:1-11.

Krieger, D.T., Kreiger, H.P. 1967. Circadian pattern of plasma 17-hydroxycorticosteroid: alteration city anticholinergic agents. Science 155:1421-1422.

McKhann, G., Drachman, D., Folstein, M. et al 1984. Clinical diagnosis of Alzheimer's disease: Report of the NINCDS-ADRDA work group under the auspices of Department of Health and Human Services Task Force on Alzheimer's Disease. Neurology 34:939-944.

Mohs, R.C., Davis, B.M., Johns, C.A. et al 1985. Oral physostigmine treatment of patients with Alzheimer's disease. Am. J. Psychiatry 142:28-33.

Perry, E.K., Gibson, P.H., Blessed, G., Perry, R.H., Tomlinson, B.E. 1977. Neurotransmitter enzyme abnormalities in senile dementia. J. Neurol. Sci.34:347-265.

Rosen, W.G., Mohs, R.C., Davis, K.L. 1983. A new rating scale for Alzheimer's disease. Am J Psychiatry 141:1356-1364.

Rosen, W.G., Mohs, R.C., Davis, K.L. 1984. A new rating scale for Alzheimer's disease. Am J Psychiatry 141:1356-1369.

Rossor, M.N., Iversen, L.L. 1986. Non-cholinergic neurotransmitter abnormalities in Alzheimer's disease. Br. Med. Bull. 42:70-74.

Sitaram, N., Nurnberger, J.I., Gershon, E.S. et al 1980. Faster cholinergic REM sleep induction in euthymic patients with affective illness. Science 208:200-202.

Summers, W.K., Majovski, L.V., Marsh, G.M., Tachiki, M., Kling, A. 1986. Oral tetrahydroaminoacridine in long-term treatment of senile dementia, Alzheimer's type. N. Engl. J. Med. 315:1241-1245.

Wettstein, A., Spiegal, R. 1984. Clinical trials with the cholinergic drug RS 86 in Alzheimer's disease (AD) and senile dementia of the Alzheimer's type (SDAT). Psychopharmacology 84:572-573.

4-AMINOPYRIDINE (4-AP) - DERIVATIVES AS CENTRAL CHOLINERGIC AGENTS

P.G. Waser, S. Berger, H.L. Haas, and A. Hofmann
Institute of Pharmacology, Univ. of Zürich, Gloriastr. 32, CH-8006 Zürich, Switzerland

There is hope that the stimulant action of 4-AP on transmitter release might lead to an improvement in some CNS diseases such as the senile dementia and the Alzheimer's disease.
4-AP acts mainly presynaptically on the cholinergic synapse by blocking the transient K^+ current, mobilizing Ca^{2+} ions and liberating Acetylcholine (Ach) into the synaptic cleft. We have synthesized several 4-AP derivatives with different substituents in position three of the pyridine nucleus (Fig.1) and investigated their activity in peripheral and central nervous preparations.

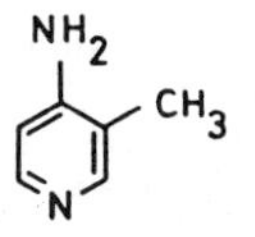

3-Methyl-4-aminopyridine
(3-Me-4-AP)

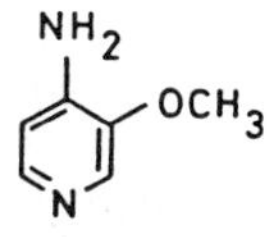

3-Methoxy-4-aminopyridine
(3-MeO-4-AP)

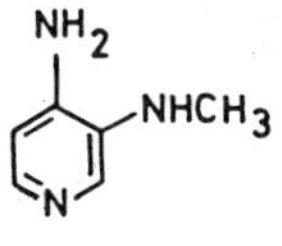

3-Methylamino-4-aminopyridine
(3-NMe-4-AP)

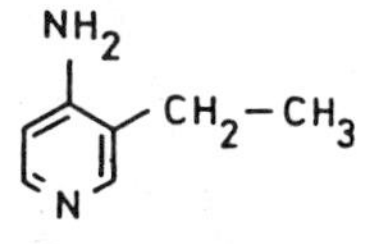

3-Ethyl-4-aminopyridine
(3-Et-4-AP)

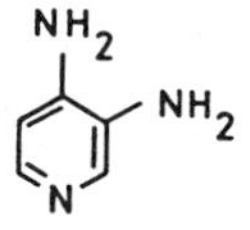

3 4-Diaminopyridine
(3 4-DAP)

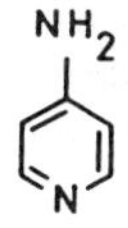

4-Aminopyridine
(4-AP)

4-AP DERIVATIVES IN RAT HIPPOCAMPAL SLICES

Method:
Transverse hippocampal slices (n=15) were cut with a McIlwain tissue chopper, about 450 µm thick from the hippocampi of young adult Wistar rats and transfered to a perfusion chamber. The slices were perfused continously with artifical cerebrospinal fluid pH 7.4, containing mM Na^+ 150.0, K^+ 6.25, Mg^{+2} 2.0, Ca^{2+} 2.5, Cl^- 138.0, HCO_3^- 26.0, HPO_4^- 1.25 and glucose 10.0, additionally superfused with O_2 95% / CO_2 5%. Glass micropipets for extracellular recording were filled with 3 M KCl, having tip resistances between 20-60 MOhm and inserted in the appropriate locations (Fig.2). In all the figures presented positivity is always downwards.

Results:
- Fibre potentials
Fibre potentials are broadened and their refractory period is markedly prolonged. This is shown for all compounds by recording the summed action potential of afferent fibres after stimulating in the Schaffer collateral/commissural fibres with an interstimulus interval of 5 msec. In the presence of $5 \cdot 10^{-4}$M drug concentration the amplitude of the second fibre volley is diminished to 15% by 3,4-DAP, 73% by 3-Me-4-AP, 76% by 3-MeO-4-AP, 79% by 3-NMe-4-AP and 87% by 3-Et-4-AP of the control.
Traces of responses elicited at various interstimulus intervals show, that the recovering of the amplitude of the second stimulus is prolonged under the effect of the drugs (Fig.3).

Schematic representation of the hippocampal structure

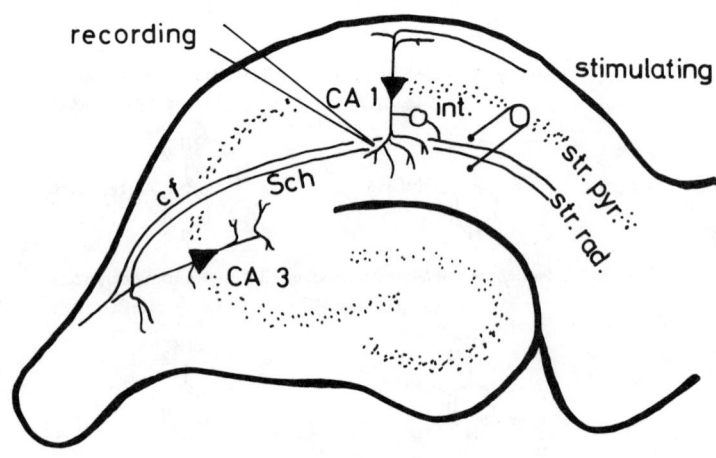

- Extracellular field potentials
Extracellular postsynaptic field potentials occurring in the apical dendrites of CA 1 pyramidal cells after orthodromic stimulation in the stratum radiatum with low intensities, which reveal no population spike, are increased and prolonged. 3-Me-4-AP is the most effective of the new synthesized derivatives with an increase of 63.7% and after washout the effect is still prominent (Fig.4).

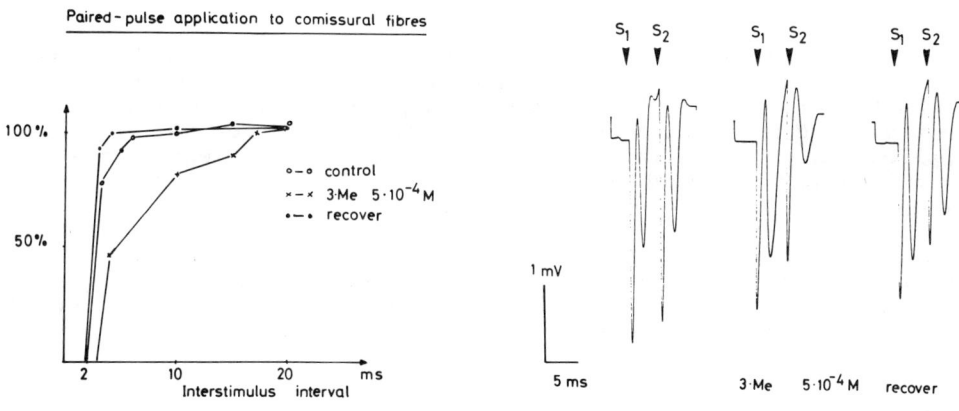

Hippocampal field potentials

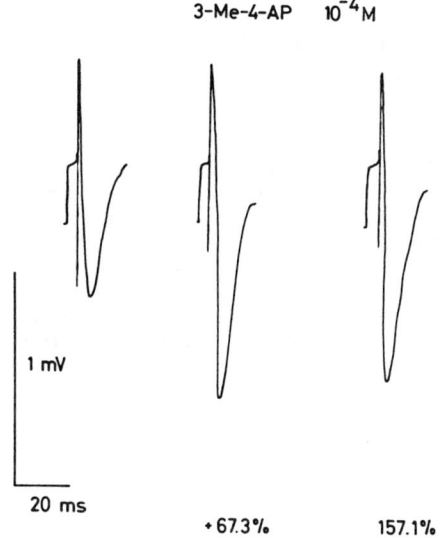

The order of potency is found to be:
3,4-DAP > 3-Me-4-AP > 3-MeO-4-AP > 3-NMe-4-AP > 3-Et-4-AP, which is also the slowest in onset with reaching its maximum effect after 17 Min.
If the neurons are stimulated twice in rapid succession with an interval of 40 msec., the amplitude of the second impulse is augmented compared to the initial response: Facilitation at the synapse occurs. The 4-AP derivatives show a clear reduction of this facilitation.

4-AP DERIVATIVES AT THE MOUSE NEUROMUSCULAR JUNCTION

Method:
Hemidiaphragms with their phrenic nerves were removed from male albino mice ICR strain, weighing between 20-30 g and mounted in a organ bath containing Krebs' solution pH 7.4. The preparation was kept at 18°C and continously gassed with O_2 95%/ CO_2 5%. After the application of a resting tension of 2.0 g the nerve was stimulated with supramaximal rectangular pulses of 1.0 msec. duration at a frequency of 0.1 Hz. Alternately a direct stimulation of 10 msec. duration was applied to the muscle. 20 min. after drug application a tetanic stimulation, 2.5 sec. train of 50 Hz and 0.2 msec. duration, was delivered to the preparation to reveal effects at the presynaptic site. The isometric tension output was recorded with a strain gauge force displacement transducer on a Grass polygraph.

Results:
All compounds increase indirectly and directly elicited twitches and show a tension fade during tetanic stimulation (Fig.7). The lipophilic 3-Me-4-AP is the most effective derivative, but it does not reach the potency of 4-AP. By stimulating the phrenic nerve it increases the amplitude to 100% of the control.
The order of potency is found to be:
4-AP > 3,4-DAP > 3-Me-4-AP > 3-MeO-4-AP > 3-NMe-4-AP > 3-Et-4-AP
The effect on the directly elicited twitches is less pronounced and takes a longer time to develop.

PARTITION COEFFICIENT

Method:
The partition coefficient was determined in a 1-Octanol/Krebs' buffer pH 7.4 system. 10 mg recrystallized substance was dissolved in 100 ml buffer and 3 ml of this solution was shaken with the same volume 1-Octanol for 2 hrs at 37° C. The two layers were separated and the concentration was determined spectrophotometrically in the water phase.
Hydrophilicity is expressed by log P, in which P is the ratio of the concentration in the organic phase to the concentration in the water phase.

Results:

	log P
3,4-DAP	− 1.442
3-NMe-4-AP	− 0.848
4-AP	− 0.761
3-Me-4-AP	− 0.513
3-Et-4-AP	− 0.162
3-MeO-4-AP	− 0.061

STRUCTURE-ACTIVITY RELATIONSHIP ANALYSIS

Method:
Stepwise regression analysis was performed with the physicochemical parameters for the hydrophilicity log P and the steric parameter MR (molar refractivity, which is defined by the density, the refraction and the molecular weight of a molecule) as variables. The F criterion was set at the 95% level for each variable entering the multiple regression.

Results:
The effect of the 4-AP derivatives in rat hippocampus is highly related, R-SQ= 0.9616, to the hydrophilicity and the steric volume and can be described by the equation:
Effect= − 6.67 MR − 16.66 log P + 97.93.

DISCUSSION

4-AP derivatives have an essentially similar action in the peripheral and central nervous system. They increase transmitter release mainly by a presynaptic blockade of K^+ channels.
From the derivatives so far synthesized 4-AP still is the most potent. It seems that its structure is ideal for the access to the site of action: The molecular volume is small and it has a high lipophilicity, so that it can easily pass through the membranes and enter the central nervous system. Further investigations have to elucidate the selectivity of the effect with focusing on the central cholinergic pathway.

SUMMARY

New derivatives of 4-aminopyridine were investigated in different organ preparations [isolated mouse phrenic nerve-hemidiaphragm, isolated guinea-pig atria and ileum (Berger, 1988), and in rat transversal hippocampal brain slices]. The presynaptic liberation of acetylcholine was observed in potentiation of twitch contraction and fading of tension during tetanic stimulation of stimulated hemidiaphragms. Fibre potentials and extracellular field potentials of the hippocampal tissue were markedly strengthened. Lipophilicity and chemical structure were related to activity and passage through the blood-brain-barrier. The mainly presynaptic activation might be of value in treatment of Alzheimer's disease.

RELEVANT REFERENCES

Buckle, P.J. and Haas, H.L. 1982. Enhancement of synaptic transmission by 4-aminopyridine in hippocampal slices of the rat. J. Physiol. 326:109-122.

Berger, S. 1988. Pharmacology of 4-aminopyridine and some new derivatives. Doctoral thesis. ETH Zurich.

Wesseling, H. 1984. Effects of 4-aminopyridine in elderly patients with Alzheimer's disease. New Eng. J. Med. 12(310/15):988.

FROM PHYSOSTIGMINE TO PHYSOSTIGMINE DERIVATIVES AS NEW INHIBITORS OF CHOLINESTERASES

M. Brufani[1], C. Castellano[1], M. Marta[1], F. Murroni[1], A. Oliverio[2],
P.G. Pagella[3], F. Pavone[4], M. Pomponi[4], and P.L. Rugarli[4]

[1]*Dipartimento di Scienze Biochimiche, Universita "La Sapienze," Rome*
[2]*Istituto di Psicobiologia e Psicofarmacologia del CNR, Rome*
[3]*Mediolanum Farmaceutici, Milan*
[4]*Istituto di Chimica e Propedeutica Biochimica,
Universita Cattolica S.C., Facolta di Medicina, Rome, Italy*

Summary

In this paper the authors report a series of physostigmine analogues, in which the methylcarbamic group has been substituted with monoalkylcarbamic, dimethyl- and diethylcarbamic groups. These inhibitors were prepared in order to investigate their possible future use in the treatment of Alzheimer's type senile dementia. The new alkaloids are competitive inhibitors of acetylcholinesterase. The percentage of anticholinesterase activity in vitro and in vivo, the acute toxicity and some behavioural effects were also evaluated for selected derivatives. The reactivation constant k_3, in vitro, supports the view that the derivatives described would be more suitable for therapeutic use than physostigmine.

Many natural alkaloids which have been identified in several plant extracts present pharmacological activities on the central nervous system (Marini Bettolo 1986).

Table 1. CNS Activity Plant Extracts.

Analgesics	PAPAVER somniferum
Antipsychotics	RAUWOLFIA serpentina
Hallucinogens	RIVEA corymbosa, IPOMAEA violacea ARGYREIA nervosa
Anesthetics	ERYTHROXYLON coca
Stimulants	STRICHNOS nux vomica
Antimuscarinics	HYOSCYAMUS niger SCOPOLIA carniolica
Anticholinesterases	PHYSOSTIGMA venenosum

The plant Physostigma venenosum is a climbing vine growing in the tropical neighbourhood of the Niger and Calabar Rivers.

The tribes living in this area employed the seeds (eserè) extract for judiciary procedures (Fraser 1863).

Physostigmine (eserine) is a competitive acetylcholinesterase (AChE, EC 3.1.1.7) inhibitor extracted from Physostigma ven. seeds. From a therapeutical point of view physostigmine is an old drug previously used for its peripheral effects (Karczmar 1970).

Since the 70's there has been a new interest in physostigmine, but this time for its central effects. This alkaloid has been shown to improve memory and to reverse scopolamine-induced dementia (Drachman 1977) and to date is the most popular cholinomimetic that is used to augment cholinergic neurotransmission in Alzheimer's disease, (AD), (Drachman 1978).

NEW INTEREST IN PHYSOSTIGMINE

1972-73 Methylphenidate-induced psychostimulation in humans is rapidly antagonized by physostigmine, but not by neostigmine. (Janowsky et al.).

1973-78 Physostigmine, but not neostigmine, causes dramatic antimaniacal effects (Janowsky et al. (Davis et al.).

1976 Effects of physostigmine on cognition (Davis et al.).

1977 Physostigmine reverses the congitive effects of anticholinergic drugs (Drachman).

1978 Physostigmine improves cognitive tests in young adults (Davis et al.).

1979 Physostigmine s.c., with lecithin, improve verbal learning in patients with AD (Peters and Levin).

1981 Physostigmine i.v. gives a significant increase in picture recognition in patients with AD (Christie et al.).

1982 Physostigmine i.v. significantly increases recognition memory in AD patients (Davis and Mohs).

1983 Oral physostigmine gives a significant improvement of verbal memory in AD patients (Thal et al.).

In fact among the three basic pharmacologic strategies for augmenting cholinergic neurotransmission in AD, the drugs which have been most studied are the cholinesterase inhibitors, and specifically physostigmine. The limits of this therapeutical application of physostigmine are, inter alia (Davidson et al. 1986):

    a) the half-live of physostigmine is merely several minutes long.
    b) its low safety index.
    c) the high incidence of untoward, peripheral effects.

In 1983 the authors considered the possibility of modifying physostigmine in order to obtain less toxic compounds, which like physostigmine would still not be curative and whose ameliorative effects are likely to be limited.

FROM PHYSOSTIGMINE TO NEOSTIGMINE

1926-29   Stedman (Edinburgh) elucidates the chemical basis of physostigmine's anti-ChE properties. He also postulates that the two linked pyrrolidine rings may not be essential for anticholinesterase activity (Figure 1).

1931   As a result of the structure-activity analysis of physostigmine performed by the Edinburgh group, Aeschlimann and Reinert develop neostigmine. Due to its quaternary ammonium group neostigmine cannot penetrate cell membranes readily, it is poorly adsorbed from the gastrointestinal tract and it is excluded from exerting significant action on CNS by the blood-brain barrier.

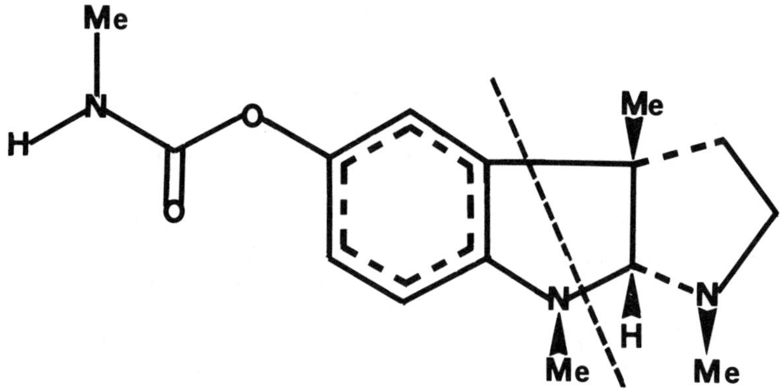

Figure 1. Stedman's Hypothesis

FROM PHYSOSTIGMINE TO PHYSOSTIGMINES (Brufani et al. 1984).

From the Stedman's hypothesis we can see from Figure 1 that the non-essential moiety, for physostigmine to have peripheral effects was of great interest to us as physostigmine is a tertiary amine in equilibrium with its own cation at physiological pH and as tertiary amine physostigmine crosses blood-brain barrier, in contrast to the quaternary anti-AChE drugs. The result of the Stedman's hypothesis was neostigmine, which meant also the removal of any specific central effects. Consequently, in order to show central effects, the carbamyl moiety of the physostigmine needs to be shuttled by the eseroline moiety, that is to say that the shuttle cannot be changed structurally, but the carbamyl moiety can (Figure 2).

In order to obtain a safer derivative with fewer peripheral effects, we have decreased the aqueous solubility of the drug by lengthening the alkyl group of the carbamyl chain. Moreover it was known from literature that the hexyl chain on the nitrogen produces a more stable carbamylated AChE (Wilson et al. 1961).

Last but not least, as the dominant type of forces involved in reactions with the second site of pseudocholinesterase (BuChE, E.C. 3.1.1.8) are Van der Waals forces, in contrast to the Coulombic attractions which favour interactions with AChE, it was expected that the more lipophilic the inhibitor, the stronger the inhibition of BuChE.

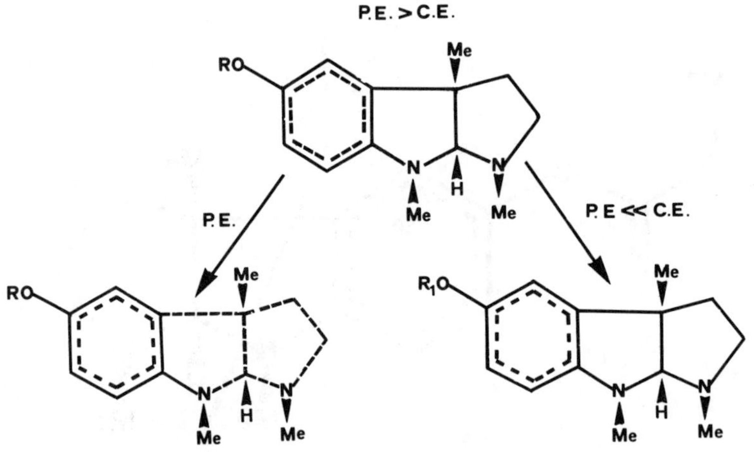

Figure 2. From Stedman's Hypothesis to Physostigmine Derivatives. CE: central effects. PE: peripheral effects.

From literature it was also known that dimethyl carbamic compounds were more stable than monomethylcarbamic compounds and in fact neostigmine has greater stability than physostigmine, even if dimethylcarbamic inhibited enzymes are, in vitro, less stable than the corresponding monomethyl carbamic inhibited ones (Wilson et al. 1961). For this reason we also needed to test dimethylphysostigmine.

ACETYLCHOLINESTERASE INHIBITION IN VITRO

Some of the new compounds were found to be significant potent inhibitor of AChE (Figure 3). The activity was slight in diethylphysostigmine, quite weak in propyl- and in ethyl-, isopropyl- and butylphysostigmine, terzbutylphysostigmine, but strong in heptyl-, nonyl- and dimethylphysostigmine. No activity was observed with phenylphysostigmine. Homologues superior to undecylphysostigmine were too insoluble in the buffer employed for the study of their inhibition parameters. Acetylcholinesterase activity in vitro was determined by the method used by Ellman et al. (1961).

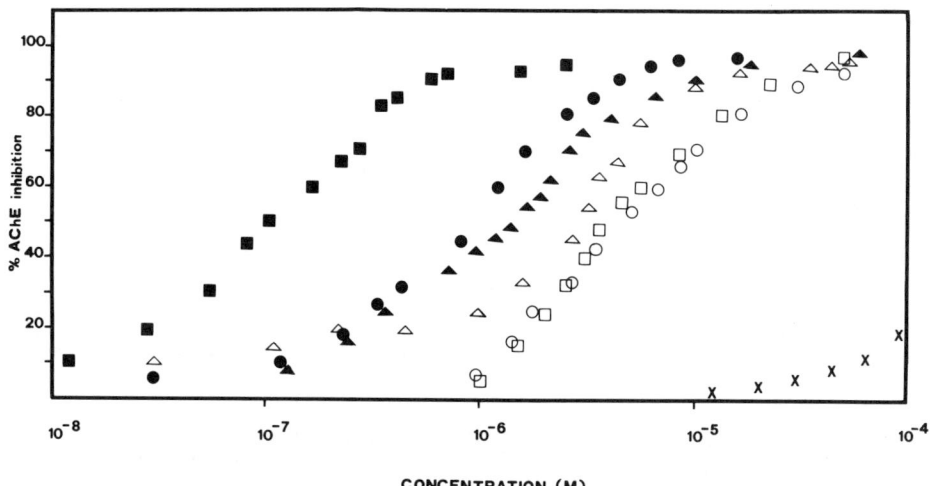

Figure 3. Percentage Inhibition of AChE from electric eel by Physostigmine (■), Heptylphy. (●), Dimethylphy. (△), Nonylphy. (▲), Butylphy. (○), Propylphy. (□) and Diethylphysostigmine (x).

All the new carbamates are competitive inhibitors of acetylcholinesterase and came to equilibrium slowly (Brufani et al. 1986).
One of the major advantages of the derivatives presented is

the value of the reactivation constant k_3 for AChE.

Carbamylated acetylcholinesterases (AChEOc) undergo reactivation according to the following scheme (Wilson et al. 1960):

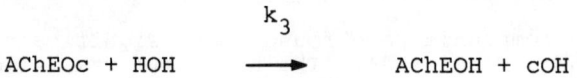

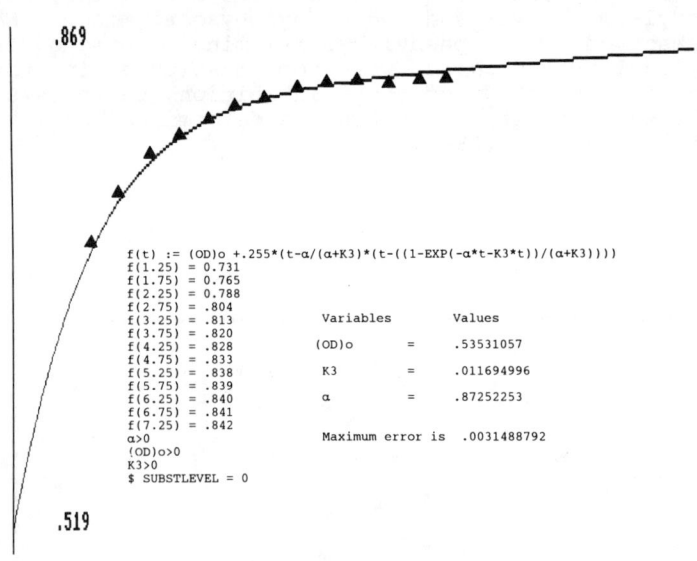

HEPTYLPHYSOSTIGMINE

ANALYSIS OF PROGRESS-CURVE FOR THE INHIBITION OF AChE IN THE PRESENCE OF SUBSTRATE.
Buffer phosphate 0.1 M, pH 8.0. Substrate acetylthiocholine iodide 4.7 x 10^{-4}M. Dithiobisnitrobenzoic acid (DTNB) 3.1 x 10^{-4}M. Heptylphysostigmine 2.2 x 10^{-5}M. AChE from electric eel 0.07 units. t = 25°C. λ 412 nm.

Figure 4. Example of computer plot demostrating fit of polynomial equation to raw inhibiton data (Brufani et al. 1985)

The reactivation time is ≈ 20 min for physostigmine while for the heptylderivative it is much longer, that is ≈ 100 minutes (Figure 4); it is under 20 min. for the dimethyl derivative; in this case the longer inhibition observed in vivo may be due to the higher dose (7.5 times greater that that of physostigmine). When acetylcholinesterase reacts with physostigmine the rate limiting step is invariably k_3. From the constant k_3 is possible to obtain the half-life of the inhibited enzyme.

ACETYLCHOLINESTERASE INHIBITION IN VIVO

In Table 2 is reported the enzymatic inhibition in brain and serum. The subjects were tested at about 90 days of age. Male rats (n=100) belonging to the Wistar strain and weighing between 180-200 g, were used for AChE inhibition. Rats were sacrificed and the whole brain rapidly removed. The brains from the rats were homogenized in 1ml of phosphate buffer (0.1 M pH 8.0) containing 1% Triton. The mixture was centrifuged and the supernatant was used for subsequent determinations of anticholinesterase activity. Acetylcholinesterase activity was determined by the procedure of Ellman et al. (1961).

Table 2 indicates that treatment with physostigmine and other selected derivatives resulted in a dose dependent 'reversible' inhibition of brain AChE activity. Thus it is evident that higher doses of heptyl are necessary in order to produce levels of AChE inhibition comparable with those of physostigmine.

Table 2. AChE Inhibition in the Brain and in Serum after i.p. Administration, at variou Time Intervals after Dosing. The Enzymatic Inhibition in Brain was expressed in arbitrary Units.

INHIBITOR	mg/Kg	Inhibition in arbitrary units S.E. ± (n = 4)					
		10 mins.		60 mins.		120 mins.	
		brain	serum	brain	serum	brain	serum
Physostig.	0.2	1.00 ±0.16	1.00 ±0.05	0.23 ±0.05	0.17 ±0.13	0.06 ±0.04	0.05 ±0.03
Heptyl	1.0	0.93 ±0.03	1.11 ±0.04	0.43 ±0.18	0.77 ±0.08	0.42 ±0.05	0.40 ±0.04
Dimethyl	1.5	0.98 ±0.13	1.38 ±0.10	0.40 ±0.10	1.31 ±0.13	0.24 ±0.04	1.22 ±0.10
Diethyl	3.0	0.64 ±0.05	1.62 ±0.02	0.71 ±0.05	1.62 ±0.08	0.62 ±0.02	1.42 ±0.05

The dimethyl- and the diethylphysostigmine show minor central action. With equivalent initial inhibitions, the duration of the inhibitory activity of heptyl, dimethyl and diethyl was longer than that of physostigmine. It is interesting to note that after two hours the inhibitory properties of physostigmine almost completely disappeared, while the effects of the other inhibitors were still evident; moreover the action of the diethyl derivative remained pratically constant. At any rate physostigmine and its heptylderivative penetrate into the central nervous system more easily than the dialkyl derivatives tested (Brufani et al. 1987).

TOXICITY

The LD_{50} for the heptyl is of the same order (35 mg/Kg in mice) of that reported for tetrahydroaminoacridine (Summers et al. 1980). The dimethyl derivative is about twenty times less toxic (LD_{50} = 11.5 mg/Kg; LD_{100} = 14.0 mg/Kg) than physostigmine. The main result emerging from the present study is that the compounds tested are significantly less toxic in comparison with physostigmine (Table 3).

Table 3. Toxicity Activity for some selected Derivatives. Phy: physostigmine; a. Quaternary salt of physostigmine obtained with the heptyl iodide. This compound does not modify the behavioral activities of the treated mice. DM-Phy: dimethylphysostigmine; DE-Phy: diethylphysostigmine.

Drug	LD_{50} (mg/Kg)	LD_{100} (mg/Kg)
Physostigmine	0.64	---
Heptyl-Phy	35.0	45.0
Nonyl-Phy	50.0	---
C7I-Phy[a]	8.5	11.0
DM-Phy	11.5	14.0
DE-Phy	38.0	---

EXPLORATORY ACTIVITY

At the behavioural level the action of heptyl-, dimethyl- and diethylphysostigmine on the brain results in an antagonism of the stimulating effects of scopolamine on the locomotor activity of mice. The latter (Table 4) was tested against the antagonism of the stimulating effect produced by scopolamine a centrally active anticholinergic agent for three selected derivatives (1.0 mg/Kg). These findings indicate that increased cholinergic function (due to inhibition of ACh hydrolysis) antagonizes the behavioral stimulation due to the anticholinergic effect of scopolamine (Oliverio and Castellano 1974).

Table 4. Exploratory Activity: Mean Nnumber of Crossing with Treatment/Mean Number of Crossing with only saline Solution (n_t/n_s). $P < 0.05$ (ANOVA One-Way and Duncam multiple Range Test); a: Data calculated from (Brufani et al. 1987).

TREATMENT	mg/Kg	nt/ns
Heptyl-derivative	3.0	0.88a
Dimethyl-derivative	0.5	1.0
	1.0	0.80
	1.5	0.49
Diethyl-derivative	5.0	0.92
	8.8	0.80
	13.2	0.41
Scopolamine	1.0	2.53
Heptyl+Scopolamine	3.0 + 1.0	1.66a
Dimethyl+Scopolamine	1.0 + 1.0	1.73
Diethyl+Scopolamine	8.8 + 1.0	2.40
	13.2 + 1.0	1.81

For exploratory activity male mice belonging to the DBA/2 strain were used. In this species the LD_{50} and the LD_{100} were also assesed. The locomotor activity was measured with a toggle floor box (24.5 cm x 9.0 cm) as previously described (Oliverio 1966). The test session lasted 25 min. long. Different groups of eight mice were tested 5 min. following drug and vehicle administration. The quaternary salts at nitrogen atom N-7 we have tested, produced poor effects.

PASSIVE AVOIDANCE

Table 5. Passive Avoidance Behaviour. Percent Increase of Mean Step-through Latencies after 48 hr for Heptylphysostigmine (i.p.) referred to Vehicle (*) and to Physostigmine (** .03mg/KG) $P < 0.01$ (ANOVA One-Way and Duncan multiple Tange Test). Physostigmine after 48 hr gave an Increase of about 33%.

Dose (0.5mg/Kg)	x 1	x 2	x 5	x 10
(*)	8.9	113.9	126.0	175.7
(**)	0.3	3.5	3.8	5.3

Table 5 indicates that the step-through latencies of mice injected posttrial with different doses of heptylphysostigmine were significantly modified. Longer step-through latencies were evident 24 and 48 hours following the posttrial injection of heptylphysostigmine, which indicated that the drug acts on consolidation mechanisms facilitating memory (Castellano et al. 1984). One of the most interesting results in terms of behavioural modification was probably the effect of heptylphysostigmine on passive avoidance learning (Castellano et al. 1984). Due to the stability of the dimethyl derivative with respect to the heptyl, it still proved to be a potentially useful drug being more than three times as active (Table 4) although three times as toxic. Moreover Table 5 shows that physostigmine gave an increase of 33% after an i.p. administration (0.03 mg/Kg) the heptyl derivative gave an increase of 175.7%.

Finally it seems reasonable to postulate that some of the new compounds are expected to be useful for the treatment of cholinergic disorders. Although little is known about the pharmacokinetics of heptylphysostigmine (or its central nervous system penetrance) a constant and long-lasting presence of the drug would nevertheless be desirable.

REFERENCES

Aeschliman, J. A. and M. Reinert 1931 J. Pharmac. 43, 413-444.
Aldridge W.N. and E. Reinert 1972 In: Frontiers of Biology Vol. 26 eds. A. Neuberger and E.L. Tatum. Enzyme Inhibitor as Substrates. Interactions of Esterase with Esters of Organophosphorus and Carbamic Acids. North Holland Publishing Co. Amsterdam.
Brufani, M., C. Castellano, M. Marta, A. Oliviero, F. Pavone

and M. Pomponi 1984. CNR Patent 47780 A84; U. S. Patent Appl. n. 705,009. 1986; Japanese Patent Appl. n. 38994/1985.

Brufani M., C. Castellano, M. Marta, A. Oliverio et al. 1987. Pharmacol. Biochem. Behavior 26 : 625-629.

Brufani, M., S. Lippa, M. Marta, A. Oradei and M. Pomponi Ital. J. Biochem. 35 : 328-340.

Brufani, M., Marta, M. POMPONI 1986. Eur. J. Biochem. 157 : 115-120.

Castellano, C., U. Filibek and F. Pavone 1984. Eur. J. Pharmacol. 104 : 111-116

Christie, J.E., Shering, A., Ferguson, J. and Glen, A.I.M. 1981. Br. J. Psychiatry 138 : 46-50.

Davidson, M., R.C. Mohs, E. Hollander et al. 1986. Psychopharm. Bull. 22 : 101-105.

Davis, K., Berger, P.A., Hollister, L.E. and De Fraites E.G. 1978. Arch. Gen. Psychiatry 35 : 119-222.

Davis, K.L. and Davis, B.M. 1979. In: Brain Acetylcholine and Neuropsychiatric Disease, eds K.L. Davis and P.A. Berjer, 445-458. Plenum Press, New York.

Davis, K.L. and Mohs, R.C. 1982. Am. J. Psychiatry 139 : 1421-1424.

Davis, K.L., R.C., Mohs, J.R. Tinklenberg, L.E. et al. 1978. Science 201: 272-274.

Drachman, D.A., 1977. Neurology (Minneap.) 27 : 783-790.

Drachman, D.A. 1978. Alzheimer's disease: Senile dementia and related disorders, eds Katzman R. and Bick K. 141-148, Raven Press, New York.

Ellman, G.L., K.D., Courtney, V. Jr., Andres and M. Featherstone, 1961. Biochem. Pharmacol. 7 : 88-95.

Fraser, T.R. 1863. Edinburgh Med. J., 9 : 36-56.

Janowsky D.S., M.K., El-Yousef, and J.M., DAVIS 1973. Am. J. Psychiatry 130 : 1370-1376.

Janowsky D.S., K., El-Yousef, Davis, J.M. and M.I. Sekerke, M.I. 1973. Arch. Gen. Psychiatry 28 : 542-547.

Karczmar A.G., 1970. Anticholinesterase Agents. In International Encyclopedia of Pharmacology and Therapeutics 1, eds Karczmar A.G., Usdin, E. and Willis, J.H., 1-44, Pergamon Press.

Marini Bettolo G. B. 1986. In Advances in Medicinal Phytochemistry. eds. D. Barton and W.D. Ollis 103-115. John Libbey London.

Oliverio, A. 1966. J. Pharmacol. Exp. Ther. 154 : 350.

Oliverio, A. and C. Castellano 1974. Psycopharmacologia 39 : 13-22.

Peters, B.H. and Levin, H.S. 1979. Ann. Neurol. 6 : 219-221.

Stedman, E. and Barger, G. 1925. J. Chem. Soc., 127 : 247-258.

Stedman, E. 1926. Biochem. J., 20 : 719-734.

Summers, W.K., K. R. Kaufman, F. Jr. Altman and J. M. Fischer 1980. Clin. Toxicol. 16(3) : 269-281.

Thal, L.J., P.A. Fuld, D.M. Masur and N.S. Sharpless, 1983. Ann. Neurol. 13 : 491-496.

Wilson, I.B., Hatch, M.A. and Ginsburg, S. 1960. J. Biol.Chem. 235 : 2312-2315.

Wilson, I.B., M.A. Harrison and S. Ginsburg 1961. J. Biol. Chem. 236 : 1498-1500.

Part VI

Neurochemical Correlates of Alzheimer's Disease:

Neural and Non-Neural Markers

Part VI

PERIPHERAL CHOLINERGIC CHANGES AND PHARMACOLOGICAL ASPECTS IN ALZHEIMER'S DISEASE

L. Ravizza[1], P. Ferrero[2], C. Eva[3], P. Rocca[1], L. Tarenzi[2], and P. Benna[2]
[1] Psychiatric Clinic, [2] Neurologica Clinic, [3] Dept. of Pharmacology,
University of Turin, Italy

Research into the pathogenesis of Alzheimer's Disease (AD) has increased steadily over recent years, with efforts focused on the differences between alterations of neural activity in normal aging and dementia. Brains from people with pre-mortem symptoms of AD characteristically exhibit major reductions in cortical choline acetyl transferase (CAT) suggesting parallel loss of the long axon cholinergic neurons that run from the basal forebrain and septum to the neocortex and hippocampus (Davies and Maloney, 1976). Other neurotransmitters or neuropeptides may also be deficient in such brains: somatostatin levels in the cerebral cortex may be markededly reduced, as may there of noradrenaline in the locus coeruleus (Davies et al., 1980; Gottfries and Roos, 1973). However the magnitude of the cholinergic defects, its ubiquity in affected brains and the frequency with which cholinergic defects is the only major neurochemical change observed, all suggest the possibility that a pharmacological therapy that potentially restores the cholinergic functions might provide the basis for a useful treatment strategy for this progressive and otherwise fatal disease.

Since biopsies cannot usually be performed in AD patients, the investigation of brain material must generally be done post mortem. However autoptic data are mostly from terminal patients and may reflect changes also due to the terminal pathology and not to the disease.

Only in vivo examinations, in which biochemical and clinical factors can be assessed together, might elucidate the relationships of early changes of central neurotransmitter function to clinical symptoms.

In the living patients the investigation of cerebrospinal fluid (CSF) can give valid informations about the metabolism of brain tissue. CSF investigations in AD patients have focused on studying biochemical markers reflecting the brain metabolism of many neurotransmitters systems. It is difficult on the basis of CSF studies, however, to make conclusion about Acetylcholine (Ach) metabolism in brain because there are some methodological problems in studying cholinergic brain functions via CSF investigations (Haber et al., 1980).

Available informations show that Acetylcholinesterase (AchE) enzyme activity is present in the CSF of mammalian including human.

Several Authors have reported lowered CSF AchE levels in AD patients, but the validity of the findings has been disputed because it cannot be concluded that this enzyme activity really reflects the brain cholinergic metabolism (Soininen et al., 1981; Huff et al., 1986; Elble et al., 1987; Ruberg et al., 1987).

PERIPHERAL STUDIES

Although the clinical and pathological findings in AD are restricted to the brain, several studies have reported abnormalities attributed to the disorder in peripheral tissues, including blood cells and fibrobalsts (Peterson and Goldmann, 1986; Babey et al., 1986; Kessler, 1987). The investigation of cholinergic activities in blood of AD patients is of considerable interest. Plasma acetylcholinesterase (AchE), as opposed to pseudo-cholinesterase, has received scant attention due partly to the difficulties in measuring low levels of AchE against a background of high pseudocholinesterase (Felstoff and Fernandez, 1981; Perry et al., 1982). AchE is also present in blood cells membrane, although its function is completely unknown. Recently it has been observed that human lymphocytes are one of those blood cells which possess considerable AchE activity besides erytrocytes (Bartha et al., 1987). The presence of AchE in lymphocytes is of special interest since several data suggest that lymphocytes, like brain cells, possess a nerve type of cholinergic information, that is based upon the stimulating effect of Ach. Augmentation of the cytotoxic responses of thymus-derived (T) lymphocytes, increase in cyclic cGMP levels, increase in proteins and RNA synthesis, as well as enhanced lymphocytes motility in the presence of cholinergic agonists have been reported (Strom et al., 1974; Schreiner and Unanue, 1975; Masturzo et al., 1985). It most of above instances, the cholinergic responses have have been diminuished or blocked by the specific muscarinic antagonist, atropin. Investigators have previously used radioligand binding techniques to characterized the lymphocyte cholinergic muscarinic receptor system (Gordon et al., 1978; Maloteaux et al., 1982; Zalcman et al., 1983; Adem et al., 1986). Recently Rabey et al. (1986) have suggested that the binding of the Ach antagonist ^{3}H-quinuclydinyl benzylate (^{3}H-QNB) to human lymphocytes might be a way of studying changes in the central cholinergic systems as in the case of Alzheimer's dementia. They demonstrated a lower density of ^{3}H-QNB binding sites on lymphocytes from AD patients. There are, however, a number of methodological problems in identifying lymphocytes cholinergic receptors and conflicting results have been reported by the different authors which used ^{3}H-QNB as a ligand. The controversy, which is most evident when intact lymphocytes instead of lysed lymphocytes membrane preparations are used, might arise from the hydrophobic properties of ^{3}H-QNB. The ligand could be enter and non specifically bound into the cells. As matter of fact, from literature, ^{3}H-QNB binding to lymphocytes muscarinic receptors in both intact or broken cells, shows a high non specific component and kinetic property which do not apply properly to Scatchard analysis.

^{3}H-N-methyl scopolamine binding (^{3}H-NMS) to human lymphocytes. A model to study cholinergic dysfunction in Alzheimer's Disease.

In attempt to more accurately characterized cholinergic receptors of human lymphocytes we studied the binding of another muscarinic antagonist, the ^{3}H-NMS which is hydrophilic and has much lower partition coefficient than ^{3}H-QNB. Thus ^{3}H-NMS would have less capacbility than ^{3}H-QNB to be trapped into the cells and be more suitable for studying muscarinic receptors in viable lymphocytes. We therefore compared the binding characteristics of ^{3}H-NMS to the traditionally used ^{3}H-QNB on both viable lymphocytes and crude lysed membrane preparation. Details of method have been previously reported (Ferrero et al., 1987). Although both radioligands showed specific binding, only the ^{3}H-NMS displayed saturation capacity. In both intact and broken cells ^{3}H-QNB binding saturation curves were biphasic. The first component showed a transient plateau up to 20 nM; in the second component the specific binding showed a linear increase with the ligand concentration and no saturation could be reached up to 100 nM ^{3}H-QNB.

Conversely ^{3}H-NMS specific binding was **saturable in** both lymphocytes and lysed lymphocytes membranes. Scatchard analysis indicate a single class of binding sites for both preparation with comparable Bmax and Kd values (Tab. I).

TAB. I - CHARACTERISTICS OF ^{3}H-QNB AND ^{3}H-NMS SPECIFIC IN LYMPHOCYTES

Lymphocytes preparations	^{3}H-QNB S.B. Bmax (fmol/mg prot)	Kd (nM)	^{3}H-NMS S.B. Bmax (fmol/mg prot)	Kd (nM)
Intact	not saturable		130±21	18
Membrane (P$_1$+P$_2$ fractions)	not saturable		140±20	16

Despite the ^{3}H-NMS binding sites on lymphocytes display a lower affinity with respect the Kd for ^{3}H-NMS in other tissues, they behave like true muscarinic receptors. The muscarinic agonists, carbamylcholine and oxotremorine, and the selective antagonists (M$_1$) piranzepine are significantly less potent than classical muscarinic antagonists (atropine, metylscopolamine) to compete with ^{3}H-NMS binding. The Kds for carbamylcholine and piranzepine were 3 log less than the Kds for atropine and scopolamine and the Hill plots coefficients was significantly lower than 1 (Tab. II).

TAB. II - BINDING PARAMETERS FOR ^{3}H-NMS DISPLACEMENT BY VARIOUS DRUGS.

Displacer	IC$_{50}$ (uM)	K$_i$ (uM)	n Hill
Atropine	0.26±0.015	0.15±0.0086	0.91
Methylscopolamine	0.53±0.076	0.30±0.043	0.81
Pirenzepine	9.3 ±0.92	5.3 ±0.53	0.23
Oxotremorine	7.0 ±0.64	4.0 ±0.36	0.54
Carbamylcholine	25 ±1.8	14 ±1.00	0.56
d-Tubocurarine	>1000	>1000	N.D.

Values are the mean ± SEM from 3 experiments.

Taken together these data suggest the presence of either M$_1$ or M$_2$ subtypes of muscarinic receptors. The nicotinic antagonist d-tubo curarine was not able to displace the tracer at all, further indicating the muscarinic nature of these binding sites. Based on this initial observation we therefore studied ^{3}H-NMS binding in lymphocytes from AD patients in order to assess whether it might be useful to evidentiate possible cholinergic defects in such patients. Moreover to examine the specificity of possible receptors changes we also studied the ^{3}H-NMS binding in lymphocytes from age matched non demented patients with Parkinson's Disease (PD). In both AD and PD the diagnosys were made according to established clinical criteria. Compared to the normal age matched control group, ^{3}H-NMS binding was significantly reduced in lymphocytes from AD patients, but not in lymphocytes from PD cases (Tab. III).

TAB. III - ^{3}H-NMS S.B. BINDING ON LYMPHOCYTES FROM PATIENTS WITH ALZHEIMER'S DISEASE, PARKINSON'S DISEASE AND CONTROLS (SCATCHARD ANALYSIS)

DIAGNOSIS	AGE (years)	^{3}H-NMS S.B. Bmax (fmol/mg prot)	Kd (nM)
Controls	28.5+5.8	§66.4	18.5
	40.3+6.6	89.6	17.0
	64.3+6.9	§*126.7	18.5
Alzheimer's Disease	62.3+6.9	*74.4	19.0
Parkinson's Disease	62.5+7.1	130.9	19.0

*p<0.01 when AD are compared to older controls
§p<0.01 when older controls are compared to younger controls
(Student's t test).

Scatchard analysis in AD group indicates a significant deacrease of the receptors number with no changes in their affinity.

It would be expected that the PD patients, in which abnormalities of cholinergic systems have been frequently reported, behave like AD patients. Conversely in this group the density of lymphocytes muscarinic receptors was similar to that of the controls. The central cholinergic defect appears more variable in PD than AD and also affects different cholinergic pathways. This might be especially true for the PD cases with pure extrapyramidal symptoms as those of the present study.

Although the physiological significance of muscarinic receptors in lymphocytes is not yet known the decrease observed in AD might indicate that the immune system is primary or secondary affected. A connection between lymphocytes and Alzheimer should be taken into consideration as AD patients are known to have abnormal immune responsiveness and laboratory signs of altered immune reactivity (Miller et al., 1981). AD afflicts primarly older people, significantly increasing in incidence with age. Aging is also associated with deterioration of immune systems; with advancing age there is increased incidence of autoantibodies, response to T cells mitogens declines, suppression functions have been reported diminuished (Burnet, 1970; Kay and Makinodan, 1976; Weksler et al., 1978).

Because the abnormal cholinergic activities on lymphocytesof AD patients might be expression of some unknown mechanisms related to aging itself we have therefore compared the ^{3}H-NMS binding in lymphocytes of AD patients to that observed in both age-matched elderly and younger controls. Tab. III shows that AD patients had a significant reduction in the number of their lymphocytes cholinergic receptors only in comparison to the older controls, but not to the younger controls. In turn the older controls exhibited a significant higher number of muscarinic lymphocytes recognition sites than the younger controls.

The age dependent increase in ^{3}H-NMS binding sites was further characterized in a larger populations of healthy volunteers of increasing age. The ^{3}H-NMS binding increased linearly with advancing age. Scatchard analysis indicate selective increases of Bmax values without changes in the affinity (Tab. III).

The interpretation of the apparently increase of lymphocytes cholinergic activity with aging is intriguing especially when we compared these results with the findings of cholinergic hypofunctions observed in AD. The more simplest interpretation is that in Ad it appears defective the physiological regulation of the lymphocytes cholinergic functions induced by aging itself. What this means it remains unexplained from our present findings. This might be a result of specific alterations in cholinergic reception by the lymphocytes or, alternatively, a result of changes in other membrane constituents which may in turn lead to alterations in the general physical properties of the lymphocytes membrane. To reach a possible interpretation of the data, it seems necessary to identify functional mechanisms and transducer systems to which lymphocytes cholinergic receptors might be coupled.

PHARMACOLOGICAL ASPECTS

While evidence for impairment of the cholinergic system in AD, and its specific relationship to memory and cognitive disorders in dementia is convincing, the etiology and mechanisms by which cholinergic impairment occur in this disease remain obscure. Recent attempts to reverse the cholinergic deficiency in AD by pharmacologic manipulations, as well as others investigative approaches, have provided some insights into the underlying mechanisms of the cholinergic dysfunction. It must be emphasized, however, that there exist no effective pharmacologic therapies for the treatment of patients with AD or age-related memory loss.

A principal aim of therapeutic trials has been the restore of brain Ach functions. Accordingly two major strategies have been suggested: inhibition of the degradation enzyme AchE and stimulation of the synthesizing enzyme CAT by raising substrate. Suitable central AchE inhibitors are not available. Therefore several groups have tried the "increase substrate availability strategy", namely with choline and lecithin (Haudrich et al., 1974; Cohen and Wurtman, 1975). Given that the available evidences indicate normal functioning of the postsynaptic cholinergic system, it might be reasonable to employ loading strategy in order to increase brain Ach levels in these patients. However results of choline and lecithin to patients with AD and age-related memory loss have not been encouraging, although several subgroups of patients demonstrated mild benefits (Boyd et al., 1977; Etienne et al., 1978).

The approach with pharmacologic agents which augment cholinergic neurotransmission is not without uncertainties. If cholinergic neuronal elements are lost, no currently available treatment could restore these elements or, indeed produce a noticeable improvment in their functions. Because there is some plasticity in brain, however some benefits could result, probably temporarly, from cholinergic enhancement. It must be recognize that safe, orally active and long lasting cholinomimetics are not available at present. Thus a major problem for clinical investigation is having effective drugs. Another criticism is that cholinergic function is not the only significant biochemical disorder in AD. Several other neurotransmitters and neuroregulators deficits have been widely reported in AD. For each of these substances a speculative theoretical rationale could be developped.

Treatment of Alzheimer's Disease with L-Acetyl Carnitine.

The possibility that increasing availability of acetyl coenzyme A might augment cholinergic function has received relatively little attention. However, an approach based on increasing either the concentration on this compound, or its transport from mitochondria to the cholinergic terminals, might also be pursuing. Several compounds have been proposed to act as a carriers of acetyl groups from the intramitochondrial acetyl-CoA to the extramitochondrial space. Among them cytrate, acetate and acetyl-carnitine appear the most important (see Tucek, 1978 for review). Carnitine/acetyl-carnitine represent a biological system involved in energy metabolism acting as the essential carrier of acyls to intramitochondrial sites of beta-oxidation, as well as in structural metabolism of the phospholipid neuronal membrane compound (McCaman et al., 1966). Acetyl-carnitine is ubiquitously distributed in peripheral organs and brain. The mechanism by which acetyl-carnitine affects neural functions is not known, but two possibilities have been suggested: an indirect action involving synthesis of Acetyl-CoA followed by the acetylation of choline to form Ach or a direct action as a transmitter substrate (Janiri and Tempesta, 1983; Onofri et al., 1983). Sterri and Fonnum (1980) observed that the degeneration of cholinergic nerve terminals in a brain region is associated with a marked decrease in the activity of carnitine acetyl transferase which suggest that acetyl-carnitine might play a specific role in the synthesis of Ach. Moreover the effects of either direct application of acetyl-carnitine on the cerebral cortex or of acute administration of L-acetyl-carnitine (LAC), which passes through the blood brain barrier, are suggestive with cholinergic muscarinic mechanisms of action.

The efficacy of LAC administration in improving the cognitive and behavioral function in Ad has been test by a limited numbers of pilots uncontrolled studies (Testa and Angelini, 1982; Loeb et al., 1986; Bergamasco et al., 1986). In our investigation we examined 20 patients with presumptive AD (age, $M \pm SD$, 60.9 ± 4.8 years) of recent onset (<3 years). The diagnosys was made according to DSM III criteria, exclusion of toxic and methabolic disorders and neuroradiological examinations including CT scanning. Hachinsky Ischemic Score was used to exclude vascular or mixed dementia. All the patients received comprehensive neuropsychological battery including Mini Mental State Examinations (MMSE), Blessed dementia and information, memory, concentration rating scales (BD and Bi.m.c.), Rey's Test (short and long term), WAIS Digit Span (forward and backward) and Gibson Spiral Maze. Depression was excluded based on the Hamilton Scale. Furthermore during the course of LAC treatment each patient was rated on the Clinical Global Impression Scale according to the relative appreciation of impairment-improvement.

A preliminar neurophysiological evaluation has been also carried out in ten patients. We studied cerebral evoked potentials, including visual evoked potentials (VEP), somatosensory evoked potentials (SEP) and brain-stem auditory evoked potentials (BAEP). Initial results indicated that LAC treatment induced an improvement of the VEP waveshape components in 6 out of 10 patients with an amplitude increment of the P_1N_2 deflection. Furthermore an increment of SEP cortical components amplitude was evident in 4 of these patients.

TAB. IV - NEUROPSYCHOLOGICAL BASELINE PROFILE

Test	Score (mean S.D.)
Mini Mental State	18.1 $\pm$ 7
Blessed Dementia score	8.3 $\pm$ 2.7
Blessed i.m.c. score	22.3 $\pm$ 7.2
Rey's test short-term	10.3 $\pm$ 3.4
Rey's test long-term	3 $\pm$ 0.9
Digit Span forward	4.4 $\pm$ 0.9
Digit Span backward	2.3 $\pm$ 2.2
Gibson Spiral Maze	7.7 $\pm$ 4.2

Tab. IV summarized the neuropsychological baseline profile and the relative score before LAC treatment. Neuropsychological evaluation was performed in each patient prior to treatment and every month during the LAC treatment.

All the patients received LAC (2 mgs p.o. daily) for six months. Prior to starting treatment a drug washout period was performed in every patient in order to avoid overlapping effects of other administrated psychotropic drugs. Only in one patient a side effect was observed, with excited behavior suppressed by the reduction of daily dose. The non parametric Wilicoxon test was used to evaluate statistical significance variation of neuropsychologic test scores in pretreatment and during treatment conditions. A positive, although slight, trend of mean scores was reported for all the studied test except the Digit Span Test which instead did not show relevant modification. There was a significant variation of the behavioral profile as rated on the Blessed Dementia Scale ($p<0.01$) as well as in both MMSE and Blessed i.m.c. which suggest an improvement of orientation, attention and memory functions ($p<0.01$). An improvement of memory functions was also suggest by the Rey test scores which were also significant ($p<0.05$). The Gibson Spiral Maze test also reveals a significant increase of the score already on the 2nd month drug treatment. The impairment of the neuropsychological profile in all the 20 patients examined was already evident from the 4th month of the followed period. This data may represent an indirect evidence that all patients were undergoing the progressive deterioration of cognitive functions which is typical of AD.

In conclusion our findings show that the neuropsychological battery employed in our study can be useful to evaluate the modification of performances in AD patients. Furthermore our results suggest that some of the neuropsychological tests (MMSE, Blessed i.m.c., Rey test) could effectively monitor drug effects or short term impairment in AD, whereas other memory tests (WAIS Digit Span) could not evaluate the progression of the disease. In relation to the evaluation of the potential efficacy of LAC therapy in AD some cautions might be exercised. It appears evident from our findings that LAC improves behavioral and memory activity in the examined patients. Because of the small size of the sample and the relatively short duration of the treatment together with the absence of placebo control group, however it is impossible to make any definite conclusion about the real efficacy of LAC treatment in Alzheimer's Dementia. Our results must be therefore confirmed in larger and controlled population.

References.

Adem, A., Nordberg and P. Slanina. 1986. A muscarinic receptor type in human lymphocytes: a comparison of ^{3}H-QNB binding to intact lymphocytes and lysed lymphocyte membranes. Life Sci. 38: 1359-1368.

Bartha, E., Z. Rakcon-Czay, P. Kasa et al. 1987. Molecular form of human lymphocyte membrane-bound acetylcholinesterase. Life Sci. 41: 1853-1860.

Boyd, W.D., J. Graham-White, G. Blackwood et al. 1977. Clinical effects of choline in Alzheimer's Senile Dementia. Lancet i: 508-509.

Bergamasco, B., L. Tarenzi, D. Leotta and Scarzella. 1985. Activity of acetyl-l-carnitine in primary degenerative dementia. IVth World Congress of Biological Psychiatry, Philadeplphia, Abstracts 142.6: 81.

Burnet, F.M. 1970. An immunological aspect of aging. Lancet ii: 358-360.

Cohen, E.L. and R.J. Wurtman. 1976. Brain acetylcholine: control by dietary choline. Science 191: 561-562.

Davies, P. and A.J.F. Maloney. 1976. Selective loss of central cholinergic neurons in Alzheimer's Disease. Lancet 2: 1403.

Davies, P., R. Katzman and R.D. Terry. 1980. Reduced somatostatin like-immuno reactivity in cerebral cortex from cases of Alzheimer's disease and senile dementia of Alzheimer type. Nature 288: 279-280.

Elble, R., E. Giacobini and G.F. Scarsella. 1987. Cholinesterase in cerebrospinal fluid. Arch. Neurol. 44: 403-407.

Etienne, P., S. Gautier, G. Johnson et al. 1987. Clinical effects of choline in Alzheimer's Disease. Lancet i: 508-509.

Ferrero, P., C. Eva, P. Rocca et al. 1988. Studio dei recettori colinergici linfocitari. Implicazioni nella malattia di Alzheimer e nel morbo di Parkinson. Atti XV Congr. LINPE, Alba. In press.

Gordon, M.A., J.J. Choen and I.B. Wilson. 1978. Muscarinic cholinergic receptor in murine lymphocytes: demonstration by direct binding. Proc. Natl. Acad. Sci. U.S.A. 71: 1330-1333.

Gottfries, G.C., and B.E. Roos. 1973. Acid monoamine metabolism in cerebrospinal fluid from patients with pre-senile dementia (Alzheimer's disease). Acta Psychiat. Scand. 49: 257-263.

Haber, B. and R.G. Grossman. 1980. Acetylcholine metabolism in intracranial and lumbar cerebrospinal fluid and in blood. In Neurobiology of cerebrospinal fluid. ed. J.H. Wood, 345-351. Plenum Press, New York.

Haubrich, D.R., P.F.L. Wang and P. Wedeking. 1974. Role of choline in biosynthesis of acetylcholine. Fed. Proc. 33: 477.

Huff, F.J., J.C. Maire, J.H. Growdon. 1986. Cholinesterase in cerebrospinal fluid. Correlations with clinical measures in Alzheimer's disease. J. Neurol.Sci. 72: 121-129.

Kay, M.M.B., and T. Makinodan. 1976. Immunobiology of aging: evaluation of current status. Clin.Immunol.Immunopathol. 6: 394-413.

Kessler, J.A. 1987. Deficiency of a Cholinergic differentiating factor in fibroblasts of patients with Alzheimer's disease. Ann.Neurol. 21: 95-98.

Janiri, L., and E. Tempesta. 1983. A pharmacological profile of the effects of carnitine and acetyl-carnitine on the central nervous system. Int.J.Clin. Pharm.Res. III 4: 295-306.

Loeb, C., F. Iannucelli, C. Traverso and C. Albano. 1985. Evaluation of the activity of acetyl-l-carnitine in the senile dementia Alzheimer type. IVth World Congress of Biological Psychiatry, Philadelphia, Abstracts

152.6: 106.

Maloteaux, G.M., A. Gossuin, C. Waterkeyn and P.M. Laduron. 1982. 13th CINP, Jerusalem, Abstracts 2: 462.

Masturzo, P., M. Salmone, O. Nordstrom et al. 1985. Intact human lymphocyte membranes respond to muscarinic receptor stimulation by oxotremorine with marked changes in microviscosity and an increase in cyclic GMP. Febs Letters 192: 194-198.

McCaman, R.E., M.W. McCaman and L. Stafford. 1966. Carnitine acetyltransferase in nervous tissue. J.Biol.Chem. 4: 930-934.

Miller, A.E., P.A. Neighbour, R. Katzman et al. 1981. Immunological studies in senile dementia of the Alzheimer type: evidence for enhanced suppressor cell activity. Ann.Neurol. 10: 506-510.

Onofrj, M., I. Bodis-Wollner, P. Pola et al. 1983. Central cholinergic effects of levo-acetyl-carnitine. Drugs Exp. Res. 9: 161-169.

Peterson, C. and J.E. Goldman. 1986. Alterations in calcium content and biochemical processes in cultured skin fibroblasts from aged and Alzheimer donors. Proc.Natl.Acad.Sci.USA 83: 2758-2762.

Perry, R.H., M.J. Wilson, J. Robert et al. 1982. Plasma and erytrocyte acetylcholinesterase in senile dementia of Alzheimer type. Lancet i: 174-175.

Rabey, J.M., L. Shenkman and G.A. Gilad. 1986. Cholinergic muscarinic binding by human lymphocytes: changes with aging antagonist treatment and senile dementia of the Alzheimer type. Ann.Neurol. 20: 628-631.

Ruberg, M., A. Villageois, and A.M. Bonnet. 1987. Acetylcholinesterase and butyrylcholinesterase activities in the cerebrospinal fluid of patients with neurodegenerative diseases involving cholinergic systems. J.Neurol. Neurosurg.Psychiat. 50: 538-543.

Shekman, L., J.M. Rabey, and G.M. Gilad. 1986. Cholinergic muscarinic binding by rat lymphocytes:effects of antagonist treatment, strain and aging. Brain Res. 380: 303-308.

Schreiner, G.F. and E.R. Unanue. 1975. The modulation of spontaneous and anti-Ig-stimulated motility of lymphocytes by cyclic nucleotides and adrenergic and cholinergic agents. J.Immunol. 114: 802-808.

Soininen, H., T. Halonen and P.G. Riekkin. 1981. Acetylcholinesterase activities in cerebrospinal fluid of patients with senile dementia of the Alzheimer type. Acta Neurol. Scand. 64: 217-224.

Strom, T.B. , A.T. Sytkowski, C.B. Carpenter et al. 1974. Cholinergic augmentation of lymphocytes-mediated cytotoxicity. A study of the cholinergic receptor of cytotoxic T lymphocytes. Proc.Natl.Acad. Sci. USA 71:1330-1333.

Testa, G. and C. Angelini. 1981. Preliminar trial of acetyl-carnitine in dementia. Current Reports in Neurology 5: Suppl. 1: 4.

Tucek, S., K.M. Havrane and I. Ge. 1978. Synthesis of (acetyl ^{14}C) carnitine and the use of tetraphenylboron for differential extraction of (acetyl ^{14}C) choline and (acetyl ^{14}C) carnitine. Anal.Biochem. 84: 589-593.

Weksler, M.E., J.B. Innes and G. Goldstein. 1978. Immunological studies of aging: the contribution of thymic involution to the immune deficiences of aging mice. J.Exp.Med. 148: 996-1009.

Zalcman, S.J., L.M. Neckers, O. Kaayalp and R.J. Wyatt. 1983. Muscarinic cholinergic binding sites on intact lymphocytes. Life Sci. 29: 69-73.

Subject Index

Acetylcholine
   metabolism   43-52, 95-100
   CSF changes   113-122
   kinetics   15-30
   turnover   43-52
   release   247-258

L-Acetylcarnitine (Ravizza)

Acetylcholinesterase   293-302
   CSF changes   113-122

Anticholinesterase agents
   behavioral effects   15-30
   general features   15-30
   side effects   15-30
   synaptic action   15-30

4-Aminopyridine (4-AP)   199-210, 323-328

Bethanechol   301-310, 311-318

Carbachol   199-210

CNS lesions
   cholinergic and adrenergic   63-72

Choline
   CSF changes   113-122
   effect   95-100
   metabolism   95-100

Cholinergic agonists   301-310, 311-318, 319-322

Cholinergic receptors   247-258

Cholinesterase inhibitors/side effects   15-30

Clonidine
   acute effects   179-190

3-4-Diaminopyridine (3-4-AP)   199-210
   effects on memory   179-190

Diisopropylfluorophosphate (DFP)   31-42

Galanthamine   191-198, 199-210, 283-292

Glycopyrrolate   87-94

Heptyl-physostigmine   329-338

Hippocampal structures
   effect on   323-328

Huperzines   277-282

Intracerebroventricular administration
   physostigmine   113-122
   bethanechol   301-310

Lecithin   95-100, 237-246
   lecithin and THA   237-246

Lesions
   cholinergic and adrenergic   63-72

Methanesulfonyl fluoride (MSF)   293-302

Metrifonate   15-30, 269-276

Neuroendocrine changes
   effect of physostigmine   247-258

Neuromuscular junction
   effect of THA   199-210

Nicotinic receptors   247-258

Obidoxime (Toxogonin)   31-42

Organophosphates   15-30, 31-42

Oxotremorine   319-322

Phospholipids   95-100

Physostigmine
   autoimmune-related retention deficits   53-62
   acetylcholine release   113-122, 247-258
   chemistry   329-338
   clinical trials   103-112, 113-122, 123-140
   detection   173-176
   experimental trials   63-72, 179-190, 191-198
   learning/retention   53-62, 63-72, 87-94, 123-140, 141-152
   memory effects   141-152, 153-162, 179-190
   neuroendocrine changes   163-172
   recognition/memory   141-152, 153-162
   steady state infusion   73-86, 123-140

Physostigmine derivatives 329-338
    heptyl-physostigmine 329-338

Pralidoxime (2-PAM) 31-42

Pyridostigmine 87-94

R-86 319-322

Sarin 31-42

Scopolamine 87-94, 123-140, 153-162

Tetrahydroaminoacridine (tacrine, THA)

    acetylcholine release 247-258
    anesthesia (Gershon)
    anti-delirium properties (Gershon)
    anti-psychotomimetic properties (Gershon)
    cholinergic transmission 199-210
    cholinesterase inhibition 15-30,
       73-86
    chronic effect 211-216
    clinical trials 225-230, 231-236,
       237-246
    effects on memory 179-190
    efficacy 225-230
    history of efficacy (Gershon)
    morphine antagonism (Gershon)
    nerve terminal currents 191-198
    neuromuscular junction 199-210
    pharmacology (Gershon)
    side effects 225-230, 259-266